Sacred Groves of Rajasthan:
Threats and Management Strategies

Sacred Groves of Rajasthan:
Threats and Management Strategies

G. Singh
Scientist G
Division of Forest Ecology,
Arid Forest Research Institute, New Pali Road,
Jodhpur-342005, Rajasthan (India)

Published for :

Arid Forest Research Institute, Jodhpur

by :

Scientific Publishers (India), Jodhpur

Published by:
ARID FOREST RESEARCH INSTITUTE
New Pali Road,
Jodhpur - 342 005, Rajasthan (India)
www.afri.icfre.org

Printed by
SCIENTIFIC PUBLISHERS (INDIA)

Jodhpur –

5 A, New Pali Road
P.O. Box 91
Jodhpur - 342 001 INDIA

Delhi –

4806/24, Ansari Road
Daryaganj
New Delhi - 110 002 INDIA

ISBN: 978-93-86102-88-1

Printed in India

Chancellor, FRI University
कुलाधिपति, व.अ.सं. विश्वविद्यालय

Dr. Ashwani Kumar, IFS
Ph.D, D.Sc, D.Sc (h.c.), FIASc, FAFESc, AIFC
डा. अश्वनी कुमार, भ० व० से०
पीएच.डी. डी.एससी. डी.एससी. (एच.सी.) एफआईएएससी. एफआईएफसी

सत्यमेव जयते

Director General
Indian Council of Forestry Research and Education
P.O. New Forest, Dehradun
(An ISO 9001 : 2008 Certified Organisation)

महानिदेशक
भारतीय वानिकी अनुसंधान एवं शिक्षा परिषद्
डाकघर न्यू फॉरेस्ट, देहरादून
(आईएसओ 9001:2008 प्रमाणित संस्था)

FOREWORD

Holy places are not simply geographically beautiful area but are more holy than shrines. Sacred groves are virtually patch of forest or a clump of trees associated with other forms of life, and affords protection on the basis of religious practice and faith. They have existed since time immemorial as patches of densely wooded areas and venerated on religious grounds. The groves are dedicated to a deity God or Goddess and causing a simple damage to life in the grove area is taboo. Their existence today is based on ancestral worship and focus on the conservation of forest patches globally. By distribution over a wide ecosystem, sacred groves help in the conservation of rare and endemic species. Well preserved sites are store houses of biological, ecological, medicinal, ethno-cultural and religious values. The preserved rare and endemic wild plant species hold potential benefit to man in medicine, agriculture and industry. In fact, sacred groves represent the ancient Indian way of *in situ* conservation of genetic diversity. Studies on floristic and ethnobotanical aspects of sacred groves provide detail scientific account of these sacred places in India.

Despite scattered all over the country, sacred groves are associated with traditional management cultures and do not enjoy protection via federal legislation, while increased pressure from modernization and market linked economy is resulting in their degradation. The projected number of sacred groves in western India in general and Rajasthan in particular is very high (>25000), but the numbers of documented sacred groves and its associated problems are hardly 10% or less. Though the unique management and ownership of the sacred groves is one significant reason that they are so special, but present ways of liberalization, privatization and globalization lead losses of such traditional treasure on which iivelihoods of the local communities depends. Concerted efforts are required to protect and conserve the sacred values of these deteriorating life support resources for the survival of existing biodiversity in view of predicted changes in climate harnessing existing knowledge and available financial and human resources.

-2-

In the book '**Sacred Groves of Rajasthan: Threats and Management Strategies**' Dr. G. Singh, Scientist G, Arid Forest research Institute, Jodhpur has presented biological, social and economic implications of the sacred groves across the region in general and Rajasthan in particular. The author has compressively incorporated the existing reports and scientific literatures on sacred places, sacred plants and sacred groves, thoroughly supported by field observations and has presented in an implementable way so that these natural heritage can be conserved and managed for the benefits of people, society and environment as whole. Other important aspect of this book is the highlights of 123 sacred groves across the state focusing biological diversity, soil characteristics, tree dominance and their regeneration status, invasion of exotic species. This book will be useful in view of conserving and managing this resource under global climate change policies and restoring them for people livelihood and environmental preservation.

I hope this work is real contribution to society and the researchers both and it would be a useful tool for the planners, foresters, non-government organizations and society as whole engaged in protecting, restoring and conserving this ancient treasure of our civilization.

With best wishes,

(Ashwani Kumar)

PREFACE

Sacred groves are traditionally protected forest fragments. They are considered as the relics of original forests that had covered the region in past. In most of the cases these preserved forest patches are close to the human settlement and are the abodes of gods/goddesses. Groves are often part of the heterogeneous landscape mixed with agricultural land, plantation, barren land, water bodies and the villages. These sacred groves link present and past society in terms of biodiversity, culture, religious and ethnic heritage and remain with rare or endangered species and unique ecological functions. Significance of sacred groves can bc viewed in terms of its global distribution, where diverse cultures recognize them in different ways encoding various rules and traditions for their protection. The rare and endemic species supported by sacred groves are extra-sensitive, compared to the common species, and persist only in the habitats available in the groves.

Developmental activities and corresponding changes in socio-economy and land use practices have transformed the landscapes and put the sacred groves into varying stages of degradation. The sacred groves are currently facing threats like submergence, sometimes clear felling, mining, quarrying, encroachment, invasion of alien exotics and other depletive factors. Their presence in the agricultural lands, and unauthorized and unregulated cutting of plants, grazing and erosion of social and cultural values are some of the many reasons of its degradation. Due to failure of pure legal machinery to conserve biodiversity, it has become utmost necessary to formulate policies based on local people's traditional knowledge to protect, conserve and manage this vanishing natural resource. To frame out restoration, rehabilitation and management strategies there are needs to identify gaps in knowledge about the endemic and endangered species, population, regeneration status, ecology etc. of these sacred groves and socio-religious environments and climatic conditions in the region. In Rajasthan, where the groves are relatively larger in size, faith in the deities is still strong and the groves harbour a significant number of floral and faunal species, but the availability of knowledge in this regard is very scarce. This work is based on past and present researches and field studies including direct observations with the financial support of State Forest Departments, Rajasthan.

Present book encompass biological, social and economic status of the sacred encompasses groves prepared through a concerted and determined attempt. Divided into 7 chapters, this book has been initiated with the background context and methods of field observation recording, followed by an overview of the sacred places, trees and groves. Status of 123 sacred groves distributed throughout

Rajasthan, dominant and subdominant tree and shrub species; biological diversity and invasions; soil characteristics and carbon status; livelihood and threats to the sacred groves and people perception and management strategies have been included in the next chapters. This also focuses on the benefits accrued from the sacred groves to the villagers and various threats arising out of overexploitation, developmental and mining activities and encroachments of various types. Along with accommodating the desire and level of participation of the local villagers in protecting and conserving these sacred groves, it also focuses on management strategies by people participation and adopting landscape approach dovetailing the probable needs of the grove dependent communities.

This publication aims to enhance grove tree cover, resilience and livelihoods of the local population and to improve the environmental conditions of this degrading ecosystem in local, region and global benefits. It may be useful to the policy makers, forest managers, non-government organization, extension agents, environmentalists as well as researchers, who are involved in developing, conserving and managing community resources in general and sacred groves in particular.

This work has been possible through encouragement of Sh N.K. Vasu, Director, AFRI, Jodhpur, Dr T.S. Rathore, former director, AFRI, Sh. Rahul Kumar, former PCCF, Dr. S.S. Chaudhary, PCCF, Rajasthan Forest, Sri S.K. Srivastava, the then Addl. PCCF (Dev), Rajasthan Forest Department, Jaipur. Prof. G. Tripathi, Head, Department of Zoology, JNV University, Jodhpur, Late Sh. A.K. Upadhyay, Addl. PCCF, Dr Shruti Sharma, Addl. PCCF, Rajasthan Forest Department, Jaipur, provided their comments and suggestions for improvement of this work. Helps of varying kinds are also rendered by scientists, staffs and project fellows working in different programmes in the Division of Forest Ecology, AFRI, Jodhpur. Mention to some are Sh N. Bala, Scientist F, Dr. Bilas Singh, Scientist B, Dr. Smita Shukla, Research Associate, Sh Kaushal Singh, Mrs. Ritu Sharma, Km Priyanka Garhwal, Research scholars, and Deepak Mishra, Sh Jalaj Saxena and Km Versha Saxena, project fellows. I would like to thank all of them for their contribution of varying kinds.

I hope this book will be useful to both policy makers and researchers equally and help in effective planning and transferring the knowledge in protecting and conserving this ancient institution, restoring it back with effective participation of village folk and promoting grove's ecological value and rural livelihood.

Jodhpur
5th June, 2016

G. Singh
Scientist G
AFRI, Jodhpur

CONTENTS

1

BACKGROUND CONTEXT AND OBSERVATIONS

Traditionally, there are different forms of practices to worship nature. One such important tradition of nature worship followed throughout the world is providing protection to patches of forests designated as sacred groves dedicated to deities or ancestral spirits. In general sacred groves are being used to dedicate by the local communities to their deities. These groves may consist of a cluster of trees, multi-species or multi-tier primary forest, depending on the history of the vegetation. A decision was established in 2007 on the incentives under the United Nations Framework Convention on Climate Change (UNFCC) for Reducing Emissions from Deforestation and Forest Degradation (REDD). This is an effort to create a financial value for the carbon stored in forests or other related ecosystems, offering incentives for developing countries to reduce emissions from forested lands and invest in low-carbon paths to sustainable development. The REDD+ goes even beyond the deforestation and forest degradation, and includes the role of conservation, sustainable management of forests and enhancement of forest carbon stocks (UN-REDD, 2012). Thus REDD+ also provides possibilities to include the rights of local peoples and includes their involvement in the conservation and management of forest or community resources including sacred groves (Sikor et al., 2010).

1. BACKGROUND

Community forests dedicated to one or more deities and worshipped by the local communities are termed as sacred groves or forests (King et al., 1997; Dagla et al., 2007; Samati and Gogoi, 2007). These are one of the most valuable, primitive practices of nature conservation throughout the world, and Sochaczewski (1998) called them as 'Life Reserve'. Sacred groves are the relict climax communities preserved by the local people for certain beliefs. Such forests are rich in biological diversity and harbour many rare and endangered both plant and animal species. An inextricable link between present and past society in terms of biodiversity, culture, religion and ethnic heritage exists in these sacred groves (Khurana, 1998; Jyothi and Nameer, 2015). Sacred groves are distributed across the globe, and diverse cultures recognize them in different ways encoding various rules for their protection (Sheridan, 2009). These have been a source of natural wealth like fodder, fuel, timber, berries, roots and herbs and are regarded as a 'symbol of prosperity' for the community that has owned it. Moreover, sacred groves play an important role in promoting a flourishing livestock based economy and growth of livestock rearing communities.

Cultural aspects such as religion, faith and traditions bring people closer to nature and the natural environment (Khumbongmayum et al., 2005a & b). Memorial Parks and Sacred Groves are examples of two such cultural landscapes that exist today also (Pruthi and Burch, 2009). There are many examples from Asia, Africa and Europe, where natural phenomenon have also been termed as 'sacred' and worshipped. These are types of natural sites like rivers, lakes, forests, gardens, groves or even a tree, that have been institutionalized by attachment of sacred values with intensions to make a collective management easy and sustainable. The Hinduism in India has grown out of the merger of local traditions that has been intertwined with nature (Chandra and Gadgil, 1998). Deep religious reverence for nature, rather than resource scarcity is the basis for longstanding commitment for preserving these community preserved forests (Ramakrishnan, 1998; Chandran and Hughes, 2000).

Sacred groves have existed in India since time immemorial and are scattered all over the country, but unfortunately it does not enjoy protection via a federal legislation. It appears that total number of sacred groves in India may be more than 100000, but the numbers of documented sacred groves are relatively less, i.e. 14272. The situation in Rajasthan state is even worse with highest estimated number (i.e., 25000 sacred groves covering an area of about 6 lakhs hectare ranging between <1 ha and as larger as 15,000 ha Bhadariyaji Oran in Jaisalmer) in India. Of these, nearly 5370 square kilometers area under Orans is in the Thar Desert. Some are as small as a few square metres. This indicates a wide variation in reporting on the number of 'Orans' exists not only in Rajasthan, but in a district too. Literatures on sacred groves are also meager in Rajasthan, i.e. less than 1% particularly for western Rajasthan. Highest available literatures are from North-East (30%), Western Ghat (26%), Eastern Ghats (24%), central India, Western Himalayas (10%) and Gangetic plain (9%) regions.

The present status of sacred groves everywhere is a matter of deep concern as they are disappearing fast from the countryside. These are currently facing threats like submergence, clear felling, mining, quarry, encroachment and other depletive factors (Nair et al., 1997). Their presence in the agricultural lands, and unauthorized and unregulated cutting of plants, grazing and erosion of social and cultural values are some of the many reasons of its degradation. In view of this, and due to failure of pure legal machinery to conserve biodiversity, it has become increasingly necessary to formulate policies based on traditional knowledge of local people. Therefore, there is an urgent need not only to protect sacred forests, but also to revive and reinvent such traditional ways of nature conservation and biodiversity management (Khumbongmayum et al., 2004). Several encroachments have taken place, and worse are being regularized by the Governments. Lack of effective legislation promotes illegal activities of the offenders. Further, lack of faith in the younger generation is a major problem to loss of this valuable land resource and the biological diversity residing in it (Patnaik and Pandey, 1998; Vinothkumar et al., 2011).

There is need to reverse this trend in benefits of the local population (Hamilton, 2002; Debabrata et al., 2014). There is lack of documentation and inventorisation of the sacred groves of Rajasthan, whereas the available data are scanty and sporadic.

United Nations Convention on Biological Diversity (UNCBD) emphasizes on the development of inventory of organisms within each country for their proper use and sharing benefits which arise out from it. In view of the above facts it is beneficial to inventories the biological diversity of the sacred groves Rajasthan for development of management strategies based on the importance of the species types existing in a particular sacred grove or region (Khoshoo, 1994; Khanwalkar, 2005). Realizing the fact and need of the day documentation of sacred groves of Rajasthan has been done and assessment of biological diversity of some of them has been carried out for improved management and people livelihoods under the funding of Rajasthan Forest Department, Jaipur. This is to develop an inventory on biological diversity and soil carbon storage as well as to develop management plan for revival of these sacred groves for their effective conservation and management. Hence the objectives of the present compilation are: (i) to document and inventories floral diversity of the important 'sacred groves' in different districts of Rajasthan; (ii) to identify and record faunal diversity countered in the above-mention 'sacred groves'; and (iii) to suggest suitable management strategies to increased diversity and productivity for improved local benefits and climate change adaptation.

2. GEOGRAPHICAL LOCATION

Located in the north-west of India, Rajasthan is the country's largest state covering about 10.4% of the total geographical area of the country. It lies between latitudes 23°3' and 30°12' North and longitudes 69°30' and 78°17' East. Rajasthan is bound on the west and northwest by Pakistan, on the north and northeast by the States of Punjab, Haryana and Uttar Pradesh, on the east and southeast by the States of Uttar Pradesh and Madhya Pradesh, and on the southwest by the State of Gujarat. It covers about 3, 42,239 sq km area (Table 1.1).

Table 1.1 Total geographical area and district-wise distribution of population, forest area and rainfall in Rajasthan.

SNo.	District	TGA (km^2)	Population (Nos)	Forest area (km^2)	% of TGA	Rainfall (mm)	
						Normal	2003-2012
1	Ajmer	8,481	2,584,913	61310	7.23	601.8	493.85
2	Alwar	8,720	3,671,999	178495	21.30	657.3	666.28
3	Banswara	4,536	1,798,194	123667	24.55	950.3	976.88
4	Baran	6,955	1,223,921	223962	32.20	873.8	814.88
5	Barmer	28,387	2,604,453	62722	2.21	265.7	342.63
6	Bharatpur	5,044	2,549,121	43494	8.59	663.9	641.81
7	Bhilwara	10,455	2,410,459	77876	7.45	683.2	630.43
8	Bikaner	30,382	2,367,745	124906	4.58	243.0	286.14
9	Bundi	5,500	1,113,725	156678	28.23	773.4	613.73
10	Chittorgarh	7,880	1,544,392	182019	16.77	841.5	829.16
11	Churu	13,792	2,041,172	7122	0.42	354.7	413.17
12	Dausa	3,420	1,637,226	28263	8.24	561.0	644.22
13	Dholpur	3,009	1,207,293	63845	21.05	744.5	633.88

SNo.	District	TGA (km^2)	Population (Nos)	Forest area (km^2)	% of TGA	Rainfall (mm)	
						Normal	2003-2012
14	Dungarpur	3,770	1,388,906	69273	18.37	728.9	772.59
15	Hanumangarh	9,580	6,663,971	94566	6.72	273.5	296.57
16	Jaipur	11,061	672,008	58129	1.51	563.8	554.1
17	Jaisalmer	38,401	1,830,151	45261	4.25	185.5	225.81
18	Jalore	10,640	1,411,327	134979	21.70	370.0	496.14
19	Jhalawar	6,219	2,139,658	40535	6.84	844.3	822.48
20	Jhunjhunu	5,928	3,685,681	24303	1.06	405.1	483.72
21	Jodhpur	22,250	1,458,459	180206	32.59	313.7	302.19
22	Karauli	5,039	1,950,491	131082	24.07	670.7	647.61
23	Kota	5,204	3,309,234	24093	1.36	732.4	736.47
24	Nagaur	17,718	2,038,533	96358	7.78	311.7	370.52
25	Pali	12,357	868,231	140704	34.18	424.4	491.96
26	Pratapgarh	4,360	1,158,283	39658	8.32	-	-
27	Rajsamand	4,636	1,338,114	93772	8.91	567.8	599.51
28	Sawaimadhopur	5,221	2,677,737	63935	8.27	873.4	673.26
29	Sikar	7,881	1,037,185	163865	31.91	440.3	469.82
30	Sirohi	5,136	1,969,520	63344	7.93	591.2	883.05
31	SriGanganagar	11,604	1,779,650	23946	1.89	226.4	254.78
32	Tonk	7,200	1,421,711	33597	4.67	668.3	590.35
33	Udaipur	11,761	3,067,549	414170	29.83	645.0	763.54
	State	342,327	68,621,012	3270135	9.56	575.1	663.26

Source: Directorate of Economics and Statistics, Government of Rajasthan, Jaipur.
TGA: total geographical area

2.1 Physiography of the region

The main physiographic feature of Rajasthan is the Aravalli Range, which runs across the state from southwest (Mount Abu) with peak height of 1,722 m to Khetri in Jhunjhunu district situated in the northeast running for almost more than 850 km. This range divides Rajasthan into 60% in the North West of the lines and 40% in the southeast. The northwest tract is sandy and unproductive with little water but improves gradually from desert land in the far west and northwest to comparatively fertile and habitable land towards the east. The western area includes the Great Indian (Thar) Desert, which is generally sandy and dry and extends into adjoining portions of Pakistan.

The Aravalli Range intercepts the moisture-giving southwest monsoon winds off the Arabian Sea, leaving the northwestern region in a rain shadow. This region receives less than 400 mm of rain in an average year. The Godwar, Marwar and Shekhawati regions lie in the northern region of the state and constitute major part of Thar Desert. The hilly Vangad region lies in southernmost Rajasthan and bordering Gujarat. With the exception of Mount Abu in Sirohi district, Vangad is the wettest region in the state, and is the most heavily forested. North of the Vangad lies the

Mewar region, home to the cities of Udaipur and Chittorgarh in Rajasthan. The Hadoti region lies to the southeast, on the border with Madhya Pradesh covering Bundi, Kota, Baran and Jhalawar districts. North of Hadoti and Mewar is the Dhundhar region, the home to the capital city Jaipur. Mewat is the easternmost region of Rajasthan, bordering Haryana and Uttar Pradesh and relatively better in climate and vegetation.

To understand implications and role of climatic and edaphic resources in agricultural and allied sectors, Rajasthan has been divided into ten agro-climatic zones (Plate 1.1, Table 1.2) covering threes agro-ecological regions (Sehgal and Abrol, 1994).

Table 1.2. Agroclimatic zone and their distribution in Rajasthan

S.No	Agro-climatic Zones	Rainfall (mm) range	Districts
1	IA-Arid Western plain	200-370	Barmer, Jodhpur
2	IB-Irrigated North Western plain	100-350	Sriganganagar, Hanumangarh
3	IC-Hyper-arid partly Irrigated Zone	100-350	Bikaner, Jaisalmer, Churu
4	IIA-Internal Drainage Dry Zone	300-500	Nagaur, Sikar, Jhunjhunu
5	IIB-Transitional Plain of Luni Basin	300-500	Jalorc, Pali, Sirohi
6	IIIA-Semi arid Eastern Plain	500-700	Jaipur, Ajmer, Dausa, Tonk
7	IIIB-Flood Prone Eastern Plain	500-700	Alwar, Dholpur, Bharatpur, Karauli and Sawai Madhopur
8	IVA-Sub humid Southern Plain	500-900	Bhilwara, Rajsamand, Chittorgarh
9	IVB-Humid Southern plain	500-1100	Dungarpur, Udaipur, Banswara and Pratapgarh
10	V-Humid Southern Eastern Plain	650-1000	Bundi, Kota, Baran, Jhalawar

Source: http://agropedia.iitk.ac.in/content/agro-climatic-zone-rajasthan.

2.2 Climate of Rajasthan

Because of its location in the western part of India and varying topography, Rajasthan exhibits varying climate. For example, the rocky Aravalli, the western arid plains, the eastern fertile plains etc experiences different climatic conditions. The weather pattern in the state can be divided into pre-monsoon, monsoon, post-Monsoon and winter. **Pre-monsoon is** summer and the most parched and hot season of the year that precedes the monsoon and extends from April to June. The temperature ranges from 32°C to 45°C. The desert lies in the west and the north west of the Aravalli region becomes drier due to the scorching heat of the sun and the highest temperature is recorded sometime about 47-48°C during April/May. The single hill station of the state – the Mount Abu records the lowest temperature. Temperature however falls steeply at night in the arid regions. Heavy winds blow from the south-western boundary and bring dust storms, which are generally known as 'Andhi' or 'Kali-Pili Andhi'. **Monsoon** brings relief to the sultry and sun-baked terrain of this state during the month of June in the eastern region and mid- July in the western arid regions. The temperature drops from 40 to 35 degree Celsius. With the fall in temperature, humidity increases. The state receives 80 to 90 per cent rainfall during this period. There is a second phase of monsoon that continues from July to September. **Post-Monsoon** season is followed by the monsoon that commences from mid-September and continues till November. This season is once again characterized by increased air temperature, when average maximum temperature ranges between 33°C and 38°C. The average minimum temperature is between 18°C and 20°C. **Winter** in Rajasthan extends from December to March and January is the coldest month of the year. Temperature varies throughout the state and the lowest temperature reaches even below 0°C especially in the Churu district. There is slight rainfall in the north and north eastern region of the state. During this period, relative humidity ranges from 50% to 60% in the morning and 25% to 35% during noon time. On the average basis, winter temperatures range from 8°C to 28°C.

2.3 Rainfall Pattern

Rainfall is very low highly erratic and variable throughout the Rajasthan plain mean Annual Rainfall is from 100 mm to 400 mm on eastern edge of Indo-Pak border. Rainfall decreases from East to West and from South West to North East. There is a very rapid and marked decrease in rainfall west of the Aravalli range making the western Rajasthan an arid tract. The average annual rainfall in western part of Aravalli range ranges from less than 100 mm in north-west part of Jaisalmer to 200-300 mm in Sriganganagar, Bikaner and Barmer regions, 300-400 mm in Nagaur, Jodhpur, Churu and Jalore regions and more than 400 mm in Sikar, Jhunjhunu and Pali regions and along the western fringes of the Aravalli range. On the eastern side of the Aravalli range, the rainfall ranges from 550 mm in Ajmer to 1020 mm in Jhalawar regions. In plains, Banswara (920 mm) and Jhalawar (950 mm) districts receive the maximum annual rain. However, highest rainfall in the state, i.e. 1638 mm occurs in Mount Abu in Sirohi district.

However, rainfall varies widely across districts and for the same district it varies widely from year to year. But over a 16 year period, the rainfall shows a declining trend for all the districts with the fall being the steepest for Jaipur and Jodhpur (Fig 1.1). Trends in annual rainfall of 1901-1935 vs 1971-2005 indicates a decrease by 14.3% at Jaisalmer but no change in Jodhpur (Khaladkar et al., 2009). Extreme northwest region of India, the Thar Desert received record rainfall of 55 cm during August 16-25, 2006 (Jayanthi et al, 2006). Rainfall pattern during 2003 to 2012 indicates an increasing trend in IB, IC, IIA, IIA and V bioclimatic zones, whereas a decreasing trend has been observed in IIB, IVA and IVB zones. Others zones show almost constant rainfall during this period (Fig 1.1).

2.4 Land use pattern

Compared to Indian averages, Rajasthan has slightly better proportion of total cropped area and net sown area (Table 1.3). The significance of the region is increase in area under irrigation even in desert region, where people are growing many cash crops by irrigating them with brackish ground water.

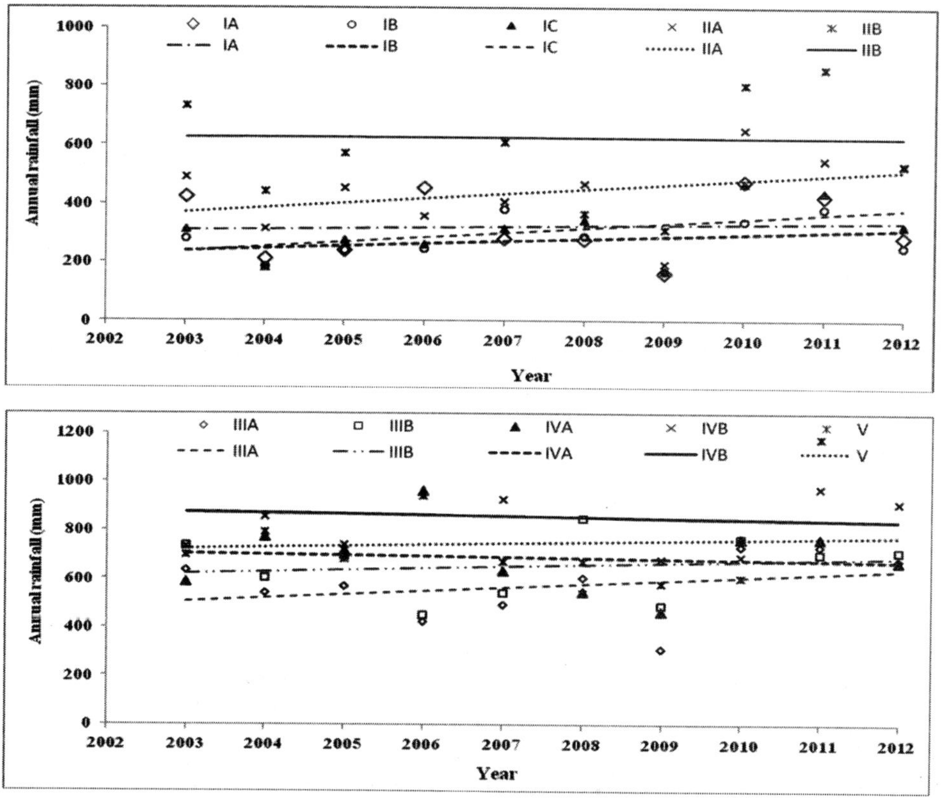

Figure 1.1. Pattern of rainfall in different agroclimatic zones of Rajasthan during 2003-2012.

Table 1.3. Land used statistics in India and Rajasthan.

Land use	India	Rajasthan
Reporting area for land utilization statistics	305611	34270
Forests	70042	2735
Not available for cultivation	42954	4268
Permanent pastures & other grazing lands	10149	1697
Land under misc. tree crops & groves	3351	17
Culturable waste	12857	4475
Fallow lands other than current fallows	10484	2048
Current fallows	15753	2055
Net area sown	140022	16974
Total cropped area	192197	21745
Area sown more than once	52175	4770

Source: GoI (2011-12; http://data.gov.in/dataset-export-tool?nid=5914) (in x10^3 ha)

Forest cover: Forests are an important part of any ecosystem and contribute by providing humans with food, fodder and fuel and by enriching and aiding the nutrients to the ecosystem too. In Rajasthan, coverage of forests is 9.54% of total geographic area as compared to the country's average of 20.6%. This is because nearly 13 districts of western Rajasthan are under desert conditions. There are only two zones namely, the Humid Southern and the Humid Southern Eastern Plain that is endowed with substantial proportion under forests, i.e. one-fourth of the total reporting area is under forests. In rest of the zones it varies from 6 to 15 per cent.

Natural Vegetation: In the western region, sparse vegetal cover comprises mainly xerophytic plants, which are thorny and have stunted growth. The dominant species are *Acacias* and *Prosopis* spp. In depressions particularly in saline areas, halophytes are available. The eastern region has vegetation ranging from mixed deciduous forests to subtropical evergreen forests, but it has been adversely affected by reckless cutting and grazing. Forests in the eastern sector comprise *Anogeissus* spp. (Dhav), *Tectona grandis* (teak). *Boswellia serrata* (Salar), *Lannea comandelica* (Gurjan), *Cassia auriculata* (Aunwal), *Diospyros melanoxylan* (Tendu), *Acacia catechu* (Khair), *Azadirachta indica* (Neem), *Madhuca indica* (Mahuva), *Syzygium* spp. (Jamun), *Acacia nilotica* (Babul) etc. Among grasses, dominant species are *Apluda mutica, Cenchrus ciliaris, C. setigerus, Chrysopogon montanus, Chloris barbata, Cymbopogon martini, C. jwarncussa, Cynodon dactylon, Dichanthium annulatum, Heteropogon contortus, Iseilema laxum, Sehima nervosum, Themada quadrivelvis* etc.

Area under non-agriculture uses: This category belongs to the land put to non-agriculture uses such as residential, roads/paths, water bodies etc. The share of such land use is only about 5 percent of the reporting areas. In different climatic zone, it varies from 3% to 8%.

Barren and Unculturable Land: This category of land is considered as non-suitable for agricultural operation. About 7% of the total area of Rajasthan is categorized as barren and un-cultural waste land. In different zones it varies between 10% and 20% area. However, in irrigated North Western Plain, this category is negligible, i.e. <1%.

Grazing Lands: This is one of the most important categories of land use. The availability of permanent pasture and grazing land determines the status of livestock economy in the regions. It constitutes about 5% of the total area in Rajasthan. In irrigated North Western Plain, the grazing land is negligible, whereas in eastern parts it constitutes about 4 to 7%.

Land under Misc. Tree Crops and Groves: Area under fruit crop falls under this category of land use. In Rajasthan, it is less than 1%. In certain regions, area under fruit crop is absolutely missing.

Culturable Waste Land: This is also one of the major categories of land use covering about 13% of the total reported area of Rajasthan. Agriculture operations are possible on these lands. In Hyper - arid partial Irrigated Zone, land under this category is in substantial proportion, i.e. 40% followed by sub humid southern and Irrigated North Western Plain zones with 15% and 12%, respectively. In rest of the zones it varies from 2 to 8 per cent.

Fallow land: There are two types of fallow land, i.e. current fallow and long fallow. Lands with suspended agriculture operation for one to five years fall under this category. At the state level, it covers about 11% area of Rajasthan.

Agriculture: About half of the total reported area is under agriculture operation. The irrigated northern-western region and internal drainage dry zones leading ahead as compared to other zones in bringing larger proportion of area under agriculture that depends upon the availability and access to irrigation facilities, status of aquifer and geophysical features of the regions.

3. POPULATION DYNAMICS

3.1 Human population

Population of Rajasthan in 1951 was 15.97 million persons, which has been increased by 4.3-fold in 2011 (Fig 1.2). According to census 2011, the population of Rajasthan is 68.62 million persons with a density of 201 person km^{-2}. It is well below the national average of 324 person km^{-2}.

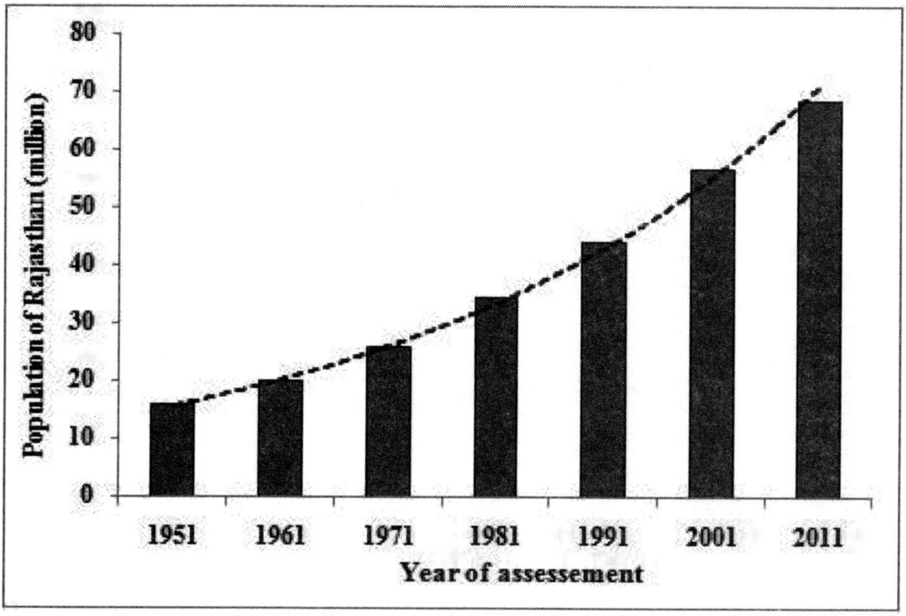

Figure 1.2 Changes in human population during 1951 to 2011 in Rajasthan

According to it Rajasthan is the eighth-most populous state in India and constitutes 5.67% of India total population. Population density ranges between 7 persons km^{-2} in Jaisalmer to 595 person km^{-2} in Jaipur. Rural population constitutes about 75.1% population of the state. Nearly 30 percent of the total population belongs to the vulnerable communities - Scheduled Castes comprising 19.2 percent and Scheduled Tribes 13 percent. Decadal growth is 21.4% during 2001-2011. However, it is observed higher in tribal and desert regions. Among the districts, the highest decadal growth is in Barmer followed by Jaisalmer, whereas it is lowest in Ganganagar followed by Pali district. The decadal rate of urban population has been observed higher (29.7%) as compared to rural population (18.9%) in 2001-2011 periods. In arid (including Ajmer now) and semiarid regions, the population densities in the respective region are 139.7 persons km^{-2} and 305.7 persons km^{-2} according to census 2011 (GoI, 2011a).

3.2 Livestock population

Livestock population in Rajasthan comprises large and small ruminants. Cattle and buffaloes are the main large animals, while sheep and goats are the major small ruminants. There is 2.3-fold increase in total livestock population between 2051 and 2012 in Rajasthan. According to estimates of Livestock Census of 2012, the share of cattle population is about 23 per cent and buffaloes constitute about 22.5% of the total livestock population. Share of cattle, sheep and camel in the total livestock population show a declining trend from 42.3%, 26.1% and 1.4% in 1951 to 23.1%, 15.7% and 0.56% in 2012, respectively (Fig 1.3). However, shares of buffaloes, goats and pigs have increased from 11.9%, 21.8% and 0.2% of the total livestock population to 22.5%, 37.5% and 0.4%, respectively during the same period. Small ruminants comprising of sheep and goats account for 52.2% of the total livestock population in the state. However, a decrease in the share of small ruminants in 2012 as compared to that in 2007 (i.e., 58%) indicates an increase in the population in favour of large milching animal.

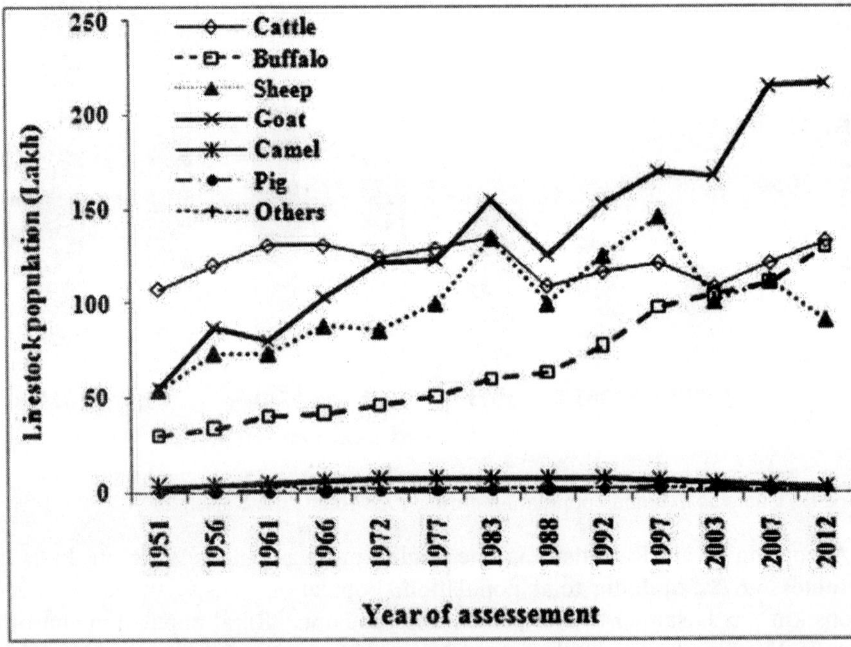

Figure 1.3 Population growths of livestock of different categories during 1951 to 2012.

4. METHODS OF OBSERVATION RECORDING

Methodology requires for biophysical forest data collection depends upon the site conditions encountered within the site and topography and available drainage line in the area. However, detailed explanations of the sampling design are necessary so that this can be repeated during revisits and therefore, allow comparison between sites. There is need of coordinates to locate the plots on the map prepared for the sampling purpose. The plots are then located on the ground using a compass and pacing the distances between the plots. The position of each plot on the ground is recorded using a GPS. A total number of 123 sacred groves have been surveyed covering all 33 districts of Rajasthan (Table 1.4).

It ranges from lowest of 3 numbers of sacred groves in most of the districts to 9 numbers in Kota district. Though a significant number of sacred groves/Dev bans/ Orans are available in the forest or in the vicinity of the forest areas, but they are relatively smaller in size. The data presented on the groves in this work cover bigger sacred groves situated outside forest area particularly in the community lands. The effective area of measurement has covered 1.25 ha (i.e., 0.25 ha × 5 plots = 1.25 ha) in each sacred grove. It is also to develop an effective management plan for the bigger groves. Positions of these sacred groves have been geo-referenced for easy access of the area in different locations (Plate 1.1A). The numbers of a particular sacred grove have been described in details in the preceding sections for their geographical position and other related information.

Table 1.4. Numbers of sacred groves studied during this programme covering different districts of Rajasthan.

SNo	District	Divisions	SNo	District	Divisions
1	Ajmer	3	18	Bhilwara	3
2	Barmer	5	19	Bundi	3
3	Bikaner	3	20	Chittorgarh	3
4	Churu	3	21	Dausa	3
5	Ganganagar	4	22	Dholpur	3
6	Hanumangarh	3	23	Dungarpur	3
7	Jaisalmer	4	24	Jaipur	3
8	Jalore	3	25	Jhalawar	5
9	Jhunjhunu	3	26	Karauli	3
10	Jodhpur	5	27	Kota	9
11	Nagaur	4	28	Pratapgarh	4
12	Pali	5	29	Rajsamand	3
13	Sikar	3	30	Sawaimadhopur	3
14	Alwar	5	31	Sirohi	3
15	Banswara	3	32	Tonk	3
16	Baran	3	33	Udaipur	8
17	Bharatpur	2	**Total**		**123**

4.1 Stratification and sample size

In general, area of the forests or sacred groves is divided into different strata or zones and then in each strata the plots of suitable size are laid out. The laying of the plots should be 20 m to more than 200 m from the forest/grove edge to minimize the edge effects and the human interferences generally occurs in the groves (Baimas-George, 2012). Number of sampling plots depends upon the number of strata and size of each stratum, though it should not be less than three. For stratification of the study area in Rajasthan, administrative forest divisions are considered as the strata and in each forest division 3-9 sacred groves are selected for the detail study as given below:

Rajasthan →Forest division → Sacred grove (3-9 sacred grove in each division).

4.2 Area surveyed and field observation design

Design of the survey and size of the plots vary depending on the conditions and resource availability. Sometimes, three concentric and nested plots of 10 m, 3 m and 1 m radius are used in data recording. In this tree, defined as woody plants of at least 10 cm diameter at breast height (DBH) data are collected in the 10 m radius plots, whereas saplings (i.e., woody plants of diameters between 2.5 and 10 cm) data are collected in the 3 m radius plots. Data on all other vegetation and seedlings are collected in the 1 m radius plot (Ostrom, 1998; Kemeuze et al., 2015). However, rectangular plots, strip plots (10 m × 50 m) or square plots (25 m × 25 m, 31.5 m × 31.5 m or 50 m × 50 m) are used in vegetation study particularly for the trees. For shrubs and tree saplings nested plots of 5 m × 5 m and for herbaceous vegetation and seedlings plots of 1 m × 1 m are used (Induchoodan, 1996; Kulkarni and Kulkarni, 2013; Singh, 2014). Further, clustered plots of 5 or 9 with nested sampling plots for smaller vegetations are also in use (ICFRE, 2013). In the study carried out in Rajasthan, all thirty three districts of Rajasthan have been covered during survey so that spatial variability in the types and structures of the vegetation in these sacred groves could be captured. For detailed study a cluster design of fives plots of 0.25 ha (50 m × 50 m) each have been followed to make observations on the tree species and their population in 1.25 ha area in total area coverage of about 5 ha of the sacred grove (Fig 1.4). However, in case of smaller size sacred grove, main plots of smaller size (20 m 20 m) with same design are followed.

This is only to have knowledge on frequency of distribution of a particular species under study and to calculate the importance value index (IVI) of each species for a sacred grove. In this design, four plots of above-mentioned area are distributed at the four corners of a central plot. Distance between the corner of the central plot and the corner of the other plots are 50 m. The smaller plots for recording observations on shrubs and herbs/grasses were nested in the above-mentioned plots (Fig 1.4). The survey study was conducted during March 2013 to February 2014.

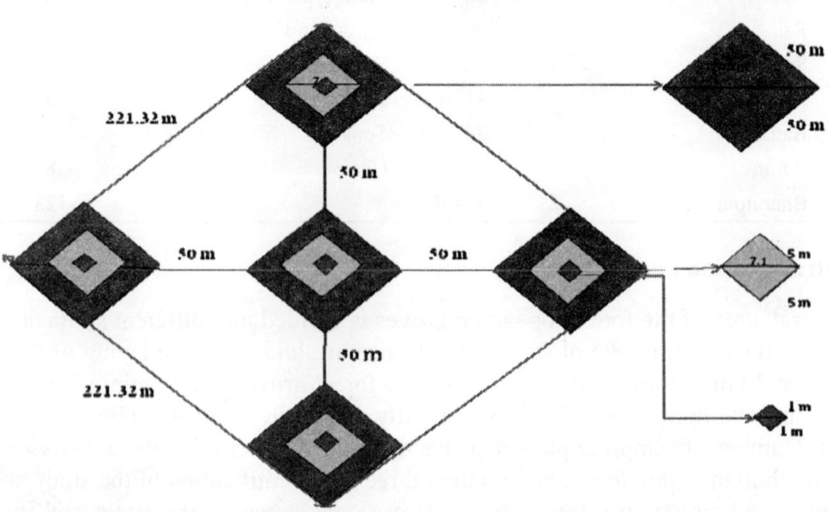

Figure 1.4. A cluster design of plots nested with the smaller plots for recording observations on trees, shrubs and herbs, respectively.

4.2.1 Vegetations and growth measurements

All trees species having girth at breast height (GBH, at 1.37 m height) >10 cm in all plots of 0.25 ha are measured for girth and total height (Plate 1.2). In each 0.25 ha plot, nested plots of 5 m × 5 m are laid out for recording growth variable of shrub species. Based on the measured diameter at breast height trees are further categorized into >30cm dbh and 10-30 cm dbh to have a knowledge about saplings recruitments. All shrubs are counted and measured for collar girth at 15 cm above soil surface and height in the plots of 5 m × 5 m size nested in the above-mentioned tree plots. Numbers of tillers are also counted and measured for their length for the shrubs. In case of less population, all shrubs are counted and measured in the plots. Observations on herbaceous vegetation and grasses available in the areas are recorded in plots of 1 m × 1m size nested in the plots laid out for recording observation on shrub species (Fig 1.4). Types of herbaceous/grass species and their population are recorded in each sacred grove depending upon the availability.

Regeneration of tree species is also recorded in the plots of shrubs (5 m × 5 m size) to determine natural regeneration patterns and changes in species compositions across mature and regenerating phases of trees and seedling recruitment of tree species under the influence of climatic conditions (Jayakumar and Nair, 2013). In this tree saplings of >10 cm tall and girth of 3 to < 10 cm are counted and measured for collar diameter and total height.

Identification of tree and shrubs species have been made on the basis of people interaction (local name), morphological and floral (if available) characters as well as by use of 'Flora of Rajasthan', flora and working plan of concerned districts and help of Botanical Survey of India (BSI, Jodhpur). Morphological characters are noted from the specimens. The plant specimens are brought to the laboratory at AFRI, Jodhpur, where microscopic characters are studied and identification of the specimen is made with the help of standard references (Hooker, 1897; Bor, 1960; Shetty and Singh, 1993; Bhandari, 1990) and related literatures comparing the characters with the other related species.

4.2.2 Mapping and general information of sacred groves

Sacred groves are identified based on the interactions with the local people as well as the forest field staff working in the districts or region. Observations are recorded on the general geo-morphological factors related to the groves and status of soil moisture conditions, litter cover, soil erosion status etc. Details regarding the existing management regimes are also recorded through interviews with the community people.

4.2.3 Recording of disturbances

Details about weed infestation, invasion by species from nearby plantations, soil erosion, relative distance from the roads, division of sacred groves by roads and railway lines, lack of protection, cattle grazing etc. have been collected. Likewise types and extent of encroachment has also been captured during the survey.

4.3 Socio-economic survey

Contribution of local socio-economy is an important aspect in maintaining sacred groves in a healthy condition. While poverty is recognized generally as a major factor hindering the conservation of sacred groves, socio–cultural and religious functions of the groves boost the tourism sector if it can be properly harnessed by the community as well as the government (Oyelowo et al., 2014). The sacred groves of India represent a long-held tradition of community management of forests for cultural reasons. To determine local attitudes toward the sacred groves, elements of sacred grove management including restrictions on resource use, as well as ceremonies associated with sacred groves there is need to use social science research methods (Ormsby, 2013). For this purpose a well designed questionnaires has been developed for interacting with the residents of the nearby area of the sacred groves and recording observations (Plate 1.3). Questionnaire is used for recording the views of the local people about general management, protection, social acceptance of the sacred groves and their opinion for developing management plan of these sacred groves. A total number of 615 people have been interacted with, where they belong to Hindu (95.4%), Muslim (3.7%) and Sikh (0.8%) religions. The respondents are also categorized depending upon number of family member and income categories, land holding size and occupation.

Maximum numbers of respondents during the survey have been observed in the family size of 6-10 persons per house hold, i.e. 55.3%. It has been followed by 5 and less persons per house hold. Greater than 10 persons per house have been recorded for only 6.5% respondents. Likewise, income categories are categorized into Rs <5001 per month, Rs 5001-10000 per month and Rs >10000 per months. In each category there have been 28%, 54.3% and 17.7% respondents, respectively (Table 1.5). This indicates that highest number of respondents fall into Rs 5001-10000 per month income group category, which is relatively low as compared to state average.

Table 1.5. Number of family members and the income categories of the respondent in sacred grove surveyed area of Rajasthan

SNo.	Family members	Frequency	Percent	Income (Rs/month)	Frequency	Percent
1.	1-5 members	235	38.2	<5001	172	28.0
2.	6-10 members	340	55.3	5001-10000	334	54.3
3.	>10	40	6.5	>10000	109	17.7
	Total	615	100.0	**Total**	615	100.0

According to land holding size, about 11.4% respondents residing in the nearby villages of the sacred groves are landless and work as labours on the farmlands of the other, migrate to nearby city or depend upon common lands for their livelihood. About 58.5% respondents are marginal in category, whereas 21.3% respondents are small farmers in category. About 8.8% respondents are of bigger in land holding size (Table 1.6). This indicates that people residing in the villages nearest to sacred groves are marginal and poor in category. The mostly depend upon common lands or the nearby sacred groves for their livelihood and use to collect fuelwood, fodder, and minor forest produce for their day to day requirements.

While considering occupation of the respondent, about 78.1% respondents are farmers, 5.7% respondents are in business and 3.6% respondents are in government service. About 9.4% respondents are labourer in category. Only 1.6% respondents are herdsmen and the rests are of mixed in categories (Table 1.7).

Table 1.6. Categories of respondents based on the land holding size in studied area of Rajasthan

SNo.	Group of land holding	Frequency	Percent
1.	Land-less	70	11.4
2.	Marginal Farmer (< 11 bigha)	360	58.5
3.	Small Farmer (11-30 bigha)	131	21.3
4.	Big Farmer (> 30 bigha)	54	8.8
	Total	615	100.0

Table 1.7. Categories of respondents based on occupation in studied area of Rajasthan

SNo.	Occupation	Frequency	Percent
1.	Agriculture	480	78.1
2.	Business	35	5.7
3.	Govt. Job	22	3.6
4.	Labour	58	9.4
5.	Agriculture and Labour	6	1.0
6.	Agriculture and shop	4	0.7
7.	Herdsman	10	1.6
	Total	615	100.0

5. CHARACTERIZATIONS OF SOIL PROPERTIES

Because of being well protected, the sacred groves are contributing towards carbon sequestration also. However, these well protected sacred groves are so far overlooked from carbon sequestration point of view (Cardelús et al., 2013). An estimate of soil carbon stocks and stock changes in tree biomass are necessary for reporting to the United Nations Framework Convention on Climate Change (UNFCCC) and may be useful for Kyoto Protocol reporting and funding under REDD+ (Olsson and Ouattara, 2013). The climatic variables like temperature, precipitation, and soil variables such as clay fraction and soil pH are all important regulators of SOC and SOC stock changes in forests, sacred groves or agro-ecosystems. There is in general positive relationship between carbon stocks and biodiversity, and both are high in tropical moist forests (Hicks et al., 2014). Likewise, relationships of SOC stocks in home garden and coffee agroforestry systems and tree species diversity are also strong, highlighting the possibilities to increase carbon stocks if proper tree species assemblies can be maintained (Hombegowda et al., 2016). Though soil samples are collected up to 1 m soil depth (Singh, 2015), the studies conducted across Rajasthan, where soil samples have been collected from the centre point of the central plot of the cluster in 0-30 cm soil layer using iron core sampler. After sampling soil samples are put in polythene bags and transported to laboratory for analysis work. Soil samples are then air dried and passed through a 2 mm mesh sieve so that stony part (gravel fraction) could be separated from the soils and actual soil fraction (fine earth fraction) could be determined as:

Soil fraction = 1- G (gravel fraction) .... Eq 1

5.1 Bulk density measurement

An iron core cutter of fixed volume placed with a steel dolly on top is hammered adequately and cautiously so that complete core should penetrate into the intact soil. Care should be taken not to rock the core cutter. Core samples are dug out taking care not to damage the core. End of the core level with the ends of the cutter is trimmed by means of a spatula and steel knives. Core samples have also been collected for 0-30 cm soil layer, respectively. In laboratory, core samples are weighed to the nearest gram and put into dry to constant weight at 110°C (for a constant weight) in a hot air Oven after putting the sample into a pre-weighed glass beaker for 48 to 72 hrs. Dry weights of the samples are recorded in g and volume of the core is recorded in cubic centimeters from its dimensions measured to the nearest 0.5 mm. Bulk density is calculated as equation provided below and it is reported as g/cm^3 (equivalent to g/cm^3) on an oven-dry basis to the nearest 0.01 g/cm^3.

$$\text{Bulk density (BD) } (g/cm^3) = \text{Mass of dry soil (g)/ Volume of Core } (cm^3) \text{....... Eq 2}$$

5.2 Soil pH and organic carbon

Soil pH, organic carbon and other soil nutrients vary with dominance of species in a particular grove, diversity extent of degradation etc. For example, Oak forests are characterized by high SOM, soil moisture, water holding capacity, and soil fertility (Gosain et al., 2015), thus management options can be designed accordingly. Soil pH has been monitored in 1:2 soil-water ratio, whereas organic carbon (SOC) has been determined using standard procedures (Jackson 1973; Walkley and Black, 1934). Walkley-Black method (1934) of soil organic carbon (SOC) determination provides an accurate estimate of SOC with 100% recovery (Wang et al., 2012) and has been used in carbon analysis. In this potassium dichromate ($K_2Cr_2O_7.5H_2O$) is used as the oxidant and ferrous ammonium sulphate ($FeSO_4$ $(NH_4)_2SO_4.6H_2O$) as the reducing agent (IPCC, 2007).

$$\% \text{ organic carbon} = 1.334 \times [(V_{blank} - V_{sample})/Wt] \times 0.003 \times M \text{ .... Eq 3}$$

Where V_{blank} is volume of ferrous ammonium sulfate required to titrate the blank (ml), V_{sample} is the volume of ferrous ammonium sulfate required titrating the sample (ml), Wt is weight of the soil sample (g), and M is the molarity of ferrous ammonium sulfate solution.

5.3 Soil nutrients

Available nitrogen (NH_4-N and NO_3-N) is determined using different UV spectrophotometer like Model Shimadzu-1650PC in this study after 0.5 M K_2SO_4 extraction. Extractable phosphorus is determined by the Olson's extraction method (Jackson, 1973) and by using above-mentioned UV spectrophotometer.

5.4 Data calculation and Statistical analysis

5.4.1 Diversity variables and forest type assessment

The diversity variables like number of species, species richness in terms of Menhinick diversity index, Shanon-Weiner Index of diversity (H'), species evenness (*e*') and species dominance (D) have been calculated following standard literatures

(Magurran, 1988; Shannon and Weiner, 1963; Simpson, 1949; Pielou, 1966; Misra, 1968). Importance value index (IVI) is calculated as below to know the dominant trees/shrubs and to identify the dominant species either trees or shrubs in a particular sacred grove.

Importance Value Index (IVI) = RD + RF + RDom .... Eq 4

Where RD is relative density, RF is relative frequency and RDom is relative dominance.

Simpson's Index (D) measures the probability that two individuals randomly selected from a sample will belong to the same species. Formula for calculating D is as below:

$$D=\Sigma\ (n/N)^2 \quad \text{.......... Eq 5}$$

Where n is the total number of a particular species and N is the total number of all species. The value of D ranges between 0 and 1. With this index, 0 represents infinite diversity and 1, no diversity. That is, the bigger the value of D, the lower the diversity. This is neither intuitive nor logical, so to get over this problem, we calculated Simpson's Reciprocal Index, i.e. 1/D. The value of this index starts with 1 as the lowest possible figure, i.e. a community containing only one species. The higher the value, the greater is the diversity. The maximum value is the number of species in the sample. For example if there are five species in the sample, then the maximum value is 5. Likewise the number of equally-common species that give a particular value of an index is called the "**effective number of species**" (ENS). This is the true diversity of a community in question and calculated in general using Shannon-Wiener index and known also as Shannon entropy (Table 1.8). For examples, ENS is exponential of Shannon-Wiener index.

Table 1.8. Conversion of common indices to true diversities. (Source: Jost, 2006; Chao et al., 2014).

Index	Formula	Diversity in terms of x:	Diversity in terms of p_i :
Species richness	$x \equiv \sum_{i=1}^{S} p_i^0$	x	$\sum_{i=1}^{S} p_i^0$
Shannon entropy	$x \equiv -\sum_{i=1}^{S} p_i \ln p_i$	exp(x)	$\exp(-\sum_{i=1}^{S} p_i \ln p_i)$
Simpson concentration	$x \equiv \sum_{i=1}^{S} p_i^2$	1/x	$1/\sum_{i=1}^{S} p_i^2$

5.4.2 Endemic species population and regeneration analysis

The Red list of threatened species, prepared by the International Union for Conservation of Nature (IUCN), has listed 132 species of plants and animals as Critically Endangered, the most threatened category, from India. Plants appeared to be the most threatened life form with 60 species being listed as critically endangered and 141 as endangered. Trees and shrubs are categorized into different categories of IUCN Red data list, criterion of which are provided below:

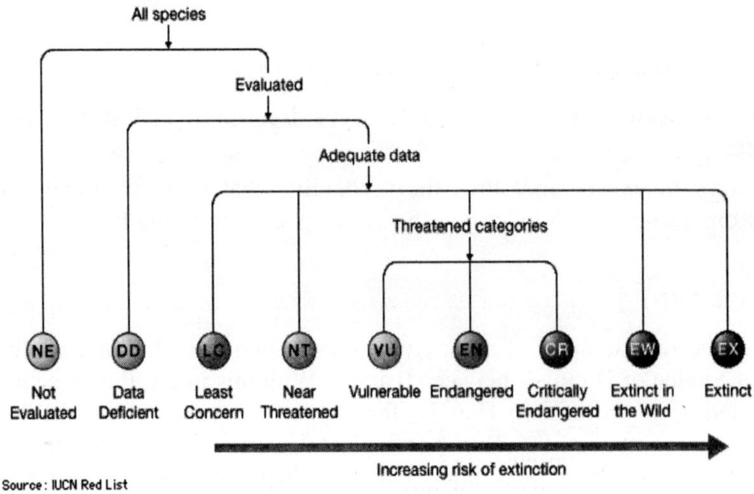

Source: IUCN Red List

5.4.3 Calculations of soil carbon density and stock

For calculation of oil organic carbon (SOC) or soil inorganic carbon (SIC) stock or carbon stock per unit area, there is need of quantity of carbon available in a particular soil sample, volume of the unit under calculation like length × width × soil depth and bulk density of the soil under investigation. Moreover these soil carbon stock are calculated using the following equation (Batjes, 1996):

$$Qi = CiDiEi\,(1-Gi) \times 10000 \quad \text{.......... Eq 6}$$

Here Qi (tons or Mg C ha^{-1}) is soil organic/inorganic carbon content in a soil layer i, Ei is soil depth in meters), Ci is carbon content in g C g^{-1} soil, Di is bulk density in Mg m^{-3} and Gi is volume fraction of coarse (gravel or stones of >2 mm size) elements defined as SOC in preceding sections.

5.5 Statistical Analysis

Data are analyzed statistically using different statistical package like SPSS version 8.0 for accuracy of the data. Since the data on soil samples for bulk density, SOC, soil nutrients and growth parameter and diversity variables have been obtained from 103 sacred groves covering 33 districts of Rajasthan, these data have been analyzed using one way ANOVA considering dominant species as the main factor and above-mentioned parameters as the dependent variables. Wherever necessary the data are transformed to reduce heteroscenesdity (Sokal and Rolf, 1981). Duncan Multiple Range Tests (DMRT) have also been applied to group homogeneous subsets of districts at the $P < 0.05$ levels. To obtain relations among rainfall, SOC and biomasses/carbon in different components of the ecosystem, Pearson correlation coefficient is calculated. Regression relations are also observed to relate rainfall, vegetation diversity, soil carbon stock and soil properties to draw inferences about the trend of these data

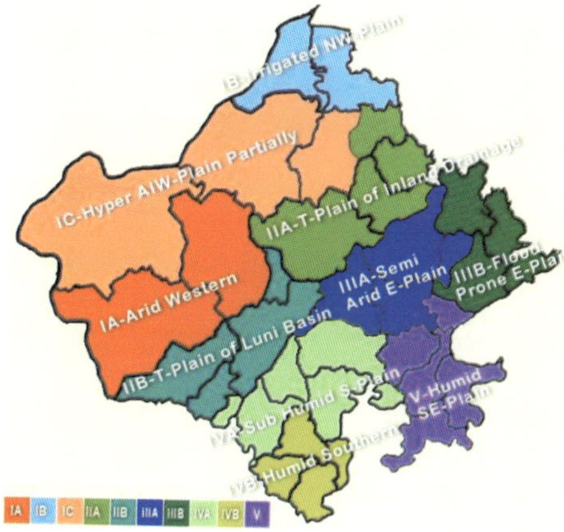

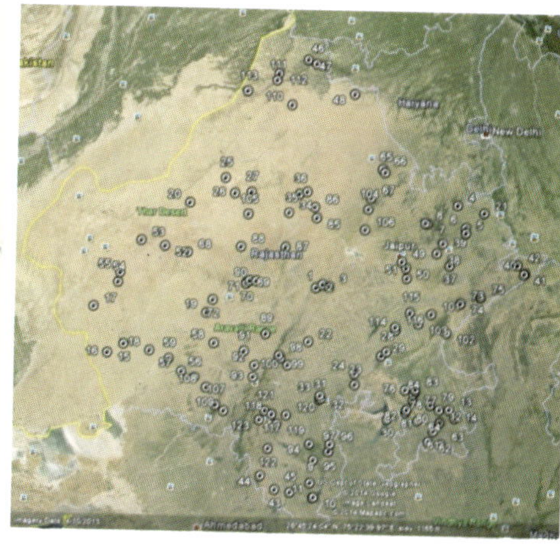

Plate 1.1. Different agro climatic zones (left) and distribution of sampling plots in Rajasthan (right).

Plate 1.2. Measuring *A. nilotica* (left) and *C. pendulus* (right) of 98 cm girth on *P. cineraria*.

Plate 1.3. People interaction (left) and stem growth of an old *Clerodendrum phlomidis* (right)

2

SACRED PLACES, TREES AND GROVES: AN OVERVIEW

The sacred is the object of veneration and admiration. It is a Latin word '*sacer*' means restricted or set off. A place, a person, a tree or an object may be designated as sacred, which is regarded as extraordinary or unique. The holy places are not only simply geographically beautiful but are more holy than shrines. These places are not commercialized rather sacred to whom the greatest respect is shown. They are used for the generation of our people, the continuum of our life: because that's where we begin and that's where we return (Hubert, np). Among primitive people the concept of nature as a totality was unknown and only individual natural phenomena like stars, rain, and animals- have been comprehended as natural objects or forces that influence them and are thus in some way admirable of being venerated or conciliated.

1. SACRED PLACES

There are sacred places or area too that encompass various individual sites and phenomena as an integral part of the whole or a sacred landscape (Hemmat et al., 2012). These are springs, caves, meadows on a mountain, rivers or its origin place like Gangotri and Jamunotri in Uttarakhand, Amarkantak in Madhya Pradesh etc. Sites connected by a river, legends or stories, the histories of individuals or groups, and/or a pilgrimage routes like the centuries old Way of Lord Rama toward Lanka or similar other are also considered sacred. These sacred natural sites are important areas for environmental conservation (Lee and Wauchope, 2003). Certain sacred places have persisted for centuries or even millennia and are attracting thousands or even millions of pilgrims and visitors. Mention to some are Bodh Gaya in Bihar, Ajanta, Ellora and Elephanta caves in Maharashtra, Amritsar in Punjab, Haridwar, Allahabad, Nasik and Ujjain are famous for hosting the holy Kumbh fair in different parts of India, where people make holy dip; whereas Ramdeora in Jaisalmer, Pushkar in Ajmer, Baneshwar Dham in Banswara in Rajasthan are other sacred sites. Sita Mata Sanctuary in Pratapgarh district and Ghotiya Amba in Bansawra have some mythological religious relations with 'Lord Rama'. While Sita Mata sanctuary is associated with Maharshi Balmiki Ashram and the birth place of Devi Sita's two sons 'Lav' and 'Kush", Ghotiya Amba is associated with Rama, who visited Ramakunda as well as Pandava, who visited Bhim Kunda during their exile, where the latter has used a tunnel as their passage during rainy season that connect Bhim Kunda with Ghotiya Amba. In contrast, there are sacred places where humans are excluded or access is strictly limited to a special class of individuals such as ritual specialists, healers, or elders. Sacred places are considered as a complex phenomena that can be viewed usefully as varying along several continua ranging from natural to socio-cultural; prehistoric to historic, recent, or newly created; secret or private to public; single culture (or religion) to multicultural (or multi religious); intrinsic to extrinsic in value; uncontested to contested; and protected to endangered.

2. SACRED PLANTS

Plants are the medium that connects the living and nonliving things in the environment. There exists a unique relationship between trees and human beings (Choudhary et al., 2008; Gorshunova, 2008; Donovan et al., 2013). For example both trees and human produce the gasses that enable the other to exist. Trees produce oxygen as byproduct that we need to breathe, and we produce carbon dioxide which trees utilize to make food, i.e. carbohydrates. Trees have played important roles in many of the world's mythologies and religions, and are given deep and sacred meanings throughout the ages. Trees are symbols of physical and spiritual nourishment, transformation and liberation, sustenance, spiritual growth, union and fertility. In addition to the places of birth and death, trees are also used as sacred shrines and places of spiritual pilgrimage, peace, ritual, ceremonies and celebrations (Dafni, 2007; Fowler-Smith, 2009). Vatsyayan (1992) informs the sacredness of trees seen in every part of India. Deodar (*Cedrus deodara*) is considered the abode of the gods, Sal (*Shorea robusta*) is venerated in Uttar Pradesh, Bihar and Madhya Pradesh and Coconut (*Cocos nucifrea*) in south India. Similarly Rudraksha (*Elaeocarpus ganitrus*), Bel (*Aegle marmelos*), Ashok (*Saraca asoka*), Kadam (*Anthocephalus chinensis*), Pipal (*Ficus religiosa*), Kalpavriksha (*Adansonia digitata)* and 'Sami' or 'Khezri' (*Prosopis cineraria*) are considered sacred in Rajasthan. The best example in Rajasthan is worshiping of two revered trees for more than 800 years old, i.e. 'Kalpavriksha' (*Adansonia digitata*) - a wish-fulfilling divine tree in Hindu mythology situated at Mangaliyawas near Ajmer, Rajasthan on an Amavashya (new moon night) day in the Hindu month of Shravan (Plate 2.1). There is belief that sage '*Durvasa*' meditated under the Kalpavriksha and Lord Shiva's daughter Ashok Sundari was created from Kalpavriksha tree by '*Parvati*', to alleviate her loneliness.

Banni (*Accacia ferruginea*) is also considered sacred because it was in one such tree that Arjuna, as stated in the Mahabharat, had hidden his Gandiva bow and arrows while in his one year *agyatavasa*, (one year of total anonymity), when the Pandavas were exiled to the forest (Murthy, 2006). Sacrifice of life in protecting trees can only be observed in the history of Rajasthan, when the King of Jodhpur ordered his soldiers to cut the 'Khezri' (*Prosospis cineraria*) tree in village Khezadali in 1730 AD. The soldiers together with the workers did not accept the goddess's interdiction, and began to cut the trees anyway, telling the villagers: "If you want to save the trees, then pay a tax; if not we will cut the trees." In response Amrita Devi - a lady said in her rhymed couplet:

Dam diya, dag lage, tukro na dev a, dan;
Sir sante runkh rahe to bhi sasto jan.

With this rhyme the villagers wrapped themselves around the trunks of the trees, and the labourers with their axes chopped them down. In this way, one by one, 363 people were martyred in order to save these trees. Other examples of such faith are the alpine plants *Saussurea obvallata* and *Delphinium vestitum* locally called Dongar and Loshkar, which have been protected through local customs and beliefs by linking these plants with local deity and not allowing any one from the village or outside to pluck the flowers except the village priest during specific day and time only (Meenakshi et al., 2011). Sacred trees are not only related to Hindu tradition but also to Shamanic, Egyptian, Sumerian, Toltec, Mayan, Norse, Celtic and Christian traditions and thus elaborate mythology of trees exists across a broad range of ancient cultures throughout the world (Table 2.1). The World Conservation Monitoring Centre (1982) states that the tradition is thought to date back several thousand years or at least the pre agrarian period of the hunter-gatherer societies. Being largest plant on earth, the tree has been a major source of stimulation to the mythic imagination and have been conferred on in all cultures with a dignity unique to their own nature, and tree cults, in which a single tree or a grove of trees is worshipped, have flourished at different times almost everywhere in the world including Europe and Asia (Gorshunova, 2008). Christianity and Islam treat the worship of trees as idolatry and this lead to their destruction in Europe and most of West Asia.

Table 2.1. Some important sacred trees of the world and belief behind these trees.

Local name	Botanical name	Belief
Willow Tree	*Salix alba*	In Japanese tradition, it is believed that wherever a willow is present, a ghost will appear.
Ash tree	*Fraxinus* spp.	It is associated with healing and enchantment, and in particular to the Welsh magician/god Gwyddion, who had an ash wand. This tree is also associated in Norse mythology to the god Odin.
Sacred Oak	*Quercus muhlenbergii*	Native Indian legend claims that the tree possesses the power to heal. It is looked upon as the shrine tree of the Delaware Indians, who use to pray for it whenever they need help.
Italian Cypress	*Cupressus sempervirens*	Known as "The Mournful Tree", and has been associated with death and mourning for the past 2,000 years.
The Bodhi Tree	*Ficus religiosa*	Lord Buddha had gained his enlightenment under this tree, which became a symbol of his presence and an object of worship after Buddha's death
Witch Tree	*Thuja occidentalis*	Sacred among the Chippewa Indian tribe. People are now prohibited to come near the tree.
Banyan Tree	*Ficus benghalensis*	Due to its shape, the tree represents eternal life. In 'Bhagavad Gita' Krishna uses the banyan tree as a symbol to describe the true meaning of life to 'Arjuna'.
Lebanese Cedar	*Cedrus libani*	It holds a religious Christian significance and mentioned in the bible 75 times and a symbol of national pride of Lebanon.
Baobab Tree	*Adansonia digitata*	Due to its massive size, fascinating shape, and long aging, people believe that baobab holds the spirits of the dead and magical properties, and help in making wise decisions.
Christmas Tree	*Abies alba* 'pyramidalis'	The most renowned sacred tree and was known as Yule tree in pagan tradition and has been used in decoration as an outdoor live tree with hanging candles.

India is country in the world, where many communities practice different forms of worship of nature (Gold and Gujar, 1989). Worship of stones and hills, worship of plants and trees, worship of animals, worship of the elements, worship of the heavenly bodies and worship of man are some of them (Shah and Patel, 2011). Tradition of nature worship by providing protection to trees or patches of forests designated as sacred groves dedicated to deities or ancestral spirits is date back to third and fourth millennia BC, i.e. during highly evolved Harappan Civilization (Bhatla et al., 1984; Haberman, 2013). One such example is carving of stylish *Pipal* tree (*Ficus religiosa*) with two heads of unicorn emerging from its stem on the seals of Mohenjodaro. Even during Vedic period, tree worshiping has occupied a very important place, where Lord *Soma* has been described as the presiding deity of trees. Hindu religion is replete with the events of celebration involving several trees and herbaceous plants (Table 2.2). Many of the trees of different species have special associations with particular deities. For example, the Lord Vishnu is associated with the 'Pipal' (*Ficus religiosa*), 'Bargad' (*Ficus benghalensis*), and 'Gular' (*Ficus glomerata*); the Lord Shiva with 'Bel' (*Aegle marmelos*) and 'Maulashri' (*Mimusops elengi*); the Lord Dattatreya with 'Gular' (*Ficus glomerata*); 'Bel; (*Agele marmelos*) and 'Rudraksha' (*Elaecarpus ganitrus*) are associated with Lord Rudra (an incarnation of Lord Shiva). *Acacia ferruginea* is the most feared and respected tree because it represents the dangerous planet Saturn, and Agni, the powerful fire god.

Table 2.2. Religious festivals in India and the associated sacred trees.

Festivals	Month of festival	Species associated
Maha Shivaratri	February	*Aegle marmelose*/ *Zizyphus* spp. *Datura* spp.
Sheetla Asthami	March	*Azadirachta indica*
Dasha Mata	March/April	*Prosopis cineraria*
Neem Saptami	April	*Azadirachta indica*
Vat Savitri	May	*Ficus benghalensis*
Bilva Mangal	May – June	*Aegle marmelos*
Sawan ke Somvaar	Mid July – Mid August	*Aegle marmelos*
Kadii Vrat	September	*Musa paradisica*
Jeevit Putrica	September/October	*Sida cordata*
Aonla Ekadashi	October/November	*Emblica officinalis*
Chhauth Vrat	November	*Saccharum officinarum*
Somvari Amavasya	15th of all months	*Ficus religiosa*

Every festival in India is virtually associated with a plant species and foods derived from different species. A harvest festival is an annual celebration that occurs around the time of the main harvest of a region. Given the differences in climate and crops around the world, harvest festivals can be found at various times at different places. Annual cherry blossom festivals in Japan or Hanami are a special feature of Japanese life, when all ages spend time outdoors, enjoying the beauty of the cherry blossoms by day and by night, with their family, friends, and workmates. Festival activities include eating seasonal foods, such as bamboo shoots, rice cakes with red beans, and wild vegetables, playing games, listening to musical instruments, and singing (Primack and Higuchi, 2006). A harvest festival 'Baisakhi' is celebrated on the thirteenth day of April according to the solar calendar, the day marks the beginning of the Hindu solar new-year in India. In fact this day is celebrated all over the country as new-year day under different names. It is celebrated in North India, particularly in Punjab and Haryana, when the 'Rabi' crop is ready for harvesting. This can be seen as one of the ways of celebrating biodiversity of nature. Use of Amaranth and Buckwheat on the occasion of observing a fast is common practice in India. This reveals that our ancestors knew for ages that Amaranth (*Amaranthus* spp.) and Buckwheat (*Fagopyrum esculentum*) are not cereals. For example, out of the 34 plant species considered sacred by the Dimasas community in Assam, India, 13 species are related to worshiping, 21 species involved in naming of areas/villages and eight species are associated with the naming of the Clans (Medhi and Borthakur, 2013).

3. SACRED GROVES

A community based repositories of biological diversity sacred groves are segments of landscape with typical geographical features. It is virtually a patch of forest or a clump of trees associated with other forms of life, and affords protection on the basis of religious practice or faith (Colding and Folke, 2001). The groves are dedicated to a deity God, Goddess and it's a taboo even to cause a simple damage to life in the grove area. The groves are considered to be the fragments of original ecosystem. They constitute unique examples of *in situ* conservation of genetic resources and serve as refuge for many threatened and endemic species (Pandey, 1999; Ormsby and Bhagwat, 2010). Most often felling of trees and cutting branches or twigs from the groves is strictly prohibited. Even lifting of leaf litter is not allowed. Many a time the groves are also associated with cremating or burial grounds.

Hughes and Chandran (1998) have presented an overview of the distribution of sacred groves around the world. This indicates that sacred grove concept of biodiversity conservation has been adopted by various indigenous communities worldwide, like aboriginals of Australia, Caucasus Mountains community, ancient Slavic people, German tribes (Tokarev, 1989), Greek and Romans, Kikuyu of Africa (Huges, 1990), and Mbeere tribe of East Africa (Gowda, 2006). Before the spread of Christianity and Islam the sacred groves had covered much of the Middle East and Europe. Frazer (1915) has also provided illustrations of sacred groves which were earlier prevalent or are so even now in Europe, India, Australia, America, Africa, China and Japan. Among the Celts, the Oak-worship of the Druids is quite well known. At Uppsala, the old religious capital of Sweden there is a sacred grove in which every tree is regarded as divine. Among the Finnish-Ugrian tribes, heathen worship is used to take place within sacred groves, which are fenced and protected. The African continent-predominated by the tribal communities is also replete with sacred forests and sacred groves (Little and Brokensha, 1987; Castro, 1990; Byers et al., 2001; Campbell, 2005; Sarfo-Mesah et al., 2010).

4. SACRED GROVES IN INDIA

Sacred groves have existed in India since time immemorial and India has the highest concentration of sacred forests in the world. Agglomeration of over 40,000 endogamous groups together with heterogeneity in the Indian society in terms of religious beliefs, customs, taboos, culture, language and livelihood availability of sacred groves vary widely in size and composition (Malhotra, 1998, Malhotra et al., 2007). Further, high flexibility in Hinduism and potential of absorbing numberless local cults including many tribal ones, instead of vanquishing them as done by other major religions, a large number of sacred groves are still available in India. These sacred groves are patches of densely wooded segments of landscape containing vegetation and other forms of life and geographical features that are delimited and protected by human societies to keep them in a relatively undisturbed state on religious grounds (Kosambi, 1962; Chandran et al., 1998; Malhotra et al., 2001; Anthwal et al., 2010). These sacred groves have been dedicated by local communities to their ancestral spirits or deities. These groves are protected by local communities, usually through customary taboos and sanctions with cultural and ecological implications. It consists of a multi-species, multi-tier primary forest or a cluster of trees, depending on the history of the vegetation in the region. Sometimes a grove is composed of single tree species, whereas a single tree is also considered as a grove at places because of its size and sacredness in the region, i.e. a tree of *Ficus bengalensis*. Existence and importance of sacred groves have also been described by the first Inspector General of Forests in India, Sir D. Brandis (1897) as:

"Very little has been published regarding sacred groves in India, but they are, or rather were, very numerous. I have found them in nearly all provinces. An instance I mention the Garo and Khasi hills...... the Devara Kadus of Coorg and all the hill ranges of Salam district in Madras Presidency.... In the dry region sacred groves are particularly numurous in Rajputana...... In Mewar they usually consist of Anogeissus pendula......in Pratapgarh and Banswara....... the sacred grove , here called Malwan, consist of a variety of trees.....These.....as a rule, are never touched by the axe, except when wood is wanted for the repair of religious buildings........"

Existing information in terms of association of sacred groves with different ethnic groups indicates that the sacred groves are found among both tribal and non-tribal communities indicating variation at the regional level in terms of ethnic association. However, the association of sacred groves with castes of different Varnas is still unclear. Likewise, in the states like Bihar, Madhya Pradesh, Orissa, West Bengal, etc., where both tribal and non- tribal exists, the existence of groves in the non-tribal areas is also not clear. In fact different deities are worshipped in the sacred groves and each grove is named after the deity dwelling in the respective sacred grove. A total of 19 deities have been observed worshipping by the local people in sacred groves in Panchmarhi Biosphere Reserve, Madhya Pradesh, India (Kala, 2011). More important is that various traditional customs associated with these sacred groves have also been reported earlier from the Himalayas, North-east India, highlands of Bihar, Orissa, Madhya Pradesh, Andhra Pradesh, Karnataka, Tamil Nadu and Kerala etc (Kala, 2005; Bhakat and Sen, 2008; Jain, 2011).

According to an estimate there are numbers of sacred groves lying between 100,000 and 150,000 covering about more than 100, 000 ha area (Malhotra et al., 1999; Kala, 2010a & b) as compared to the documented ones (Table 2.3). These sacred groves are located in a variety of habitats ranging from resource rich forested landscape of Western Ghats and North-East to the extremely resource poor desert conditions in north-western and central India. In Himalayan region, Himachal

Pradesh has largest number with about 5000 documented sacred groves (Sharma, 2000). East Khasi Hill district of Meghalaya and Karbi Anglong district in Assam in north east harbour substantial number of sacred groves of varying size. These districts, both inhabited by indigenous tribal populations represent different contexts of conservation of sacred groves in the region (Khiewtan and Ramakrishna, 1989; Ramakrishnan, 1998 & 2001; Sinha and Maikhuri, 1998; Tripathi et al., 1995; Jamir and Pandey, 2002). In Mizoram, each village has its own sacred grove, though their size varies from village to village. The Western Ghat through the states of Maharashtra, Goa, Karnataka, Tamil Nadu and Kerala - one of the world's biodiversity "hotspots" have substantial numbers of sacred groves, where Karnataka alone have nearly 1,500 documented sacred groves. Though highly populated, population of sacred groves is also high in Uttar Pradesh, where Singh (2011) has reported 32 sacred groves in Devipatan region only. Tree cutting or other resource extraction is strictly prohibited in smaller groves, while larger groves often function as "resource forests," in which people are allowed to gather products for sustenance. There are 2,000 well documented sacred groves in Kerala, where more than hundreds are dedicated to snakes, which protect agricultural crops by controlling insect and rodent populations. Rajasthan is another region, where numbers of sacred groves are more than the number of villages (Chaudhary et al., 2002; Krishna, 2013). During exhaustive survey of common resources that includes Oran, guachar and other in Jodhpur district with 1846 villages indicates 3278 number of Orans, gauchar, Agore and Parat Bhumi in combine and 3497 numbers of water bodies (Anon, 2004). However projected number of sacred groves in Rajasthan is 25000 based on the ratio of 1027 recorded groves and number of villages of Rajasthan (Pandey, 2000).

Table 2.3. Documented sacred groves in India with their local names and sources.

SNo.	State	Local name	Nos	References
1	Andhra Pradesh	Pavithravana	761	WWF -A.P. (1996); Naidu et al. (2015)
2	Arunachal Pradesh	Gumpa Forests	65	Chatterjee et al. (2000); Chakrabarti, (2010)
3	Assam	Than, Madaico	40	Chakrabarti (2010)
4	Bihar	Sarana	-	-
5	Chhattisgarh		600	Patnaik and Pandey (1998); Pandey (2000)
6	Delhi		1	Baviskar (2014).
7	Goa	Devarais	55	Kakar (2013)
8	Gujarat	Deorai, Pann	65	Gupta et al. (2000); Patel and Patel (2013a&b); Patel et al. (2014)
9	Haryana		248	NAEB (1995)
10	Himachal Pradesh		5000	Sharma (2000)
11	Jammu & Kashmir		96	Kumar et al. (2011); Gupta and Sharma (2014); Sharma (2014)
12	Jharkhand	Sarana	21	Jayapal et al. (2014)
13	Karnataka	Devara Kadu	1424	Kalam, 1996, Gokhale, 2000

SNo.	State	Local name	Nos	References
14	Kerala	Kavu, Sara Kavu	2000	Rajendraprasad (1995)
15	Madhya Pradesh		275	Srivastava (1994)
16	Maharashtra	Devrai, Devrahati, Devgudi	2808	Deshmukh et al., 1998
17	Manipur	Gamkhap, Mauhak	365	Chakrabarti (2010).
18	Meghalaya	Ki Law Lyngdoh, Ki Law Kyntang, Ki Law Niam	79	Chakrabarti, (2010); Tiwari et al. (2013)
19	Mizoram	Mawmund, Ngawpui	Many	Tiwari et al. (2013)
20	Orissa	Jahera, Thakuramma	322	Malhotra et al. (1997); Debabrata et al. (2015)
21	Puduchery	Kovil Kadu	15	Sambandan and Dhatchanamoorthy (2012)
22	Rajasthan	Orans, Kenkris, Jogmaya	3164	Gold and Gujar (1989), Singh and Sisodia (2003).
23	Sikkim	Gumpa forests	56	Chakrabarti (2010)
24	Tamil Nadu	Swami shola, Koilkadu	448	Amrithalingam (1998)
25	Uttarakhand	Deo Bhumi, Bugyal	55	Gokhale and Pala (2011)
26	Uttar Pradesh	-	36	Singh (2011), Garg and Singh, (2013); Malik (2015)
27	West Bengal	Garamthan, Harithan, Jahera, Sabitrithan, Santalburithan	670	Deb et al. (1997)
	Total		18669	

The sacred groves in India are known by different names at different places like '*Devray*' in Maharastra, '*Devarkand*' and '*Siddarvanam*' in Karnataka, '*Orans*', '*Kenkari*', '*Malvan*' and '*Yogmaya*' in Rajasthan and '*Saranya*' in Bihar (Ramchandran, 1999; Joshi and Shrivastava, 2000), and *Jaher* (Xaxa, 1991) in tribal region of Jharkhand and Orissa. There is no generic name for sacred groves in Haryana, though there are about 248 sacred groves in Kurukshetra district (Silza et al., 2008). In Mizoram, sacred groves are called Ngawpui. Each village has its own sacred grove of varying sizes. No tree is allowed to be cut in the Sacred Grove, except for those trees that show signs of age and decay.

5. TYPES OF SACRED GROVES

Sacred groves have been categorized depending upon the deities, types of protection or sacredness like Mandir groves, Samadhi groves, Ashram groves and Tirath groves (Yadeo et al., 2010). Moreover, there are three types of sacred groves and these are Dev van or bani, Kaker bani and Oran. Pandey and Singh (2000) have divided sacred groves of the southern Aravalli ranges and Vindhyanchal ranges into three major categories.

The first type of sacred groves are developed and managed by tribes, and are located in forests, near streams or on hills and is known as Dev Bans. The second type is devoted to Shankara. They are located in watershed areas. The third type consists of single trees like banyan (*Ficus benghalensis*), pipal (*Ficus religiosa*), and so on. Institutional structures in management of these sacred groves have been used to assure by linking them to religious sentiments. However, sacred groves are also categorized based on their legal status and management, which vary among regions and individual villages. Some groves are associated with and managed by separate caste groups within a community, some by a village as a whole, and some by neighboring districts within a larger geographical area. There are also large "pan-Indian" groves that involve people from many parts of the country. In some groves, all forms of resource extraction are strictly prohibited, while in others people may collect material such as fallen branches and leaves from the forest floor or fruit from the trees. Most of the sacred groves fall under following three categories:

Control of state forest departments;

Control of revenue and other government departments; and

Privately owned sacred groves.

The unique management and ownership of the sacred groves is one of the significant reasons to make them so special. In most cases, the sacred groves are owned and managed by local communities, and sometimes by one or more families or clans (Ormsby, 2011; Ormsby and Ismail, 2015). Most sacred groves like that in Maharashtra are either under the control of the Forest Department and/or under the control of the Revenue Department (Godbole et al., 1998). Many groves associated with temples in western Maharashtra have been brought under the scrutiny of the government by forming the Paschim Maharashtra Deosthan Prabodhan Samiti in 1960s (Gadgil and Vartak, 1976). In Rajasthan, Orans are managed by Grama Panchayat, whereas sacred groves in Meghalaya are under the control of District Councils (Tiwari et al., 1998). However, management of sacred groves are also done by temple committee consisting of members of different villages, i.e. Haryali grove in Garhwal (Sinha and Maikhuri, 1998), by clan elders, i.e. Mahadeo Kolis of Pune district, village elders, i.e. Kunbis of Kolhapur district, two clans of the Proja Tribe in Kantabanshini Thakurma sacred grove in Koraput district, Orrisa. Clan-based management is widespread practice among the Santhal, Oraon, Milnda, Kharia and other tribes of central, eastern and north-eastern India. In Kerala, several sacred groves are owned by a family (i.e, Ollur Kavu sacred grove), a group of families (i.e., S.N. Puram Grove), a caste, or a trust (i.e., Iringole Kavu sacred grove) (Chandrashekara and Sankar, 1998). It is clear that encroachments are not permitted or tolerated rather it can only be used for open grazing; there is no commercial use of the resources extracted from the grove; dry wood extraction is permitted only for religious function in the grove; green felling, open toilets and hunting in these groves are totally not allowed. The guilty party is punished and has to provide grain at the local village court and is also fined with a sum of money.

6. IMPORTANCE OF SACRED GROVES

Various live forms in sacred groves serve a variety of ecosystem functions including biodiversity conservations, removal of atmospheric pollutants, oxygen generation, microclimate regulation, stabilization of soil, ground water recharge, prevention of soil erosion and carbon sequestration. Likewise scared groves possess a great heritage of diverse gene pool of many forest species having socio-religious attachment and possessing high values including medicinal ones (Anderson, 1996; Sharifi et al., 2000; Joshi and Sringi, 2014). For example, 78 species in two small sacred groves in Kuchh region of Gujarat has been recorded (Patel et al., 2014). In this region about 75% of tree species of arid region of Gujarat state are available (Patel and Patel, 2013a & b). Major services provided by these sacred groves are: provisioning services focusing on directly supplying food and non food products; regulating services, which are obtained from regulation of ecosystem processes; supporting services, which provide support to habitats and ecosystem functioning; and cultural and amenity services relating to recreation and human inspiration (Table 2.4).

Table 2.4. Various services provided by the sacred groves.

Services	
Provisioning services	➢ Soil and water conservation
	➢ Water supply for drinking and irrigation
	➢ Sources of NTFP (medicinal plants, fruits, fuel wood, fodder etc.)
Regulating services	➢ Maintenance of hydrology and water flow regulation
	➢ Mitigation of natural hazards, i.e. pollutants
Supporting services	➢ Carbon sequestration and climate change mitigation
	➢ Pollination and seed dispersal
	➢ Amelioration of microclimate for flora and fauna
Cultural and amenity services	➢ Religious, cultural tradition and ethnic identity
	➢ Landscape aesthetics, research and education

The main economic benefits of sacred groves are: place of grazing for the livestock, (ii) place of drinking water, (iii) resting places, and (iii) medicine in ethno botanical form; whereas the main environmental benefits are: refugia for bio-genetic diversity, repositories of ethno-social codes of relation and regulation, venues of local and universal manifestation of aesthetic tradition and religiosity, and community resources, commons or cultural space between private domain and the rest of cosmos.

6.1 Religio-cultural

Most of the sacred groves are associated with certain deities. Villagers use to perform annual rituals and ceremonies to appease the presiding deity and ensure the well-being of the community. Sacred groves help to define the cultural identity of the communities that revere and protect them. During these rituals sacrifices of birds and other animals are being done. However, in sanskritized groves offerings of vegetable, fruits or sweats are made. These rituals are performed for the well-being of the people, animals, crops, etc. The presiding deities are believed to look after the well-being of the people, and also protect the groves by administering punishment (mostly death) to the offenders (Sisodia and Malhotra, 1963; Kalam, 1996; Roy Burman, 2014). People take vows for wish-fulfillment when there is a crisis, particularly bearing on health and offerings mostly of terracotta of animals, birds, humans, etc. are made. In some of the groves of West Bengal heaps of such terracotta offerings of elephants and horses are found. Groves are mostly managed by the local community, and owned by a family, group of families, or a clan. A second category is represented by those managed by the entire village community. The local-level groves are those where people from larger geographical areas, usually a few neighbouring districts come to worship a particular grove, i.e. Iringole in Kerala and Kantabanshini hakurma in Orissa, which area managed by the local community and/or committees. The regional-level sacred groves are those where people from several districts or states participate like that in Sabarimala in Kerala and Ramdeora in Rajasthan. Such groves are usually managed by temple trusts (http://www.thealternative.in/society/on-the-knees-to-nature/). The next higher level is Pan-Indian character where people from many parts of the country participate. Hariyali sacred grove in Garhwal Himalayas is example of Pan-Indian (Sinha and Maikhuri, 1998) and managed by temple trusts. Another category of SGs includes those that are believed to be abodes of ancestral spirits.

Sacred groves have important socio-cultural functions too. Several festivals are performed in these sacred groves (Rao et al., 2011). Annual festivals associated with the sacred grove range from one to nine days in length and vary among the regions (Ormsby, 2013). Some examples are: offerings of food and liquor in groves among the Bhils of Ratanmal once a year on the occasion of Deepavali (Nath, 1960); social gatherings on the occasion of Salui and Karam festivals, as well as wedding ceremonies among the tribes of southwest Bengal; celebration of major festivals at the sacred groves among different tribes of Bihar (Vidyarthi and Rai, 1997); socialization of the youth among the tribes of Chhotanagpur (Fernandes, 1993) and celebration of festivals like Holi, Navratri, Dev Diwali in sacred groves in Ratnagiri district of Maharashtra (Godbole et al., 1998; Waghchaure et al., 2006).

6.2 Conservation of Biodiversity

Cultural and spiritual aspects of the ecology of indigenous societies help conserve biodiversity, ecosystems, and landforms in their habitat as there are significant relationships between sacred places, biodiversity and the conservation measures (Mgumia and Oba, 2003; Chandrakar et al., 2014). Sacred groves are dynamic systems because the species composition of any plant community change naturally over time with ecological succession, even in absence of natural perturbations or human disturbances. As a result, a landscape can be a mosaic of patches of various types of plant communities, and that help sustain some of the biodiversity in a region. Because of cultural restriction of access to and interference with sacred groves to reduce the human impact, the sacred groves have been evolved as reservoirs of biodiversity and germplasm allowing the complex and diverse array of ecological processes to continue uninterruptedly over long period of time (Yelvattimath and Kotresha, 2011, Gokhale et al., 2011). The religious or cultural designation of an area as sacred promotes the conservation of its associated biodiversity (Maru and Patel, 2013). These sacred groves complement national parks, sanctuary and other protected areas (Khandal, 2014; Niraj and Gupta, 2015). Many a times these sacred groves show greater diversity in terms of number of species and their importance as compared to the nearby forest areas, which are generally due to monoculture plantation in forest areas (Singh, 2008; Vipat and Bharucha, 2014). Pushpangadan et al (1998) have recorded 722 species of angiosperm in a sacred grove in Kerala with only 1.4 sq. km as compared to 960 species have observed in 90 sq. km of the Silent

Valley forest. Many endangered, rare and endemic species and their wild relatives find their place in these sacred groves. Some of them are *Phoeba hainsiana, Rhus hookeri* and *Flacourtia cataphracta* in sacred groves of Manipur (Haridasan and Rao, 1985), *Kunstleria keralensis* (Mohanan and Nair 1981), *Belpharistemma embranifolia, Buchanania lanceolata, Syzygium travuncoricum* and *Cinnamomum quilonensis* in sacred groves of Kerala (Unnikrishnan, 1995) and *Myristica malabarica, Garcinia gummi-gutta* and wild pepper in sacred groves of Karnataka (Chandran et al., 1998). Some of the rare plants observed in orans of Rajasthan are *Acacia catechu, Acacia nilotica, Anogeissus sericea, Blepharis sindica, Boswellia serrata, Caralluma edulis, Ficus benghalensis, Ficus glomerata, Ficus religiosa, Glossonema varians, Haloxylon recurvum, Commiphora wightii, Tribulus rajasthanensis, Zizyphus* spp. etc. (Singh, 2008; Meena and Singh, 2012; Rathore and Shekhawat, 2012).

Interesting to see are the facilitative effects of trees or shrubs to live together in these sacred areas (Plate 2.2A). Byers et al. (2001) have highlighted the importance of sacred groves in terms at least 50% lower deforestation than in their secular counterparts, where some 133 species of native plants is recorded in the sacred forests, but are under variously threatened, endangered, or extirpated categories in Zimbabwe. Mgumia and Oba (2003) have recorded a greater woody species richness and taxonomic diversity in a sacred grove than in the state managed forest Reserve despites of relatively small area of the sacred groves. Some of the woody species are absent in the forest plots, also suggesting that groves serve as a refuge for some species. Christopher et al. (2014) has provided an account of species richness and diversity in sacred groves of different ecosystems and abiotic factors and observed significantly higher herbaceous α-diversity in the fallows than the sacred groves at the low altitude. The tree species richness has been observed higher at the low altitude compared to the high altitude with tree β-diversity increasing with altitude. Varying combinations of soil pH, total P, total K, CEC and slope per cent have been related to some of the parameters and indicates that biodiversity changes in the sacred groves are governed by biophysical drivers, while a combination of human and biophysical explain the variation in rotational fallow vegetation. Many animal species including birds that are otherwise threatened or becoming rare find a safe refuge in many a sacred groves throughout the world (Mondal and Roy, 2010; Yadeo et al., 2010; Ray et al., 2014).

6.3 Ethnomedicinal importance

Besides conservation of biodiversity, the role of sacred groves is also important as a life support system. The sacred groves help indigenous communities by providing edible fruits, leaves, fibers, gums, resins and medicinal plants (Koradia, 2000; Bhakat and Pandit, 2003; 2004; Nithyadevi and Sivakumar, 2015). The people are in opinion that the required species if not found elsewhere around their village surroundings, there are high probability of its occurrence in the sacred grove sites. Thus sacred groves act as a nursery and storehouse of many of the ayurvedic, tribal and folk medicines too (Rao and Sunitha, 2011; Sambandan and Dhatchanamoorthy, 2012; Rao et al., 2014). Most of the indigenous people residing near the groves are illiterate, but have scrupulously nurtured their traditional customs, rituals, ceremonies and a way of forest life through folk beliefs with great vitality (Sahu et al., 2013).

The fact that 'Sanjeevani' found in Himalayan region was used by Lord Hanuman to resuscitate Lakshman the brother of Lord Rama is well known to all Indians. In view of reduced human interferences, there are lesser chances of species extinction, if they are preserved in the sacred groves. These groves also serve to preserve genotypes which may be useful in forest tree-breeding programmes and as indicators of the natural productivity of the region. Ecologically valuable species like *Albizia lebbeck*, *Ficus glomerata*, *Prosopis cineraria* conserve high amount of nutrients both in soils and the plant system itself. Keystone species that contribute to the maintenance and enhancement of biodiversity and socially valued by local communities for cultural or religious reasons, are often found in sacred groves. Medicinal plant conservation is an integral part of sustainability (Pearce and Moran, 1994; King-Oliver et al., 1997; Rahangdale et al., 2014) and there is a need for protection of large number of medicinal plants as an important component of common resources including sacred groves already documented in various studies (Vartak et al., 1987; Bhandary and Chandrasekhar, 2003; Pandit and Bhakat, 2007, Rao and Sunitha, 2011).

6.4 Source of non-timber produce

People normally do not collect or harvest plants from sacred groves for domestic use (Gadgil and Vartak, 1976; Roy Burman, 1995; Pushpangadan et al., 1998; Malhotra, 1998; Swamy et al. 1998). But there are reports (Singh and Saxena, 1998; Jha et al., 1998; Godbole et al., 1998) indicating biomass extraction and deriving certain direct economic benefits in a restricted/sustainable way without affecting the local resources (Bhakat and Sen, 2012). Sacred groves provide shade, fuel wood, fodder and even food and livelihood for humans and animals. In western Rajasthan, orans also protect local livelihoods and provide minor forest produce that are collected from the Oran and used or sold. Fuel wood and fodder for animals are available in the sacred groves and are utilized by the villagers. Grasses are gathered for basket-weaving, clay for pottery and leaves for making brooms. Honey, fruits, seeds and medicinal plants are other minot forest produces available for local use or the market. Nesting birds, such as owls, parakeets and woodpeckers, build their homes in the 'Khajjur' and other large trees, assisting farmers by eliminating insects and other agricultural pests. The faunal wealth of sacred groves is also worth mentioning. Harvesting of certain species of trees (i.e., *Caryota urens*, *Mangifera indica*, *Entada phaseoloides, Diospyros melanoxylan* etc) from groves provide livelihood to many families (Gadgil and Vartak, 1976; Malhotra et al., 1997; Unnikrishnan, 1990). In some cases wood is also extracted from many groves dedicated to ancestor spirits for cremation (Mitra and Pal, 1994). Some of the important trees/plants of orans in Rajasthan provide fruits (*Salvadora oleoides*, *Capparis decidua*, *Prosopis cineraria*, *Zizyphus* spp.), oils (*Salvadora* spp., *Citrullus colosynthesis*), fodder (*P. cineraria*, *Tecomella undulata*, *Zizyphus* spp.), fuel wood (all the above-mentioned species including *Maytenus emarginata*) etc.

6.5 Sources of water supply and recharge of aquifers

Most of the sacred groves hold water resource in the form of springs, ponds, lakes, streams or rivers (Chaudhary et al., 2002; Singh, 2011). People believe, with some justification, that forests bring rain. Watersheds associated with forests have sanctity in every village. Sometimes the only source of water in a village is the grove's spring. Groves can supply fresh water in regions where other water is saline or not suitable for drinking. The sacred groves of Parinche valley in Pune, Maharashtra is present near water resources and is assumed to be responsible for water availability in lean season (Waghchaure et al., 2006). Likewise the Lum Shyllong-Nongkrim sacred groves in Meghalaya are the source of as many as eight streams that supply water to Shillong city, the capital of Meghalaya. However, the most significant contribution of sacred grove in water conservation could be witnessed in Rajasthan, where water bodies associated with Orans serve as lifeline for local inhabitants as well as livestock. In general small natural water bodies are maintained near or inside the sacred groves to take care of drinking water problem during drought (Plate 2.2B).

There are 3497 water bodies documented in Orans, pasturelands, paratlands and agores of Jodhpur district (Anon., 2004). However, there are problems of siltation of the ponds and Kunds in different regions reducing capacity of these water harvesting devices (Saha et al., 2010). In western Rajasthan, the villagers are now improving the surrounding areas of water bodies by excavation and applying concrete structure to conserve natural flowing and maintain it in a hygienic condition. In this regard the works of renovating ponds and Talabs in the Agore or sacred groves in western Rajasthan by Jal Bhagirathi Foundation (JBF), a Jodhpur based NGO is worth mentioning that has not only enhanced the availability of drinking water in many villages but has also to the adjoining villages in addition to the ground water recharge. While many Orans like Gar vaji, Adaval, Talv raksh, Kalaka, Bhar thari, Naraini Mata, Nadeshwarji, Parashar ji and other in Sariska sanctuary area of Alwar district, Rajasthan, have large perennial water springs used for irrigation and other purposes all the year-round and thus protect watersheds and/or water sources (Singh, 2011), only 7.6% of the ponds located in the Orans of Jodhpur districts are perennial and contain water for the entire year (Chaudhary et al., 2002). Orans conserve valuable water in the ponds, reservoirs, step wells, streams and springs that are an essential feature of the Oran and the availability of water is major incentive for communities to use this valuable resource in a sustainable manner. In Hadoti region, dev bani are maintained in the belief that the local deity would protect the community's water bodies and other water harvesting structures. Groves are developed on the main bund of the talab to strengthen it. A large patch of land downstream of the talab is also demarcated as part of the dev van to compensate for any losses of tree growth to submergence. One best example is Jharan Mahadeo sacred grove in Jhalawar, which is situated along the stream leading to a large tank that ensures round the year supply of water to Jhalawar city. This is the only green patch in the area harbouring some rare plants like *Bambusa hamiltonii* and *Scleichera oleosa*.

The ponds and streams in the groves are often perennial and in some cases, act as the last resorts to many of the animals and birds for their water requirements, especially during dry seasons. Thus one of the important ecological roles of these groves is to provide a more dependable source of water for the organisms living in and around the sacred groves (Puspangadan et al., 1998; Mabhal, 2013). In addition, transpiration from the sacred groves vegetation would increase atmospheric humidity and reduce temperature in the immediate vicinity and produce a more favourable microclimate for the organisms (Khiewtam and Ramakrishnan, 1989). Sometimes plants like vetiver grass (*Vetiveria zizanioides)*, *Saccharum munja*, and tree species are maintained to bind the soil thereby preventing soil erosion.

'Khadu'

It is an important traditional system of resource use in some villages of Alwar district, where each family group use to bring their livestock to graze around one particular Johad (pond), of which there are eleven in the village. Family rights to Johads are inheritable, and it is universally known to each villager that which Johad is used by which family. Management of both the watering hole and the grazing land surrounding the water point is the responsibility of the extended family group. This practice reduces the likelihood of exploitation and instead promotes wise use of the available resources. Such interlinking of ecology and kin-based history of 'Khadu' system of resource conservation and use encourages conservation and ensure welfare of future generations of a given family.

6.6 Conservation of soil and water and in nutrient cycling

Sacred groves play a crucial role in soil and water conservation. As community-aggregated specific leaf area, leaf nitrogen concentration, canopy interception, stem flow, surface runoff and soil erosion and litter biomass (Fu et al., 2009). High accumulation of litter biomass at soil surface strongly increases the maximum water-retaining capacity of litter and reduces surface runoff and soil erosion. Increased fine root biomass also reduces surface runoff and soil erosion. This indicates that high canopy characteristics in a sacred grove play a more important role in control of runoff and soil erosion. The Mawsmai sacred groves in the Cherrapunji ecosystem receive very high rainfall. With a rapid litter decomposition rate, nutrient release in the soil of these groves is very high. The soil itself has little nutrients to support a large biomass of the sacred grove. The fine roots developed on the surface layers of the soil is important for supporting the large above-ground biomass and for tight cycling of the nutrients (Lima et al., 2010; Rawat, 2012). Many microorganisms, invertebrates, fungi, etc. flourish and a vast array of species not hither to indigenous to the groves may also colonise and thrive and help in nutrient cycling (Tripathi et al., 2013). Dense root mat prevents the nutrients from leaching out and restore soil moisture favouring regeneration and growth of herbaceous vegetation. Further, fine roots act as a medium for transfer of atmospheric carbon into the soil in the form of carbon containing compounds that help enhance soil carbon sequestration. The land surrounding the sacred groves, which are devoid of necessary root mat and litter decomposition, can no longer sustain vegetation growth (Khiewtam and Ramakrishnan, 1989; Skinner et al., 2009). All of these factors indicate that the conservation of sacred groves is essential for maintaining local/regional biodiversity, the comprehensive health of a landscape, and preserving the socio-cultural integrity of local communities (Cardelús et al., 2013).

6.7 Carbon cycling and carbon sequestration

Forest preservation on religious ground (Kulkarni et al., 2010, Kulkarni and Nipunage, 2009), cultural heritage conservation (Sebastiano, 2009; Barbaro et al., (2007) and restrictions on excessive extraction of natural resources and on human settlement within the defined boundaries of sacred forests have indirectly contributed to conservation of the habitat and protected animals residing therein that help conserve carbon and mitigate climate change too (Arora, 2011). Estimation of carbon stocks and stock changes in existing tree or shrub biomass are necessary for reporting to the United Nations Framework Convention on Climate Change (UNFCCC) and which is required for Kyoto Protocol reporting also. Recent assessment indicates about 288.8 billion tons of carbon in forest living biomass in 2010 as compared to 299.2 billion tons in 1990 (FAO, 2010). Because of harbouring large biomass per unit area of land and greater diversity sacred groves make an important contribution to the global carbon cycle (Hangarge et al., 2015). Trees absorb more carbon dioxide from the atmosphere than they release under the combined effect of photosynthesis and respiration. Plants store atmospheric carbon in the form of photosynthetic products, standing biomass, leaf litter and also contribute to soil organic carbon. It has been observed that unmanaged and old growth forests have better carbon sequestration capability than plantation and managed forests (Chen et al., 2004). Thus a well maintained grove can have great potential in carbon storage due to having older trees with bigger stem bulk; presence of good number of young individuals thus ensures future carbon storage also. A study from Himachal Pradesh show higher percentage of soil carbon stock in a sacred grove in comparison to the other forest ecosystems. Furthermore, a protected sacred forest, because of its higher biomass, sequesters significantly more carbon compared to other forest ecosystems (Carret and Loyer, 2003). Hangarge et al. (2012) has studied carbon sequestration potential of tree species in Somjaichi Rai sacred grove at Nandghur village, Pune District, Maharashtra, where tree species having 4000 individuals have been recorded in 5 ha area. In this study, *Terminalia bellirica* is dominant species with 180 individuals and sequester 327.78 tons of carbon followed by *Ficus amplissima* (221 tonns). The species *Gnidia glauca* shows the lowest carbon sequestration potential, i.e. 0.0808 tons, whereas the second lowest carbon sequestration species is *Nathapodytes nimoniana* (0.182 tons). In Benin, the njangsa (*Ricinodendron heudelotii*), a taboo species, has recently been assessed as one of the trees with both the highest carbon sequestration potential and fallow improvement of cash crop tree plantations in West Africa. *Antiaris toxicaria* and the African border tree (*Newbouldia laevis*) have also proven important for the plantation in Benin (Quiroz, 2013). However, there are studies showing strong spatial patterns in soil carbon storage, declining quite sharply in the more isolated sacred groves. Tree stand density appears to be an important positive correlate of most of the bird and carbon storage service responses studied tree density, in turn, seems to decline at higher levels of site isolation (Pundee, 2007).

6.8 Sacred grove and landscape heterogeneity

Sacred grove is not an isolated system. They form significant landmarks in the landscape, both urban and rural areas (Patel and Patel, 2012). The flora and fauna observed in the sacred groves is unique to the habitat and climatic condition of the region, where the grove is located. The groves serve as both feeding and nesting grounds for innumerable bird species, which in turn help in dispersion of seeds and fertilization of soil through their droppings. Due to its presence in a mosaic of landscape elements like forests, agricultural lands, grazing lands, plantations and human settlements the groves enhance landscape heterogeneity and biodiversity. Sacred groves in combination with variety of tree cover play important role in maintaining diverse microclimate and habitat which nurture several distinct macro- and micro-flora and fauna (Bhagwat et al 2005). Sacred grove fragments offer more intra patch habitat diversity to different life forms such as epiphytes, shrubs and lianas thus increasing total pool of regional species diversity. However, fragmentation of the groves often affects seed dispersal and regeneration of rare plants which ultimately causes decline of plant population. The study of Cardelús et al. (2013) indicates significant decrease in soil carbon and nitrogen stocks in sacred groves with increased distance from forest interior indicating a decrease in soil fertility. Because of small average size, isolation from seed sources and decreasing soil status the sacred groves become more vulnerable to loss. Further, the availability of pollinators and seed dispersers is adversely affected causing decline in mutualistic relationship between trees and animals. The study from the Konkan region has also shown that larger groves have better recruitment potential (both seedling density and seedling survival) than the smaller ones with regard to *Antiaris toxicaria* and *Strychnos nux-vomica* tree species. However, it has also been indicated that apart from grove size, habitat quality, connectivity and population of seed dispersers may have a role in plant population dynamics.

6.9 Sacred groves and animal diversity

Almost all Hindu Gods are associated with animals, birds and creatures as their vehicles or vahanas. To promote harmony in nature and to maintain ecosystem balance, protection of animals and other creatures along with human beings is essential. The association of animals with Gods as Vahans (vehicles) is a deliberate attempt to weave mythology for protection of animals and plants in order to maintain ecological balance. For instance, Garuda (eagle) is symbolized as the Vahan (means of transport) of 'Lord Vishnu', Peacock to 'Muruga', Bull to 'Lord Shiva', Tiger to goddess 'Durga' and Rat to 'Vinayaka' etc. Lord Shiva uses lion skin for his attire and supports a reptile king cobra as an ornament. Serpents are also venerated on Nagpanchami day and offered milk for maintaining the natural food chain and preserving environment of sacred groves. Monkeys as a part of respect to Lord Hanuman and bears are also treated as gods as they had been a part of the army force of Lord Rama.

Though the animals found in most of the sacred groves are those which inhabit the groves like snakes, frogs, lizards and other lower organisms; a higher group of fauna who nest and den in the groves; and those who visit the grove temporarily for food, shelter, etc (Plate 2.2C). In addition sacred groves also provide shelters to many of the livestock of the adjoining village residing around the sacred groves (Ormsby and Edelman, 2010). Many sacred groves act as an abode for many rare, endemic, endangered plant and animal species, i.e. spiny tailed lizard in sacred groves of western Rajasthan. Birds constitute one of the common fauna of sacred groves and their diversity and abundance reflect ecological trends in a particular habitat (Kangah-Kesse et al., 2003). The orans in Rajasthan, in the vicinity of Bishnoi community villages provide protection to the Indian gazelle (*Gazella gazella*), blackbuck (*Antelope cervicarpa*) (Plat 2.2C) and to the migratory bird like Demoiselle crane (*Anthropoides virgo*) and the important bird Great Indian Bustard (Dagla et al., 2007).

7. THREATS AND OPPORTUNITIES

In recent age of liberalization, privatization and globalization, communities are losing their traditional bases of sustainable livelihoods. Forests, rangelands and other common land resources are being forced to enter into market system that generally regards nature as a commodity. This makes communities to deviate from the customs which were the guiding principle in protecting and maintaining sacred groves. As market-oriented social systems do not hold respectful attitude towards nature, sanctity of life has no place in this market-linked livelihood systems. Further, modern education system has no respect for local traditions. As a result this ancient and widespread institution is now weakening in terms of both cultural and biological integrity. Increasing pressure from population and livestock and encroachments are major threats. More worse is regularization of these encroachments by the Governments. Area and legal status of several sacred groves has not been clearly defined and lack effective legislation against offenders. Developmental activities, agricultural encroachment, increased resource use, cattle trampling and renovations of temples lead to damage this valuable habitat (Blicharska et al., 2013). Over-exploitation and unscientific collection of plants for varying uses, eroding community values and lack of faith in the younger generation are other threats to the sacred groves (Yadeo et al., 2010; Rao and Sunitha, 2011). Despite of such threats there are lots of opportunities too. These are responses of these sacred groves towards changing climate in terms of ecological succession and adaptability. A challenge towards devising management options with ever increasing threats of multiple uses and changing market economy.

8. CONCLUSIONS

Sacred place, sacred trees and sacred groves have been flourishing since time immemorial. Culture of worshiping each live form in India is a way of celebrating the sanctity of life and sacred groves serve as a meaningful fusion of religion and ecology. Protecting nature's biodiversity and especially the rare types of life forms through sacred groves is the astonishing response of people to conserve biodiversity. Availability of a number of plants and animals species described rare, endangered, vulnerable and threatened in sacred groves are living examples of rich and vibrant biodiversity in this cultural heritage. In addition to conserving biodiversity, sacred groves also conserve soil and water and are sources of water and ethno-medicinal plants. Besides it plays an important role protecting various life forms and sequester atmospheric carbon dioxide thus help mitigate climate change.

However, decline in traditional beliefs, westernization and market-linked economy and other socio-economic factors have undermined the status of the sacred groves. Lack of legal status and awareness, internal politics and conflicts, little or no financial support, encroachments of varying kinds and exotics invasion have led to the degradation of these resource-use areas. Grazing, lopping and removal of biomass from some of the sacred groves have also resulted in dwindling of these groves. Concerted efforts are required to protect and conserve the sacred values of these deteriorating life support resources for the survival of existing biodiversity (Sajwan and Kala, 2007). For necessary protection and to maintain their natural identity and sanctity cooperation of the local people should be extended. Likewise surrounding village communities need to be educated and guided for sustaining the sanctity of existing grove in the area and strengthening them.

Terminalia arjuna in Chandi Mata ki Bani, Pratapgarh

Salvadora oleoides in Kubad Mata Oran, Kosloo, Barmer

Adansonia digitata in Mangaliavas, Ajmer

Tamarindus indica in Hanumanji Ki bani, Baran

A devasthan under *Dolichandrone falcata* in Kota

A devasthan under *Maytenus emarginata* in Kota

Plate 2.1. Some sacred trees in Rajasthan under which people offers worship.

Plate 2.2A. Association of *Salvadora oleoides*, *Cocculus pendulus* and *Clerodendrum phlomides* in Kubad Mata Oran, Sindhari, Barmer (left) and *Salvadora oleoides*, *Maytenus emarginata* and *Cadaba fruticosa* in Mangal Das ki Bani, Alwar (right).

Plate 2.2B. Pond in Vadara Hanumanji Ka Oran, Sirohi (left) and a water body in Gosaiji Ka Oran, Nagaur (right) is utilised by villagers for varying purposes.

Plate 2.2C. Nesting of Egret and snake bird on *A. nilotica* in Raghunathji Devban, Bhiluda, Dungarpur (left) and wild life through conservation of Orans, Lohawat, Jodhpur (right).

3

DIVERSITY AND SATATUS OF SACRED GROVES IN RAJASTHAN

Sacred groves contribute to biodiversity conservation, increase religious belief and provide opportunities of income generation and ecosystem services for local community and many more... They often contain plant and animal species that have become extinct in the neighbouring areas, thus rich in genetic diversity and support human livelihood (Bhakat and Sen, 2008; Yadav et al., 2010). Sacred groves often called 'Oran' in Rajasthan are considered as a refuge for several plant and wildlife species. However, this community resource is fragile and vulnerable, and is facing severe threats from land use change, unsustainable exploitation of natural resources, desertification and climate change (Iyer, 2010). Recent studies on sacred groves depict this land use types as indigenous strategies, which help to address many local social and environmental problems (Chandrakanth et al., 2004; Jeeva et al., 2006). Conservation efforts of sacred grove indicates improvement in overall vegetation diversity and carbon stock (Carret and Loyer, 2003; Ray et al., 2012), though a base line is necessary to assess the impact of conservation and management in social and environmental benefits. Understanding the importance of plant diversity and carbon storage of these land use types in Rajasthan can be a good step towards their sustainable management for the delivery of diverse ecosystem services of this resource.

1. BACKGROUND

Being a preserver of the endemic, endangered or threatened species, the sacred groves in Rajasthan are illustrated by many names in different regions. They are name like Banis of Mewar, the Kenkris of Ajmer, the Orans of Jodhpur, Jaisalmer, Bikaner, the Shamlat Dehs and Devbanis of Alwar, Baag in south eastern Rajasthan regions and Deora, Malvan, Deorai, Rakhat Bani etc. in different localities. In Aravallis and Vindhyas areas, the sacred groves are classified into three major groups. First group of groves are those located near the village and close to a water source. These groves are also available at the top of small hillocks in Aravallis, where people worship in name of local deities like Bheruji, Bawsi and Mataji etc. For example Khanpa Bheruji, Kukawas Bheruji, Badi Roopan Mata etc. are sites available in Udaipur district (Choudhary, 2016). There is an overlap between 'gauchar' or 'roondhs', the common pastures, and 'Oran' the sacred woodland around a temple (Singh and Sishodia, 2003). Despite of climatic vagaries, sparse vegetation and limited water resources the tradition of maintaining sacred groves is still continue, and Rajasthan provides interesting insights on tribal traditions of maintaining these sacred groves. The desert region of Rajasthan supports a higher density of human and animal populations than any other desert region in the world because of the conservation practices of the people of the region (GoI, 2011). About 8 to 9% of the total area of desert accounts sacred groves called 'Oran' in western Rajasthan.

Oran word is derived from the Sanskrit word Aranya, meaning forest. These are patches of jungle preserved in the name of local deities or saints. People believe that Aranya or the forest is the place where the Aranyakas ("Forest Books"), a part of the Vedas, have been written. Development of Orans is a complex phenomena that can be viewed useful because of varying along several continua ranging from socio-cultural; prehistoric to historic, recent, or newly created; secret or private to public; single religion to multi religious; intrinsic to extrinsic in values; uncontested to contested; and protected to endangered (Rathore and Shekhawat, 2012).

Delineation of an Oran is conducted during a ceremony called 'Doodh Jal' or 'Kesar Chaanta', when boundaries of the sacred grove are marked by pouring water of sacred river Ganga or saffron milk around a specific area within the forest, which is thereafter declared as an Oran or Dev bani (God's forest). There is report that Bisnois maintained 9 sacred groves in vicinity of their villages covering about 241 hectares area (Sethi and Vishwanath, 2003). In another report Jha et al. (1998) have reported five sacred groves in Peepasar and Kherzadli village area of Jodhpur. Some of them are more than 200 year old. There are only a handful of documented sacred groves in Rajasthan, but their areas often exceed thousands of acres (Khiewtan and Ramakrishna, 1989; Tripathi et al., 1995; Malhotra, 1999; Jamir and Pandey, 2002). About 2804 sacred groves have been documented in Rajasthan against the projected number of 25000 covering an area of more than 6,00,000 hectares and are providing livelihood to the rural population (Table 3.1). These sacred groves vary from 0.1 hectares to 500 hectares (Dande, 2014). Chaudhary et al. (2002) have recorded 732 'Orans' and 443 'Gauchars' covering 334 villages in 4 tehsils of Jodhpur district that vary from less than 2 hectares to greater than 320 hectares. Interestingly number of pond availability equaled to the number of Oran and Gauchars. Some big 'orans' in Rajasthan are Bhadriyaji in Jaisalmer district that covers about 15000 hectares area and Kundla's Oran in Barmer district that covers about 7500 hectares (Khandal, 2014).

Gujjar community of Rajasthan use to plant 'Neem' (*Azadirachta indica*) and worship it as Neem-god. A Gujjar settlement normally starts near a water source or along a stream or river. Initially few houses are constructed and neem saplings procured from varying sources are planted in the enclosure around the hut, and the people worship it as the abode of God Deonarayan (Gold and Gujar, 1989). Likewise, Bishnoi community in western Rajasthan protects 'Khezri' (*Prosopis cineraria*) from cuttings and associated sacred groves for promoting wild life (Gold and Gujar, 1989). The basic philosophy of the Bishnoi community faith is that all living things have a right to live and share resources; this group has a set of abiding laws including a ban on killing animals and on felling trees, especially the most sacred khejri tree, every part of which has numerous life-sustaining properties. The '*orans*' provide a protective habitat together with pasture lands for the Indian gazelle and blackbuck (Dixit et al., 2007). In some places in south east Rajasthan sacred groves are also named as 'Baag'- an ethno-silvihorticultural garden planted near settlements for fruit, fodder, fuel wood, medicine, NTFPs and shade, and said to be the backbone of livelihoods (Pandey, 2000).

It is not the tree only but animals are also conserved through such sacred groves. One best example is the Karni Mataji Oran at Deshnok in Bikaner, where the rodents are regarded as reincarnations of the goddess's. Here one can see much deeper relation with nature on the 2,100 hectare woodland which is protected by the local communities. Thousands of rates can be seen freely roaming in the temple. Some of them are 'albino' too, which are considered to be luckier than the normal once. Villagers are following the orders of Mata Karni Devi and thousands of jhad beris (*Zizyphus nummularia*) - a fruit bearing shrub, dot the area as cattle of the whole village graze freely in the area. According to a belief, the goddess 'Karni Mata' had special affection for cows and this town was established by her. Other worshiped trees are Salar (*Boswellia serrata*) and Pipal (*Ficus religiosa*) and are prohibited from either to cut them or to burn their wood. The trees on Chavariya hill are preponderantly *Anogeissus pendula* – a species that had dominated most of the hills of Aravallies through the first half of the 20th century, i.e. before massive deforestation in the region (Gold and Gujar, 2002).

Table 3.1. Distribution of sacred groves in different districts of Rajasthan and the area covered under it.

SNo.	District	Number of Scared groves			Area (ha)	Reference
		ENVIS	CECODECON	All		
1	Ajmer	38	-	38	1424	Envis
2	Alwar	116	-	163	7459	Singh and Sisodia (2003)
3	Banswara		18	18		CECODECON
4	Baran		-	3		Gold and Gujar (1997)
5	Barmer	6	253	253	14532	CECODECON/Singh (2008)
6	Bharatpur	-	-	5		Charanji (2008)
7	Bhilwara	9	-	9	431	Envis/ Gujar and Gold (2007)
8	Bikaner	5	-	5	3142	Envis
9	Bundi	-	-	1	-	Gold and Gujar (1997)
10	Chitaurgarh	5	83	83	955	CECODECON
11	Dausa	12	-	12	149	Envis
12	Dungarpur	-	25	25		CECODECON
13	Jaipur	26	-	26	693	Envis
14	Jaisalmer	10	27	28	52743	Envis, CECODECON, CSE (2003)
15	Jhalawar	1		1		UNDP (2008)
16	Jalore	3	22	22	266	CECODECON
17	Jodhpur	-	-	1690	-	Chaudhary et al. (2002), Anon (2004)
18	Kota	-	-	2	-	Gold and Gujar (1997)

19	Nagaur	-	31	31	-	CECODECON
20	Pali	-	57	57	-	CECODECON
21	Rajsamand	-	13	13	-	CECODECON
22	Sikar	4	2	4	760	Envis/ CECODECON
23	Sirohi	2	92	92	138	Envis/ CECODECON
24	Tonk	3	-	3	331	Envis
25	Udaipur	221	46	221	6387	Envis/ CECODECON
	Total	**461**	**669**	**2804**		

Sources: ENVIS; Chaudhary et al. (2002); CECODECON; Charanji (2008); and Sewa Mandir (1997).

Sacred grove (Oran) serves four main purposes: vegetation as grazing ground for the livestock; watering place for the livestock; resting places for the livestock; and availability of medicine in ethno botanical form. It plays an important role in promoting a flourishing livestock based economy and growth of livestock rearing communities in Rajasthan. The sustainable lifestyle practiced by the desert people in the past and the common resource pool in the villages in the form of "Orans" and "gochars" have made the Thar desert one of the most livable dry lands in the whole world (The Hindu, 7th July 2007). According to a study, about 41% livestock in Barmer district of Rajasthan is dependent on these resources. Though water availability in the ponds varies widely, but most of the sacred groves have water ponds and other water body therein. Survey of all 163 existing *Orans* in Alwar district, a water body like Johad, Tank, spring, Baori, Well etc have been reported (Singh, 2010).

Sacred Forestry

'Jharan Mahadeo' sacred grove in Jhalawar is situated along the stream leading to a large tank that ensures round the year supply of water to the city of Jhalawar. This is the only green patch in the area. It is important because it protects catchment that might otherwise be silted very quickly in the absence of vegetation. Rare plants include *Bambusa hamiltonii* and *Schleichera oleosa*. It is also important because it is a de facto sanctum sanctorum of threatened plants, all of which have become extinct in the adjoining area outside the grove. The Jharan Sacred Grove is also an indicator and benchmark of forests that might have existed in the region. Today, it is a natural laboratory, a habitat island, a gene bank and a store-house of ethno-medicine. There is a perennial water spring. The Forest Department has run a forest nursery inside the groves for the last 50 years for the production of seedlings for plantations and distribution. This ensures the survival of the grove in its original condition.

UNDP (2008)

Many rituals are performed in these sacred groves. For examples, many orans host an annual mela (festival), at which communities reaffirm their commitment to the forest and the deity Presence of a Mahatma or priest is a socio-cultural phenomenon in most of the sacred groves. Somewhere between a sentry and saddhu, the mahatma resides in an Oran or asides and, in exchange for basic provisions from the community, keeps watch over the forest and its other inhabitants (Singh, 2011). However, the traditional community practices of conserving and managing these sacred groves is rapidly declining.

2. DOCUMENTATION OF GROVES

Various informations have been collected through field observations and people interaction at the sacred grove sites through a well designed format and compiled for easy access and knowledge about the particular sacred grove. Following information have been incorporated in the format together with a photograph of the site in the preceding sheets for sacred grove number 1 to 123. This includes sacred grove number for its location in the given Google map (Plate 1.1 right).

A. Basic Information about the sacred groves

Name of village

Panchayat Samiti/block

Nearest Forest block

District

Name of sacred grove

Geographical position

B. Topographical features

Erosion status

Topography

Soil depth

C. Vegetation status

Tree species (>30 cm girth) with importance value index (IVI)

Shrub species with IVI

Effective number of species (ENS): For trees (>30 cm girth) and shrubs

Herbs and grass species

D. Wild-life

Animals

Birds

E. Soil characteristics

Soil pH and electrical conductivity (EC),

Soil bulk density

Soil nutrients, organic carbon (SOC), and soil carbon density in 0-30 cm soil layer

3. TOPOGRAPHY AND SOIL EROSION

Delineation of the sacred groves has been done mostly in hill slopes or in plain land, which covered about 31% and 60% of the sacred groves, respectively. About 5% of the sacred groves are observed on hill top. Another 3 per cent sacred groves show gully or ravines due to severe erosion, whereas one sacred grove is along a river or have river passing through it (Table 3.2). About 28 (23%) sacred groves have soil depth of <25 cm. Another 16% fall into 25-50 cm soil depth. About 18% sacred groves have soil depth 50-100 cm, whereas remaining 44% sacred groves have been found with deep soil and appears to be favourable for enhanced diversity and productivity (Table 3.2).

Table 3.2. Topographical positions of the sacred groves and the soil depth therein in the sacred groves of Rajasthan.

Topography	Number of sacred groves	Soil depth	Number of sacred groves
Hill top	6 (4.9%)	Shallow (<25 cm)	28 (22.8%)
Hill slope	38 (30.9%)	Moderately deep (25-50)	19 (15.5%)
Plane	74 (60.2%)	Moderate (50-100)	22 (17.9%)
Gullied & Ravine	4 (3.3%)	Deep (>100 cm)	54 (43.9%)
Valley (riverbed)	1 (0.8%)		

Status of soil surface of the studied sacred groves indicates varying degree of erosion (Table 3.2). Out of the 123 sacred groves studied, almost 24% sacred groves indicate slight erosion and under safe condition. About 69% of the sacred groves show moderate soil erosion. This indicates that almost 93% of the sacred groves vary between slight to moderate level of soil erosion. Severity of the soil erosion has been observed in about 8% of the sacred groves, in which 4% fall into severe category, whereas another 4% fall into gullies with ravenous structure.

Table 3.3. Erosion status of the soil surface in studied sacred groves of Rajasthan.

Erosion status	Number of sacred groves	Sacred grove number
Slight	30	5-6, 8, 22-24, 27-28, 46-48, 50-51, 56, 58, 67, 75, 77, 79, 85, 88, 92, 102-104, 108, 110-112, 120
Moderate	85	1, 3, 7, 9-21, 25-26, 29-45,49, 52-53, 55, 57, 59-66, 68-70, 72-74, 76, 78, 80, 81-83, 86, 89-91, 93-100, 105-107, 113-119, 121-123
Severe	4	54, 84, 87, 109
Gullied	4	2, 4,71, 101

DATA ON SACRED GROVES OF RAJASTHAN

Sacred Grove No. 1

A. Basic information about the sacred groves

Name of sacred groves	Devnarayan Ji Ka Oran
Name of village	Devnagar
Panchayat	Pishangan
Nearest forest Block	Devnagar-khori
Land status	Community land
District	Ajmer
Area	20 bigha
Altitude (m)	473
Latitude	26° 31' 36.3 N
Longitude	74° 33' 41.5 E

B. Topographical features

Erosion status	Topography	Soil depth
Moderate	Undulating plane	Deep (>100 cm)

C. Vegetation status

Tree species with IVI

Prosopis juliflora (194), *Ficus benghalensis* (55), *Tamarindus indica* (20), *Prosopis cineraria* (17), *Acacia leucophloea* (14)

Shrub species with IVI

Lantana camara (170), *Calotropis procera* (130)

Effective Number of Species (ENS) Tree > 30: 1.38 Shrub: 2.00

Herb and Grass species

Eragrostis ciliaris, Eragrostis tenella, Eragrostis minor, Tridax procumbens, Evolvulus alsinoides, Polygala erioptera, Indigofera cordifolia, Commelina benghalensis, Dactyloctenium sindicum, Datura stramonium, Peristrophe paniculata, Phyllanthus amarus, Sida cordifolia, Tragus roxburghii.

D. Wild-life

Boselaphus tragocamelus, Canis aureus auveus, Lepus nigricolis, Pavo cristatus, Turdoides caudatus

E. Soil characteristics

pH	EC (dSm^{-1})	SOC (%)	BD (g/cm^3)	NO_3-N (mg kg^{-1})	NH_4-N (mg kg^{-1})	PO_4-P (mg kg^{-1})	C density (tons ha^{-1})
7.52	0.25	0.195	1.44	1.12	3.12	6.30	5.31

F. History and status

An ancient grove in name of Lord Devanarayanji- a locally venereated diety. It is believed that the deity protects the area from all things destructive to the region.

DATA ON SACRED GROVES OF RAJASTHAN

Sacred Grove No. 2

A. Basic information about the sacred groves

Name of sacred groves	Bheruji Ka Oran
Name of village	Naand
Panchayat	Pushkar
Nearest forest block	Naand
Land status	Community land
District	Ajmer
Area of sacred grove	70 bigha
Altitude (m)	448
Latitude	26° 28' 49.4 N
Longitude	74° 27' 21.9 E

B. Topographical features

Erosion status	Topography	Soil depth
Gullied	Undulating plane	Deep (>100 cm)

C. Vegetation status

Tree species with IVI

Maytenus emarginata (231), *Acacia excelsa* (26), *Ailanthus excelsa* (17), *Acacia leucophloea* (14), *Prosopis juliflora* (12)

Shrub species with IVI

Acacia jacquemontii (170), *Leptadenia pyrotechnica* (91), *Crotalaria burhia* (39)

ENS Tree : 1.59 Shrub: 4.70

Herb and Grass

Amaranthus viridis, Boerhaavia diffusa, Borreria articularis, Brachiaria ramose, D. sindicum, E. alsinoides, I. cordifolia, Indigofera linifolia, Indigofera linnaei, Justicia procumbense, P. paniculata, Tephrosia purpurea, Tribulus terrestris, P. amarus, T. roxburghii.

D. Wild-life

Gazella gazella, Canis aureus auveus, Boselaphus tragocamelus, Pavo cristatus, Turdoides caudatus, Francolinus pondicerianus

E. Soil characteristics

pH	EC (dSm^{-1})	SOC (%)	BD (g/cm^3)	NO_3-N (mg kg^{-1})	NH_4-N (mg kg^{-1})	PO_4-P (mg kg^{-1})	C density (tons ha^{-1})
7.28	0.19	0.060	1.58	1.60	3.10	5.00	1.80

F. History and status

People of Bhil community worship Bheruji in most of the Aravalli region of Rajasthan and many sacred groves are in this name, where this community protects the area in name of their diety.

DATA ON SACRED GROVES OF RAJASTHAN

Sacred Grove No. 3

A. Basic information about the sacred groves

Name of sacred groves	Bhuna Bai ji ka Oran
Name of village	Bhuna bai
Panchayat	Ajmer
Nearest forest Block	Muabead
Land status	Community land
District	Ajmer
Area of sacred grove	25 bigha
Altitude (m)	515
Latitude	26°29' 54.1 N
Longitude	74° 40' 19.6 E

B. Topographical features

Erosion status	Topography	Soil depth
Moderate	Hill slope	Moderately deep (25-50)

C. Vegetation status

Tree species with IVI

Prosopis juliflora (246), *Cordia gharaf* (29), *Ficus infectoria* (25).

Shrub species with IVI

Calotropis procera (125), *Jatropha gossypifolia* (109), *Grewia tenax* (23), *Lantana camara* (22), *Securinega virosa* (21).

ENS Tree : 1.31 Shrub: 4.00

Herb and Grass

Aristida funiculata , *Borreria pusilla, B. diffusa, Dipteracanthus patulus, Pupalia lappacea, Triumfetta rhomboide, Zornia gibbosa, D. sindicum, E. alsinoides, I. cordifolia, Justicia procumbense, P. paniculata, T. purpurea, Tribulus terrestris, P. amarus, T. roxburghii, Vernonia cinerea, Xanthium stramonium.*

D. Wild-life

Canis aureus auveus, Lepus nigricolis, Streptopelia copicola, Francolinus pondicerianus, Picnonotus cafer, Passer domesticus.

E. Soil characteristics

pH	EC (dSm^{-1})	SOC (%)	BD (g/cm^3)	NO_3-N ($mg\ kg^{-1}$)	NH_4-N ($mg\ kg^{-1}$)	PO_4-P ($mg\ kg^{-1}$)	C density ($tons\ ha^{-1}$)
7.23	0.09	0.105	1.55	1.25	5.12	9.00	2.98

F. History and status

As per villagers, Smt Bhuna Bai- a village lady got meditated here and protected the villagers from the ill effect of a devil. The villagers protect this area in name of Bhuna Bai.

DATA ON SACRED GROVES OF RAJASTHAN

Sacred Grove No. 4

A. Basic information about the sacred groves

Name of sacred groves	Mandal Das ji Ka Oran
Name of village	Todli
Panchayat	Ramgarh
Forest Block	-
Land status	Community land
District	Alwar
Area of sacred grove	240 bigha
Altitude (m)	262
Latitude	27° 38' 40.0 N
Longitude	76° 45' 40.6 E

B. Topographical features

Erosion status	Topography	Soil depth
Gullied	Ravine (undulating)	Deep (>100 cm)

C. Vegetation status

Tree species with IVI

Acacia nilotica (145), *Maytenus emarginata* (33), *Acacia senegal* (31), *Prosopis cineraria* (27), *Salvadora oleoides* (25), *Capparis decidua* (12), *Holoptelea integrifolia* (8), *Acacia nilotica var. cupressiformis* (8), *Acacia leucophloea* (6), *Balanites aegyptiaca* (5).

Shrub species with IVI

Ziziphus nummularia (205), *Lycium barbarum* (35), *G. tenax* (22), *Capparis sepiaria* (21), *Leptadenia pyrotechnica* (17)

ENS Tree : 5.17 Shrub: 2.28

Herb and Grass

Achyranthes aspera, Arnebia hispidissima, Cynodon dactylon, Dactyloctenium aegyptium, B. diffusa, I. linnaei, Spergula arvensis, E. minor, Cenchrus ciliaris, Saccharum munja, Argemone mexicana, Chenopodium murale, Parthenium hysterophorus, E. alsinoides, I. cordifolia, T. Purpurea.

D. Wild-life

Boselaphus tragocamelus, Hystrix indica indica, Hyaena hyaena, Lepus nigricolis, Streptopelia chinensis, Acridotheres ginginianus.

E. Soil characteristics

pH	EC (dSm^{-1})	SOC (%)	BD (g/cm^3)	NO_3-N ($mg\ kg^{-1}$)	NH_4-N ($mg\ kg^{-1}$)	PO_4-P ($mg\ kg^{-1}$)	C density ($tons\ ha^{-1}$)
7.66	0.29	0.435	1.40	2.33	3.41	6.49	17.05

F. History and status

Sh Mandal Das ji got meditated in the foothill this grove. There is a temple on the hill top at present, where people use to worship.

DATA ON SACRED GROVES OF RAJASTHAN

Sacred Grove No. 5

A. Basic information about the sacred groves

Name of sacred groves	Hanuman Ji Ka Oran
Village	Jatwada
Panchayat	Laxmangarh
Forest Block	Laxmangarh
Land status	Community land
District	Alwar
Area of sacred grove	700 bigha
Altitude (m)	226
Latitude	27° 15' 26.8 N
Longitude	76° 53' 55.5 E

B. Topographical features

Erosion status	Topography	Soil depth
Slight	Flate land	Moderate deep (50-100)

C. Vegetation status

Tree species with IVI

Salvadora oleoides (142), *Acacia. nilotica* (35), *H. integrifolia* (33), *Azadirachta indica* (31), *P. cineraria* (26), *C. decidua* (23), *Cordia myxa* (5), *A. leucophloea* (5)

Shrub species with IVI

Clerodendrum phlomidis (107), *C. sepiaria* (100), *Adhatoda vasica* (73), *Cadaba fruiticosa* (11), *Calotropis procera* (9).

ENS Tree : 4.19 Shrub: 3.54

Herb and Grass

A. aspera, C. dactylon, B. diffusa, I. linnaei, E. alsinoides, I. cordifolia

D. Wild-life

Oryctolagus cuniculus, Boselaphus tragocamelus, Hyaena hyaena, Canis aureus auveus, Columba livia, Streptopelia chinensis, Pavo cristatus, Psittacula krameri

E. Soil characteristics

pH	EC (dSm^{-1})	SOC (%)	BD (g/cm^3)	NO_3-N ($mg\ kg^{-1}$)	NH_4-N ($mg\ kg^{-1}$)	PO_4-P ($mg\ kg^{-1}$)	C density ($tons\ ha^{-1}$)
8.15	0.18	0.375	1.42	1.58	3.86	9.32	10.68

F. History and status

An ancient temple, but under renovation. People use to come and worship Lord Hanuman here. Gaint trees of *Azadirachta indica* are there in the sacred grove.

DATA ON SACRED GROVES OF RAJASTHAN

Sacred Grove No. 6

A. Basic information about the sacred groves

Name of sacred groves	Jharna Ji Ka Oran
Name of village	Lioi, Raipur, Makreta
Panchayat	Toda
Forest Block	Toda
Land status	Community land
District	Alwar
Area of sacred grove	100 bigha
Altitude (m)	232
Latitude	27° 19' 49.3 N
Longitude	76° 53' 48.0 E

B. Topographical features

Erosion status	Topography	Soil depth
Slight	Gullied area	Moderately deep (25-50)

C. Vegetation status

Tree species with IVI

P. juliflora (92), *M. emarginata* (36), *B. aegyptiaca* (34), *P. cineraria* (33), *A. nilotica* (30), *Anogeissus pendula* (24), *C. decidua* (23), *H. integrifolia* (22), *Acacia tortilis* (6).

Shrub species with IVI

Z. nummularia (150), *C. sepiaria* (63), *S. virosa* (43), *G. tenax* (31), *L. barbarum* (13)

ENS Tree: 5.90 Shrub: 3.24

Herb and Grass

B. diffusa, E. alsinoides, I. cordifolia

D. Wild-life

Boselaphus tragocamelus, Canis aureus auveus, Rattus norvegicus, Ophiophagus hannah, Lepus nigricolis, Herpestes edwardsi, Streptopelia chinensis.

E. Soil characteristics

pH	EC (dSm^{-1})	SOC (%)	BD (g/cm^3)	NO_3-N (mg kg^{-1})	NH_4-N (mg kg^{-1})	PO_4-P (mg kg^{-1})	C density (tons ha^{-1})
7.69	0.22	0.705	1.35	2.23	12.11	8.25	21.95

F. History and status

Old groves with some small temples, where people use to worship their deities. The grove is dominated by *P. juliflora*-an invasive species.

DATA ON SACRED GROVES OF RAJASTHAN

Sacred Grove No. 7

A. Basic information about the sacred groves

Name of sacred groves	Rupu Ka Vas Oran
Name of village	Godi, Arnar, Amada, Bhopada
Panchayat	Thanagari
Forest Block	Thanagari
Land status	Community land
District	Alwar
Area (bigha)	150
Altitude (m)	487
Latitude	27° 23' 17.3 N
Longitude	76° 16' 11.7 E

B. Topographical features

Erosion status	Topography	Soil depth
Moderate	Hill slope	Moderately deep (25-50)

C. Vegetation status

Tree species with IVI

A. pendula (248), *C. decidua* (52)

Shrub species with IVI

no

ENS Tree : 1.17 Shrub: 1.00

Herb and Grass

Adiantum lunulatum, Actiniopteris radiata, Aerva lanata, B. diffusa, E. alsinoides, I. cordifolia.

D. Wild-life

Boselaphus tragocamelus, Pavo cristatus, Psittacula krameri, Passer domesticus.

E. Soil characteristics

pH	EC (dSm^{-1})	SOC (%)	BD (g/cm^3)	NO_3-N ($mg\ kg^{-1}$)	NH_4-N ($mg\ kg^{-1}$)	PO_4-P ($mg\ kg^{-1}$)	C density ($tons\ ha^{-1}$)
7.66	0.18	1.140	1.32	2.56	4.68	9.64	31.15

F. History and status

An ancient sacred grove surrounding a hillock. There is a temple of Mataji on the hill top, where people use to come regularly to worship mataji. The grove is mostly dominated by *Anogeisus pendula* trees.

DATA ON SACRED GROVES OF RAJASTHAN

Sacred Grove No. 8

A. Basic information about the sacred groves

Name of sacred groves: Shiv Ji Maharaj Ka Oran
Name of village: Garh Bassi
Panchayat: Thanagazi
Nearest forest Block: Garh Bassi
Land status: Community land
District: Alwar
Area: 125 bigha
Altitude (m): 434
Latitude: 27° 23' 35.6 N
Longitude: 76° 13' 48.0 E

B. Topographical features

Erosion status	Topography	Soil depth
Slight	Foot hill slope	Moderately deep (25-50)

C. Vegetation status

Tree species with IVI

A. leucophloea (110), *A. senegal* (45), *P. juliflora* (39), *A. indica* (39), *B. aegyptiaca* (31), *H. integrifolia* (23), *A. nilotica* (10), *Moringa concanensis* (3).

Shrub species with IVI

A. vasica (73), *Z. nummularia* (55), *Dichrostachys cinerea* (34), *Rhus mysurensis* (31), *G. tenax* (29), *Commiphora wightii* (23), *Grewia flavescence* (19), *C. sepiaria* (19), *S. virosa* (10), *C. fruticosa* (7).

ENS Tree : 5.48 Shrub: 7.00

Herb and Grass

Elytraria acaulis, A. radiata, Euphorbia hirta, Pavonia arabica, T. purpurea, B. diffusa, E. alsinoides, I. cordifolia.

D. Wild-life

Picnonotus cafer, Streptopelia capicola, Pavo cristatus.

E. Soil characteristics

pH	EC (dSm^{-1})	SOC (%)	BD (g/cm^3)	NO_3-N ($mg\ kg^{-1}$)	NH_4-N ($mg\ kg^{-1}$)	PO_4-P ($mg\ kg^{-1}$)	C density ($tons\ ha^{-1}$)
7.94	0.31	0.300	1.43	2.14	1.90	11.91	5.74

F. History and status

An ancient temple of Lord Shiva exists in the centre of the groves in foothill. A fair is used to commence here in October each year.

DATA ON SACRED GROVES OF RAJASTHAN

Sacred Grove No. 9

A. Basic information about the sacred groves

Name of sacred groves	Dagia Bheruji Ka Oran
Name of village	Lakheria, Ghatol
Panchayat	Ghatol
Nearest forest Block	Lakeria
Land status	Community land
District	Banswara
Area	84 bigha
Altitude (m)	267
Latitude	23° 45' 08.5 N
Longitude	74° 24' 55.9 E

B. Topographical features

Erosion status	Topography	Soil depth
Moderate	Hill slope	Shallow (<25 cm)

C. Vegetation status

Tree species with IVI

Tectona grandis (150), *H. integrifolia* (66), *Adina cordifolia* (61), *Wrightia tinctoria* (13), *A. indica* (10)

Shrub species with IVI

L. camara (161), *Barleria prionitis* (94), *C. sepiaria* (45)

ENS Tree : 1.38 Shrub: 2.55

Herb and Grass

Elytraria acaulis, Euphorbia hirta, T. purpurea, B. diffusa, E. alsinoides, I. cordifolia.

D. Wild-life

Boselaphus tragocamelus, *Canis aureus auveus*, *Psittacula krameri*, *Corvus macrorhynchos*.

E. Soil characteristics

pH	EC (dSm^{-1})	SOC (%)	BD (g/cm^3)	NO_3-N (mg kg^{-1})	NH_4-N (mg kg^{-1})	PO_4-P (mg kg^{-1})	C density (tons ha^{-1})
6.33	0.12	0.623	1.36	2.56	2.56	10.23	9.29

F. History and status

An ancient sacred grove is here in name of Bheruji and is protected and conserved by the local Bhil community. It is dominated by *Tectona grandis* as the tree species but *Lantana camara* is dominating as the shrub species.

DATA ON SACRED GROVES OF RAJASTHAN

Sacred Grove No. 10

A. Basic information about the sacred groves

Name of sacred groves	Prathurinath ji Ka Oran
Name of village	Amlipada
Panchayat	Banswara
Nearest forest Block	Anand sagar
Land status	Community land
District	Banswara
Area	35 bigha
Altitude (m)	270
Latitude	23° 32' 51.3 N
Longitude	74° 28' 30.1 E

B. Topographical features

Erosion status	Topography	Soil depth
Moderate	Hill slope	Shallow (<25 cm)

C. Vegetation status

Tree species with IVI

T. grandis (160), *A. cordifolia* (47), *Aegle marmelos* (41), *Morinda tinctoria* (18), *Diospyros melanoxylon* (17), *Mitragyna parviflora* (17).

Shrub species with IVI

No shrub

ENS Tree : 2.39 Shrub: 0

Herb and Grass

Alysicarpus vaginalis, Cassia tora, E. acaulis, E. hirta, B. ramose, B. diffusa, E. alsinoides, I. cordifolia

D. Wild-life

Panthera pardus, Boselaphus tragocamelus, Canis aureus auveus, Corvus macrorhynchos, Psittacula krameri

E. Soil characteristics

pH	EC (dSm^{-1})	SOC (%)	BD (g/cm^3)	NO_3-N ($mg\ kg^{-1}$)	NH_4-N ($mg\ kg^{-1}$)	PO_4-P ($mg\ kg^{-1}$)	C density ($tons\ ha^{-1}$)
6.12	0.19	1.080	1.40	1.24	4.50	12.56	12.64

F. History and status

An old grove in name of Prathurinathji- a local diety exists here. It is dominated by *Tectona grandis*, but moisture loving trees are also available in the grove.

DATA ON SACRED GROVES OF RAJASTHAN

Sacred Grove No. 11

A. Basic information about the sacred groves

Name of sacred groves	Sai Baba Ka Mandir
Name of village	Anari
Panchayat	Banswara
Nearest forest block	Anand sagar
Land status	Community land
District	Banswara
Area	14 bigha
Altitude (m)	265
Latitude	23° 32' 53.4 N
Longitude	74° 27' 54.5 E

B. Topographical features

Erosion status	Topography	Soil depth
Moderate	Hill top	Shallow (<25 cm)

C. Vegetation status

Tree species with IVI

T. grandis (131), *A. leucophloea* (60), *P. juliflora* (55), *Bombax ceiba* (38), *A. indica* (16).

Shrub species with IVI

L. camara (114), *C. procera* (107), *Z. nummularia* (42), *Cassia auriculata* (37).

ENS Tree : 2.73 Shrub: 3.59

Herb and Grass

A. vaginalis, C. tora, B. ramosa, B. diffusa, D. sindicum, E. alsinoides, I. cordifolia, E. acaulis, E. hirta, Leonotis nepetifolia.

D. Wild-life

Panthera pardus, *Canis aureus auveus*, *Hyaena hyaena*, *Psittacula krameri*, *Pavo cristatus*, *Ophiophagus hannah*.

E. Soil characteristics

pH	EC (dSm^{-1})	SOC (%)	BD (g/cm^3)	NO_3-N ($mg\ kg^{-1}$)	NH_4-N ($mg\ kg^{-1}$)	PO_4-P ($mg\ kg^{-1}$)	C density ($tons\ ha^{-1}$)
7.20	0.18	0.420	1.41	1.25	3.40	12.50	7.19

F. History and status

This grove is approx. 250 years old. People believe that God Bheroji appeared from a wall, where the present statue is available. The old temple is under ruin and the new temple was built up in 1982.

DATA ON SACRED GROVES OF RAJASTHAN

Sacred Grove No. 12

A. Basic information about the sacred groves

Name of sacred groves	Devnarayan Ji Ka Oran
Name of village	Jhadoda
Panchayat	Attru
Nearest forest block	Jhadoda
Land status	Community land
District	Baran
Area	28 bigha
Altitude (m)	293
Latitude	24° 47’ 58.6 N
Longitude	76° 28’ 49.9 E

B. Topographical features

Erosion status	Topography	Soil depth
Moderate	Flat land with drainage	Shallow (<25 cm)

C. Vegetation status

Tree species with IVI

Butea monosperma (133), *P. juliflora* (40), *Tamarindus indica* (31), *M. parviflora* (30), *Phoenix sylvestris* (21), *A. marmelos* (13), *P. cineraria* (13), *A. leucophloea* (10), *Crataeva nurvala* (10)

Shrub species with IVI

Z. nummularia (139), *Flacourtia indica* (55), *C. sepiaria* (44), *S. virosa* (43), *Carissa carandas* (19)

ENS Tree : 4.09 Shrub: 4.37

Herb and Grass

Blumea eriantha, Cassia tora, E. hirta, T. roxburghii, T. procumbense

D. Wild-life

Lepus nigricolis, *Pavo cristatus*, *Psittacula krameri*, *Francolinus pondicerianus*, *Canis aureus auveus*, *Boselaphus tragocamelus*, *Corvus macrorhynchos*, *Naja naja*, Semnopithecus entellus.

E. Soil characteristics

pH	EC (dSm^{-1})	SOC (%)	BD (g/cm^3)	NO_3-N (mg kg^{-1})	NH_4-N (mg kg^{-1})	PO_4-P (mg kg^{-1})	C density (tons ha^{-1})
7.59	0.59	0.285	1.47	2.13	9.60	9.20	10.05

F. History and status

An old grove in name of Lord Devanarayan ji-a localy venerated diety of Gurjar community, to whom the people worship and maintain the grove.

DATA ON SACRED GROVES OF RAJASTHAN

Sacred Grove No. 13

A. Basic information about the sacred groves

Name of sacred groves	Musain Mata Ji Ka Oran
Name of village	Musain Mata ji
Panchayat	Attru
Nearest forest block	Musain Mata ji
Land status	Community land
District	Baran
Area of sacred grove	150 bigha
Altitude (m)	490
Latitude	24° 48' 10.3N
Longitude	76° 38' 19.8 E

B. Topographical features

Erosion status	Topography	Soil depth
Moderate erosion	Flat land	Moderate (50-100)

C. Vegetation status

Tree species with IVI

A. pendula (190), *D. melanoxylon* (32), *A. leucophloea* (27), *G. montana* (16), *Acacia catechu* (13), *Anogeissus latifolia* (12), *P. juliflora* (10).

Shrub species with IVI

Carissa carandas (186) *F. indica* (62), *C. sepiaria* (21), *Z. nummularia* (16), *S. virosa* (15)

ENS Tree : 4.22 Shrub: 2.97

Herb and Grass

E. hirta. T. roxburghii, T. procumbense, E. alsinoides, I. cordifolia.

D. Wild-life

Melursus ursinus, Panthera pardus, Axis axis, Lepus nigricolis, Pavo cristatus, Psittacula krameri, Francolinus pondicerianus, Canis aureus auveus, Boselaphus tragocamelus, Corvus macrorhynchos, Naja naja, Semnopithecus entellus, Sus scrofa.

E. Soil characteristics

pH	EC (dSm^{-1})	SOC (%)	BD (g/cm^3)	NO_3-N ($mg\ kg^{-1}$)	NH_4-N ($mg\ kg^{-1}$)	PO_4-P ($mg\ kg^{-1}$)	C density ($tons\ ha^{-1}$)
7.62	1.94	0.390	1.48	1.19	7.12	13.00	10.58

F. History and status

In ancient time the area was covered with dense forest and many wild animals lived here, roaring of panther could be heard in evening. People have started worshiping the miraculous Goddess, in name of whom this oran is conserved at present.

DATA ON SACRED GROVES OF RAJASTHAN

Sacred Grove No. 14

A. Basic information about the sacred groves

Namew of sacred groves	Kalla ji Ka Bagh (Nag)
Name of village	Tanchla
Panchayat	Chhipabadod
Nearest forest Block	Tanchla
Land status	Community land
District	Baran
Area (bigha)	86
Altitude (m)	331
Latitude	24° 41' 05.2 N
Longitude	76° 43' 39.9 E

B. Topographical features

Erosion status	Topography	Soil depth
Moderate erosion	Undulating flate land	Moderate (50-100)

C. Vegetation status

Tree species with IVI

B. monosperma (102), *P. juliflora* (47), *M. parviflora* (46), *H. integrifolia* (37), *A. leucophloea* (28), *Terminalia arjuna* (14), *A. pendula* (10), *Holarrhena antidysenterica* (9), *Syzygium cummini* (7).

Shrub species with IVI

C. sepiaria (300)

ENS Tree : 5.77 Shrub: 1.00

Herb and Grass

C. tora, E. hirta, T. procumbense, E. alsinoides, I. cordifolia.

D. Wild-life

Melursus ursinus, *Axis axis*, *Lepus nigricolis*, *Pavo cristatus*, *Psittacula krameri*, *Columba livia*, *Francolinus pondicerianus*, *Canis aureus auveus*, *Boselaphus tragocamelus*, *Corvus macrorhynchos*, Snake, *Semnopithecus entellus, Herpestes edwardsi.*

E. Soil characteristics

pH	EC (dSm^{-1})	SOC (%)	BD (g/cm^3)	NO_3-N ($mg\ kg^{-1}$)	NH_4-N ($mg\ kg^{-1}$)	PO_4-P ($mg\ kg^{-1}$)	C density ($tons\ ha^{-1}$)
7.87	0.17	0.555	1.52	1.09	9.16	9.00	16.05

F. History and status

An ancient grove in name on Kalaji – the local diety in Baran district exists here. The area is dominated by moisture loving trees like *Mitragyna parviflora*, *S. cummini* etc.

DATA ON SACRED GROVES OF RAJASTHAN

Sacred Grove No. 15

A. Basic information about the sacred groves

Name of sacred groves	Bankal Mata ji Ka Oran
Name of village	Ranigaow
Panchayat	Barmer
Nearest forest Block	Ranigaon
Land status	Community land
District	Barmer
Area	800 bigha
Altitude (m)	173
Latitude	25° 35' 57.9 N
Longitude	71° 17' 50.5 E

B. Topographical features

Erosion status	Topography	Soil depth
Moderate erosion	Flat land (undulating)	Moderate (50 100)

C. Vegetation status

Tree species with IVI

Tecomella undulata (181), *S. oleoides* (78), *P. juliflora* (41)

Shrub species with IVI

Aerva persica (249), *Z. nummularia* (51)

ENS Tree : 2.34 Shrub: 1.89

Herb and Grass

T. purpurea, Lindenbergia muraria, A. senegalensis.

D. Wild-life

Gazella gazella, *Pavo cristatus*, *Francolinus pondicerianus*, *Streptopelia decocto*, *Lepus nigricollis*, *Meriones hurrianae*, *Vulpes vulpes pusilla*, *Chamaeleo zeylanicus*.

E. Soil characteristics

pH	EC (dSm^{-1})	SOC (%)	BD (g/cm^3)	NO_3-N ($mg\ kg^{-1}$)	NH_4-N ($mg\ kg^{-1}$)	PO_4-P ($mg\ kg^{-1}$)	C density ($tons\ ha^{-1}$)
7.80	0.21	0.150	1.58	1.25	1.50	5.80	5.17

F. History and status

An old grove in name of Bankal Mataji - Kuldevi of 'Daga' and 'Ladhdha' communities in Rajasthan exists here. This grove is conserved and protected by the local communities. The grove is dominated by Marwar Teak (*Tecomella undulata*).

DATA ON SACRED GROVES OF RAJASTHAN

Sacred Grove No. 16

A. Basic information about the sacred groves

Name of sacred groves	Viratra Mata Ka Oran
Name of village	Dhok
Panchayat	Dhok
Nearest forest Block	Chohatan
Land status	Community land
District	Barmer
Area	1859 ha
Altitude (m)	234
Latitude	25° 27' 48.2 N
Longitude	71° 01' 34.4 E

B. Topographical features

Erosion status	Topography	Soil depth
Moderate erosion	Undulating plain land	Deep (>100cm)

C. Vegetation status

Tree species with IVI	Shrub species with IVI
M. emarginata (224), *S. oleoides* (58), *A. tortilis* (18).	*Z. nummularia* (92), *C. burhia* (54), *Calligonum polygonoides* (51), *Aerva pseudotomentosa* (50), *L. barbarum* (20), *A. jacquemontii* (16), *Maerua oblongifolia* (9), *L. pyrotechnica* (8)

ENS Tree : 1.65 Shrub: 5.79

Herb and Grass

Ephedra foliata, Aristida funiculata, A. hispidissima, B. erecta, D. sindicum, Mollugo cerviana, P. erioptera, T. roxburghii.

D. Wild-life

Gazela gazela, *Boselaphus tragocamelus*, *Vulpes vulpes pusilla*, *Pavo cristatus*, *Lepus nigricolis*, *Francolinus pondicerianus Chamaeleo zeylanicus*, *Varanus bengalensis*, *Aquila hastate*.

E. Soil characteristics

pH	EC (dSm^{-1})	SOC (%)	BD (g/cm^3)	NO_3-N ($mg\ kg^{-1}$)	NH_4-N ($mg\ kg^{-1}$)	PO_4-P ($mg\ kg^{-1}$)	C density ($tons\ ha^{-1}$)
7.26	0.15	0.060	1.60	1.09	1.15	5.40	2.88

F. History and status

Shree Vakrangi Mata (Vankal Mata) is the Kuldevi of Nandwana community. The Oran is covering 1859 ha area. Penalties for violations are Rs. 500 to 1000, imposed by the Oran Prabandhak Samiti, involving 7-8 villages. Earlier the Samiti was dominated by Rajputs, but now other casts have also joined and their needs have been accommodated.

DATA ON SACRED GROVES OF RAJASTHAN

Sacred Grove No. 17

A. Basic information about the sacred groves

Name of sacred groves	Mata ji Ka Oran
Name of village	Unrod
Panchayat	Shiv
Nearest forest Block	Harshani
Land status	Community land
District	Barmer
Area	2000 bigha
Altitude (m)	152
Latitude	26° 06' 18.1 N
Longitude	70° 45' 59.3 E

B.Topographical features

Erosion status	Topography	Soil depth
Moderate erosion	Plain land	Deep (>100 cm)

C. Vegetation status

Tree species with IVI

C. decidua (300)

Shrub species with IVI

Z. nummularia (127), *C. procera* (112), *L. pyrotechnica* (45), *A. pseudotomentosa* (16)

ENS Tree : 1.00 Shrub: 3.55

Herb and Grass

A. funiculata, Corchorus depressus, E. minor, Fegonia cretica, Heliotropium species, T. roxburghii.

D. Wild-life

Gazella pussill, Vulpes vulpes pussilla, Lepus nigricolis, Canis aureus auveus, Chamaeleo zeylanicus, Eryx conicus, Streptopelia decoct, Francolinus pondicerianus, Picnonotus cafer, Leptocoma zeylonica.

E. Soil characteristics

pH	EC (dSm^{-1})	SOC (%)	BD (g/cm^3)	NO_3-N ($mg\ kg^{-1}$)	NH_4-N ($mg\ kg^{-1}$)	PO_4-P ($mg\ kg^{-1}$)	C density ($tons\ ha^{-1}$)
7.78	0.18	0.135	1.59	1.50	1.60	6.56	6.05

F. History and status

This oran is protected and conserved in name of Mataji- local goddess. Local people use to worship her in the area.

DATA ON SACRED GROVES OF RAJASTHAN

Sacred Grove No. 18

A. Basic information about the sacred groves

Name of sacred groves	Khubad Mata Ji Ka Oran
Name of village	Nimbal Kot
Panchayat	Kosloo
Forest Block	Banadion ki dhani
Land status	Community land
District	Barmer
Area	210 bigha
Altitude (m)	105
Latitude	25° 31' 31.6 N
Longitude	71° 43' 52.0 E

B. Topographical features

Erosion status	Topography	Soil depth
Moderate	Plain land	Deep (>100 cm)

C. Vegetation status

Tree species with IVI

S. oleoides (174), *P. cineraria* (50), *C. decidua* (43), *T. undulata* (33)

Shrub species with IVI

C. phlomidis (138), *A. pseudotomentosa* (85), *L. pyrotechnica* (73), *Lycium barbarum* (4)

ENS Tree : 2.81 Shrub: 2.97

Herb and Grass

Cenchrus biflorus, E. minor, Heliotropium marifolium, Leucas cephalotes, Solanum albicaule, T. purpurea, T. roxburghii.

D. Wild-life

Boselaphus tragocamelus, *Gazella ussill*, *Canis aureus auveus*, *Vulpes vulpes ussilla*, *Columbia livia*, *Francolinus pondicerianus*, *Lepus nigricolis*, *Aquila hastate*, *Picnonotus cafer*, *Uromastix hardwickii*.

E. Soil characteristics

pH	EC (dSm^{-1})	SOC (%)	BD (g/cm^3)	NO_3-N (mg kg^{-1})	NH_4-N (mg kg^{-1})	PO_4-P (mg kg^{-1})	C density (tons ha^{-1})
7.90	0.15	0.090	1.58	1.23	1.23	10.55	4.27

F. History and status

It is a quite old grove dominated by *Salvadora oleoides*. There is a small temple in the grove, where local people come and worship Mataji.

DATA ON SACRED GROVES OF RAJASTHAN

Sacred Grove No. 19

A. Basic information about the sacred groves

Name of sacred groves: Mama ji Ki Selenadi
Name of village: Doli
Panchayat: Doli
Nearest forest Block: Sarvadi
Land status: Community land
District: Barmer
Area: 400 bigha
Altitude (m): 160
Latitude: 26° 05' 21.4 N
Longitude: 72° 39' 11.3 E

B. Topographical features

Erosion status	Topography	Soil depth
Moderate erosion	Plain land	Deep (>100 cm)

C. Vegetation status

Tree species with IVI

P. cineraria (101), *C. decidua* (85), *P. juliflora* (48), *S. oleoides* (66)

Shrub species with IVI

Z. nummularia (300)

ENS Tree : 3.54 Shrub: 1.00

Herb and Grass

B. diffusa, D. sindicum, Euphorbia granularis, I. cordifolia, Justicia procumbens, P. crispa, Zygophyllum simplex.

D. Wild-life

Gazella gazella, *Lepus nigricolis*, *Pavo cristatus*, *Francolinus pondicerianus*, *Chamaeleo zeylanicus*, *Varanus griseus*.

E. Soil characteristics

pH	EC (dSm^{-1})	SOC (%)	BD (g/cm^3)	NO_3-N ($mg\ kg^{-1}$)	NH_4-N ($mg\ kg^{-1}$)	PO_4-P ($mg\ kg^{-1}$)	C density ($tons\ ha^{-1}$)
7.88	0.16	0.089	1.58	1.22	1.26	9.22	4.03

F. History and status

This sacred grove area is now considered as community reserve observed in famous Dhava-Doli wild life area of Jodhpur-Barmer region.

DATA ON SACRED GROVES OF RAJASTHAN

Sacred Grove No. 20

A. Basic information about the sacred groves

Name of sacred groves	Ghanshsyam Baba Ka Ashram
Name of village	Mulla
Panchayat	Kanwara
Nearest forest Block	Kanwara
Land status	Community land
District	Bharatpur
Area	500 bigha
Altitude (m)	202
Latitude	27° 36' 50.6 N
Longitude	72° 18' 68.1 E

B. Topographical features

Erosion status	Topography	Soil depth
Moderate	Undulating plain land	Deep soil (>100 cm)

C. Vegetation status

Tree species with IVI

S. oleoides (82), *P. cineraria* (60), *A. pendula* (41), *H. integrifolia* (32), *A. indica* (23), *C. decidua* (20), *B. aegyptiaca* (16), *Ficus religiosa* (8), *Pongamia pinnata* (6), *Morus alba* (5), *Dalbergia sissoo* (4), *Cordia dichotoma* (3).

Shrub species with IVI

No shrub

ENS Tree : 6.75 Shrub: 0

Herb and Grass

B. diffusa, D. sindicum, E. hirta, I. cordifolia, J. procumbens.

D. Wild-life

Lepus nigricolis, *Pavo cristatus*, *Psittacula krameri*, *Columba livia*, *Francolinus pondicerianus*, *Canis aureus auveus*, *Boselaphus tragocamelus*, *Corvus splendens*, Snake, *Macaca fascicularis*.

E. Soil characteristics

pH	EC (dSm^{-1})	SOC (%)	BD (g/cm^3)	NO_3-N ($mg\ kg^{-1}$)	NH_4-N ($mg\ kg^{-1}$)	PO_4-P ($mg\ kg^{-1}$)	C density ($tons\ ha^{-1}$)
7.04	0.11	0.338	1.48	6.14	6.67	12.46	13.75

F. History and status

This is 25-30 years old grove. People of two villages have donated their lands for this and on demand of villagers Baba Ghanshyamdas has secured this place. Baba Ramcharn Das has constructed this temple and Baba Makhandas is now taking care of it.

DATA ON SACRED GROVES OF RAJASTHAN

Sacred Grove No. 21

A. Basic information about the sacred groves

Name of sacred groves	Adi Badri Dham
Name of village	Alipur
Panchayat	Pasopa
Nearest forest Block	Pasopa
Land status	Community land
District	Bharatpur
Area	4.5 bigha
Altitude (m)	207
Latitude	27° 32' 00.0 N
Longitude	77° 12' 18.2 E

B. Topographical features

Erosion status	Topography	Soil depth
Moderate erosion	Hill slope	Moderate deep soil (50-100)

C. Vegetation status

Tree species with IVI

M. parviflora (74), *C. dichotoma* (55), *A. indica* (54), *F. religiosa* (45), *T. indica* (24), *A. marmelos* (18), *Lawsonia inermis* (14), *A. lebbeck* (9), *Mangifera indica* (7).

Shrub species with IVI

Z. nummularia (300).

ENS Tree : 5.84 Shrub: 1.00

Herb and Grass

Not available

D. Wild-life

Lepus nigricolis,*Canis aureus auveus*, *Hyaena hyaena*, *Boselaphus tragocamelus*, *Pavo cristatus*, *Psittacula krameri*, *Columba livia*, *Francolinus pondicerianus*, *Corvus macrorhynchos*, Snake, *Macaca fascicularis*, *Acridotheres tristis*, *Passer domesticus*.

E. Soil characteristics

pH	EC (dSm^{-1})	SOC (%)	BD (g/cm^3)	NO_3-N ($mg\ kg^{-1}$)	NH_4-N ($mg\ kg^{-1}$)	PO_4-P ($mg\ kg^{-1}$)	C density ($tons\ ha^{-1}$)
7.67	0.12	0.315	1.49	7.13	3.03	8.00	11.65

F. History and status

The history is attached with lord Krishna. Nand Baba was not having a child so he prayed god for a son and assured that he would travel to all Teerth in case if a child is born. After fulfillment of his desire, Baba requested son Krishna for travel of all Teerth. Rather travelling along with all teerth, Krishna brought all Teerth at this place within a moment.

DATA ON SACRED GROVES OF RAJASTHAN

Sacred Grove No. 22

A. Basic information about the sacred groves

Name of sacred groves	Sawai Bhoj/Devnarayan Mandir
Name of village	Asind
Panchayat	Asind
Forest Block	Amesar
Land status	Community land
District	Bhilwara
Area	679 bigha
Altitude (m)	467
Latitude	25° 43' 51.9 N
Longitude	74° 20' 54.7 E

B. Topographical features

Erosion status	Topography	Soil depth
Slight	Almost plain land	Moderately deep (50-100)

C. Vegetation status

Tree species with IVI

P. juliflora (88), *Salvadora persica* (56), *A. leucophloea* (44), *P. cineraria* (31), *A. indica* (26), *A. senegal* (21), *C. decidua* (19), *B. aegyptiaca* (15)

Shrub species with IVI

Euphorbia neriifolia (113), *G. tenax* (71), *S. virosa* (61), *Z. nummularia* (31), *D. cinerea* (24)

ENS Tree : 6.16 Shrub: 4.34

Herb and Grass

A. aspera, Enicostemma axillare, A. funiculata, Apluda mutica, A. vaginalis, Barleria acanthoides, B. barbata, B. erecta, Chloris barbata, C. benghalensis, Crotalaria medicaginea, Cassia pumila, D. fovelatum, E. hirta, Heteropogon contortus, Hibiscus species, I. cordifolia, Martynia annua, Pentatropis spiralis, Sporobolus species.

D. Wild-life

Boselaphus tragocamelus, *Canis aureus auveus*, *Chamaeleo zeylanicus*, *Pavo cristatus*, *Aquila hastate*, *Streptopelia decocto*, *Francolinus pondicerianus*, *Coturnix coturnix* etc.

E. Soil characteristics

pH	EC (dSm^{-1})	SOC (%)	BD (g/cm^3)	NO_3-N ($mg\ kg^{-1}$)	NH_4-N ($mg\ kg^{-1}$)	PO_4-P ($mg\ kg^{-1}$)	C density ($tons\ ha^{-1}$)
7.33	0.20	1.080	1.42	1.78	6.30	13.34	32.91

F. History and status

This temple was built in 10th century by Sh. Sawai Bhoj on the left bank of Khari River. This temple is dedicated to Lord Deo Narain' and caters the religious need of Gujar community

DATA ON SACRED GROVES OF RAJASTHAN

Sacred Grove No. 23

A. Basic information about the sacred groves

Name of sacred groves	Devnaryan Ji Ki Bani
Name of village	Ladpura
Panchayat	Mandalgarh
Nearest forest block	Naya nagar
Land status	Community land
District	Bhilwara
Area	400 bigha
Altitude (m)	393
Latitude	25° 08' 37.8 N
Longitude	75° 06' 35.8 E

B. Topographical features

Erosion status	Topography	Soil depth
Slight	Plain gravelly land	Shallow (<25 cm)

C. Vegetation status

Tree species with IVI

A. pendula (220), *Acacia catechu* (63), *Anogeissus latifolia* (10), *P. cineraria* (7).

Shrub species with IVI

No shrub

ENS Tree : 1.56 Shrub: 1.00

Herb and Grass

A. lunulatum, B. pussilla, C. dactylon, C. tora, D. aegyptium, E. alsinoides, Funaria, Glossocardia bosvallia, H. hispida, Hibiscus species, I. cordifolia, J. procumbense, T. tennala, Themeda species.

D. Wild-life

Boselaphus tragocamelus, Canis aureus auveus, Oryctolagus cuniculus, Picnonotus cafer, Streptopelia decocto.

E. Soil characteristics

pH	EC (dSm^{-1})	SOC (%)	BD (g/cm^3)	NO_3-N ($mg\ kg^{-1}$)	NH_4-N ($mg\ kg^{-1}$)	PO_4-P ($mg\ kg^{-1}$)	C density ($tons\ ha^{-1}$)
7.42	0.22	1.155	1.41	2.18	10.67	12.86	29.48

F. History and status

Very old grove established long back and a temple was built here by Nayanagar King Mann Singh during his journey to this area. People use to come at the temple and worship here.

DATA ON SACRED GROVES OF RAJASTHAN

Sacred Grove No. 24

A. Basic information about the sacred groves

Name of sacred groves	Devnaryan ji Ka Oran
Name of village	Manpur
Panchayat	Mandalgarh
Nearest forest Block	Manpura beed
Land status	Community land
District	Bhilwara
Area	58 bigha
Altitude (m)	369
Latitude	25° 17' 55.1 N
Longitude	75° 06' 21.0 E

B. Topographical features

Erosion status	Topography	Soil depth
Slight	Flat land	Moderate (50-100)

C. Vegetation status

Tree species with IVI

A. leucophloea (98), *A. catechu* (49), *B. monosperma* (38), *A. nilotica* (37), *A. indica* (37), *F. religiosa* (32), *P. juliflora* (9).

Shrub species with IVI

S. virosa (300)

ENS Tree : 4.84 Shrub: 1.00

Herb and Grass

A. mutica, A. vaginalis, C. dactylon, C. rotundus, C. tora, E. acaulis, E. colona, E. hirta, Euphorbia indica, P. paniculata, Themeda sp.

D. Wild-life

Boselaphus tragocamelus, *Corvus macrorhynchos*, *Acridotheres tristis.*

E. Soil characteristics

pH	EC (dSm^{-1})	SOC (%)	BD (g/cm^3)	NO_3-N ($mg\ kg^{-1}$)	NH_4-N ($mg\ kg^{-1}$)	PO_4-P ($mg\ kg^{-1}$)	C density ($tons\ ha^{-1}$)
7.74	0.31	1.290	1.41	1.91	2.66	9.56	22.00

F. History and status

This grove is also maintained and conserved in Name of Lord Devnarayan ji – the main diety of Gurjar community. This grove is mostly dominated by *Acacia leucophloea*.

DATA ON SACRED GROVES OF RAJASTHAN

Sacred Grove No. 25

A. Basic information about the sacred groves

Name of sacred groves	Goga Ji Ka Oran
Name of village	Gangapura
Panchayat	Khari
Nearest forest Block	Balleri
Land status	Community land
District	Bikaner
Area of sacred grove	115 bigha
Altitude (m)	218
Latitude	27° 58' 33.9 N
Longitude	72° 53' 34.9 E

B. Topographical features

Erosion status	Topography	Soil depth
Moderate	Flat land	Deep (>100 cm)

C. Vegetation status

Tree species with IVI

Ziziphus nummularia (139), *C. decidua* (139), *P. juliflora* (22)

Shrub species with IVI

No shrub

ENS Tree : 2.51 Shrub: 1.00

Herb and Grass

Cenchrus species, C. dactylon, E. hirta, D. sindicum.

D. Wild-life

Gazella gazella, Uromastix hardwickii, Lepus nigricolis

E. Soil characteristics

pH	EC (dSm^{-1})	SOC (%)	BD (g/cm^3)	NO_3-N ($mg\ kg^{-1}$)	NH_4-N ($mg\ kg^{-1}$)	PO_4-P ($mg\ kg^{-1}$)	C density ($tons\ ha^{-1}$)
8.74	0.30	0.345	1.50	2.36	4.30	7.32	11.01

F. History and status

A quit old oran in name of Jahar Veer Gogga, who is a folk deity, worshiped in northern states of India mainly in Rajasthan, Uttarakhand, Punjab and Uttar Pradesh. He is a warrior-hero of the region and is venerated as a saint or a 'snake-god'.

DATA ON SACRED GROVES OF RAJASTHAN

Sacred Grove No. 26

A. Basic information about the sacred groves

Name of sacred groves	Karni Mata Ji ka Oran
Name of village	Mokha Charan
Panchayat	Kolayat
Nearest forest Block	Mokh Khalsa
District	Bikaner
Area of sacred grove	1800 bigha
Altitude (m)	292
Latitude	27° 45' 52.3 N
Longitude	73° 03' 36.8 E

B. Topographical features

Erosion status	Topography	Soil depth
Moderate	Flat land	Deep (>100 cm)

C. Vegetation status

Tree species with IVI

Z. nummularia (105), *S. oleoides* (68), *C. decidua* (58), *A. tortilis* (52), *P. juliflora* (17).

Shrub species with IVI

L. barbarum (300)

ENS Tree : 3.96 Shrub: 1.00

Herb and Grass

Cenchrus species, C. dactylon, E. hirta, D. sindicum, T. procumbense.

D. Wild-life

Axis axis, *Vulpes bengalehsis*, *Lepus nigricolis*, *Streptopelia decocto*, *Columba livia*.

E. Soil characteristics

pH	EC (dSm^{-1})	SOC (%)	BD (g/cm^3)	NO_3-N ($mg\ kg^{-1}$)	NH_4-N ($mg\ kg^{-1}$)	PO_4-P ($mg\ kg^{-1}$)	C density ($tons\ ha^{-1}$)
8.52	0.29	0.330	1.51	1.15	2.51	8.30	14.27

F. History and status

This sacred grove is also conserved and maintained in name of Karni Mata- whose main temple is in Desnoke. It is mostly dominated by *Zizyphus nummularia* trees.

DATA ON SACRED GROVES OF RAJASTHAN

Sacred Grove No. 27

A. Basic information about the sacred groves

Name of sacred groves	Karni Mata Ji Ka Oran
Name of village	Deshnok
Panchayat	Bikaner
Nearest forest Block	Deshnok
Land status	Community land
District	Bikaner
Area of sacred grove	2216 bigha
Altitude (m)	290
Latitude	27° 47' 07.0 N
Longitude	73° 20' 11.5 E

B. Topographical features

Erosion status	Topography	Soil depth
Slight	Flat land	Deep (>100 cm)

C. Vegetation status

Tree species with IVI

Z. nummularia (300)

Shrub species with IVI

L. pyrotechnica (141), *A. pseudotomentosa* (100), *C. burhia* (32), *L. barbarum* (15), *A. persica* (12).

ENS Tree : **1.00** Shrub: **3.06**

Herb and Grass

Cenchrus species, *C. dactylon, E. hirta, D. sindicum, T. procumbens, Lasiurus sindicus.*

D. Wild-life

Axis axis, *Vulpes bengalehsis*, *Lepus nigricolis*

E. Soil characteristics

pH	EC (dSm^{-1})	SOC (%)	BD (g/cm^3)	NO_3-N ($mg\ kg^{-1}$)	NH_4-N ($mg\ kg^{-1}$)	PO_4-P ($mg\ kg^{-1}$)	C density ($tons\ ha^{-1}$)
8.64	0.19	0.330	1.49	0.19	4.50	9.60	13.77

F. History

The famous Karani Mata temple of rat is situated in this sacred grove. According to a legend Laxman, Karni Mata's stepson, drowned in a pond in Kapil Sarovar in Kolayat while attempting to drink water from it. Karni Mata implored 'Yama', the god of death, to revive him. First refusing, Yama eventually relented and permitted Laxman and all of Karni mata's male children to be reincarnated as rats.

DATA ON SACRED GROVES OF RAJASTHAN

Sacred Grove No. 28

A. Basic information about the sacred groves

Name of sacred groves	Siya Mata ji Ki Bani
Name of village	Siya Mata ka Kheda
Panchayat	Hindoli
Nearest forest Block	Devnarayan
Land status	Community land
District	Bundi
Area of sacred grove	96 bigha
Altitude (m)	295
Latitude	25° 36' 45.9 N
Longitude	75° 36' 37.5 E

B. Topographical features

Erosion status	Topography	Soil depth
Slight	Slopey hillock	Moderately deep (25-50)

C. Vegetation status

Tree species with IVI

A. pendula (230), *P. juliflora* (32), *A. leucophloea* (29), *B. monosperma* (9)

Shrub species with IVI

No shrub (0)

ENS Tree : 3.54 Shrub: 0

Herb and Grass

Blumea eriantha, C. rotundus, E. granulata, B. pussilla, E. hirta, E. alsinoides, Mollugo cerviana, E. ciliaris, J. simplex, Digera muricata, C. benghalensis, Oropetium thomaeum.

D. Wild-life

Boselaphus tragocamelus, *Canis aureus auveus*, *Felis chaus*, *Pavo cristatus*.

E. Soil characteristics

pH	EC (dSm^{-1})	SOC (%)	BD (g/cm^3)	NO_3-N ($mg\ kg^{-1}$)	NH_4-N ($mg\ kg^{-1}$)	PO_4-P ($mg\ kg^{-1}$)	C density ($tons\ ha^{-1}$)
7.40	0.35	0.555	1.44	2.78	9.07	19.12	8.60

F. History and status

An ancient grove, which is conserved and protected in name of Siya Mataji. It is dominated by *A. pendula* tree; hovever *P. juliflora* has invaded in this grove at present.

DATA ON SACRED GROVES OF RAJASTHAN

Sacred Grove No. 29

A. Basic information about the sacred groves

Name of sacred groves	Gangrawal Ji Maharaj Ki Bani
Name of village	Chhatarganj
Panchayat	Hindoli
Nearest forest Block	Aakhlod Paandal
Land status	Community land
District	Bundi
Area of sacred grove	40 bigha
Altitude (m)	314
Latitude	25° 32' 75.4 N
Longitude	75° 30' 96.5 E

B. Topographical features

Erosion status	Topography	Soil depth
Moderate	Hill top	Moderately deep (25-50)

C. Vegetation status

Tree species with IVI

A. pendula (220), *P. juliflora* (54), *A. leucophloea* (26)

Shrub species with IVI

A. vasica (171), *C. procera* (86), *C. sepiaria* (43)

ENS Tree : 2.78 Shrub: 2.60

Herb and Grass

Acalypha indica, B. diffusa, B. ramosa, C. tora, C. benghalensis, Euphorbia chamesiana, E. hirta, J. procumbens, P. paniculata, Physalis minima, Sida cordata, Sida cordifolia, T. procumbens.

D. Wild-life

Bos taurus, Pavo cristatus, Streptopelia decocto, Psittacula krameri.

E. Soil characteristics

pH	EC (dSm^{-1})	SOC (%)	BD (g/cm^3)	NO_3-N ($mg\ kg^{-1}$)	NH_4-N ($mg\ kg^{-1}$)	PO_4-P ($mg\ kg^{-1}$)	C density ($tons\ ha^{-1}$)
7.47	0.32	0.750	1.32	2.68	7.09	14.30	12.54

F. History and status

This grove is conserved and protected in name of local diety-Gangrawalji Maharaj and dominated by *A. pendula* trees.

DATA ON SACRED GROVES OF RAJASTHAN

Sacred Grove No. 30

A. Basic information about the sacred groves

Name of sacred groves	Parbatheshwar Mahadeo Ki Bani
Name of village	Bharmal ka kheda
Panchayat	Hindoli
Nearest forest Block	-
Land status	Community land
District	Bundi
Area of sacred grove	51 bigha
Altitude (m)	303
Latitude	24° 40' 14.9 N
Longitude	75° 38' 04.3 E

B. Topographical features

Erosion status	Topography	Soil depth
Moderate	Gullied land	Deep (>100 cm)

C. Vegetation status

Tree species with IVI

M. parviflora (95), *B. monosperma* (54), *A. leucophloea* (30), *B. aegyptiaca* (25), *F. religiosa* (20), *A. nilotica* (19), *H. integrifolia* (18), *C. decidua* (16), *P. juliflora* (13), *Acacia pinnata* (7), *Maytenus emarginata* (3).

Shrub species with IVI

D. cinerea (154), *C. sepiaria* (110), *S. virosa* (36)

ENS Tree : 8.33 Shrub: 2.81

Herb and Grass species

J. procumbens, P. paniculata, Digitaria ciliaris, Triumfetta rhomboidea, Ochthochloa compressa, E. hirta, Hackelochloa granularis, E. ciliaris, E. acaulis, C. tora, Desmostachia bipinnata, B. ramosa.

D. Wild-life

Lepus nigricolis, Psittacula krameri, Funambulus palmarum

E. Soil characteristics

pH	EC (dSm^{-1})	SOC (%)	BD (g/cm^3)	NO_3-N (mg kg^{-1})	NH_4-N (mg kg^{-1})	PO_4-P (mg kg^{-1})	C density (tons ha^{-1})
8.22	0.22	0.750	1.29	2.52	4.68	12.46	13.42

F. History and status

This grove/ Bani is maintained in name of Lord Shiva as cultural and spiritual tradition of sacred spaces in belief that diety protect comunity and water harvesting systems. The dominant species in this sacred grove is *M. parviflora*.

DATA ON SACRED GROVES OF RAJASTHAN

Sacred Grove No. 31

A. Basic information about the sacred groves

Name of sacred groves	Devnarayan ji Ka Oran
Name of village	Mondalda
Panchayat	Mata ji Ki Pandoli
Nearest forest Block	-
Land status	Community land
District	Chittorgarh
Area of sacred grove	1100 bigha
Altitude (m)	439
Latitude	24° 55' 37.6 N
Longitude	74° 32' 59.0 E

B. Topographical features

Erosion status	Topography	Soil depth
Moderate	Hill slope	Shallow (<25 cm)

C. Vegetation status

Tree species with IVI

A. pendula (210), *Crateva religiosa* (28), *A. indica* (20), *Boswellia serrata* (15), *M. emarginata* (14), *M. concanensis* (13).

Shrub species with IVI

M. oblongifolia (170), *F. indica* (130)

ENS of Tree : 1.67 Shrub: 2.00

Herb and Grass

A. funiculata, A. lunulatum, Blainvillea acmella, B. pussilla, I. cordifolia, T. procumbens.

D. Wild life

Lepus nigricolis, Felis chaus, Lepus nigricolis.

E. Soil characteristics

pH	EC (dSm^{-1})	SOC (%)	BD (g/cm^3)	NO_3-N (mg kg^{-1})	NH_4-N (mg kg^{-1})	PO_4-P (mg kg^{-1})	C density (tons ha^{-1})
6.21	0.22	2.955	1.25	6.71	2.55	6.22	52.25

F. History and status

An old sacred grove and is conserved and protected in name of Lord Devanarayan- Lok Devata of Gurjar communities in Rajasthan and the adjoining states. An old temple has been established in the grove, where people use to worship their diety. It is dominated by *A. pendula*, a tree species of Aravalli region.

DATA ON SACRED GROVES OF RAJASTHAN

Sacred Grove No. 32

A. Basic information about the sacred groves

Name of sacred groves	Rishi Mangri Oran
Name of village	Paroli
Panchayat	Rohlada
Nearest forest Block	-
Land status	Community land
District	Chittorgarh
Area of sacred grove	32 bigha
Altitude (m)	445
Latitude	24° 56' 40.6 N
Longitude	74° 34' 23.6 E

B. Topographical features

Erosion status	Topography	Soil depth
Moderate	Hill slope	Shallow (<25 cm)

C. Vegetation status

Tree species with IVI

A. pendula (229), *A. leucophloea* (61), *B. monosperma* (10)

Shrub species with IVI

D. cinerea (148), *S. virosa* (125), *M. oblongifolia* (27)

ENS Tree : 1.39 Shrub: 2.36

Herb and Grass

A. aspera, B. acmella, B. pussilla, Urginea indica.

D. Wild-life

Funambulus palmarum, *Pavo cristatus*.

E. Soil characteristics

pH	EC (dSm^{-1})	SOC (%)	BD (g/cm^3)	NO_3-N (mg kg^{-1})	NH_4-N (mg kg^{-1})	PO_4-P (mg kg^{-1})	C density (tons ha^{-1})
7.29	0.18	1.290	1.31	6.35	6.42	10.46	18.71

F. History and status

It is an ancient place of worship. The temple is situated on the hill, where local people come to worship the diety. The surrounding area is conserved and protected by the local communities. The grove is dominated by *A. pendula* trees.

DATA ON SACRED GROVES OF RAJASTHAN

Sacred Grove No. 33

A. Basic information about the sacred groves

Name of sacred groves	Nahar Singhji Mata Ka Oran
Name of village	Borda
Panchayat	Chitaurgah
Nearest forest Block	-
Land status	Community land
District	Chittorgarh
Area	110 bigha
Altitude (m)	481
Latitude	24° 58' 54.8 N
Longitude	74° 33' 16.1 E

B. Topographical features

Erosion status	Topography	Soil depth
Moderate	Hill slope and pediment	Shallow (<25 cm)

C. Vegetation status

Tree species with IVI

A. pendula (277), *A. leucophloea* (46), *Diospyros cordifolia* (27).

Shrub species with IVI

L. camara (221), *A. jacquemontii* (79)

ENS of Tree : 1.20 Shrub: 1.57

Herb and Grass

B. diffusa, B. articularis, B. ramosa, D. sindicum, E. alsinoides, I. cordifolia, I. linnai, J. procumbense, T. roxburghii.

D. Wild-life

Lepus nigricolis, Axis axis

E. Soil characteristics

pH	EC (dSm^{-1})	SOC (%)	BD (g/cm^3)	NO_3-N (mg kg^{-1})	NH_4-N (mg kg^{-1})	PO_4-P (mg kg^{-1})	C density (tons ha^{-1})
7.43	0.12	0.525	1.44	6.49	5.16	8.36	3.35

F. History and status

Quite old sacred grove conserved and protected in name of Nahar Singhji Mata. There is a small temple on hill top, where people come and worship her. Most of the area is dominated by *A. pendula* trees.

DATA ON SACRED GROVES OF RAJASTHAN

Sacred Grove No. 34

A. Basic information about the sacred groves

Name of sacred groves	Junjan Dodaji Ka Oran
Name of village	Jogalsar
Panchayat	Sujangarh
Nearest forest Block	Lalgarh
Land status	Community land
District	Churu
Area of sacred grove	3500 bigha
Altitude (m)	319
Latitude	27° 30' 27.3 N
Longitude	73° 58' 33.6 E

B. Topographical features

Erosion status	Topography	Soil depth
Moderate	Almost plain land	Deep (>100 cm)

C. Vegetation status

Tree species with IVI

M. emarginata (224), *P. cineraria* (38), *C. decidua* (27), *A. catechu* (11)

Shrub species with IVI

C. procera (108), *Aerva persica* (87), *L. pyrotechnica* (62), *Z. nummularia* (43)

ENS Tree : 1.72 Shrub: 3.32

Herb and Grass

B. diffusa, B. articularis, B. ramose, D. sindicum, E. ciliaris, E. alsinoides, I. cordifolia, I. linnai, J. procumbense, T. purpurea, T. terrestris, Pedalium murex, P. amarus, T. roxburghii.

D. Wild-life

Axis axis, Lepus nigricolis

E. Soil characteristics

pH	EC (dSm^{-1})	SOC (%)	BD (g/cm^3)	NO_3-N (mg kg^{-1})	NH_4-N (mg kg^{-1})	PO_4-P (mg kg^{-1})	C density (tons ha^{-1})
6.71	0.08	0.180	1.58	2.92	6.43	14.24	8.36

F. History and status

About 500 years old grove along with a temple, which has been conserved and protected in name of Dodaji- local diety, where local people worship. A fair is generally organized on the occasion of Vijayadashmi here.

DATA ON SACRED GROVES OF RAJASTHAN

Sacred Grove No. 35

A. Basic information about the sacred groves

Name of sacred groves	Gogaji Ka Oran
Name of village	Manpura
Panchayat	Badi gatyal
Nearest forest Block	Manpura
Land status	Community land
District	Churu
Area of sacred grove	200 bigha
Altitude (m)	302
Latitude	27° 48' 25.6 N
Longitude	79° 16' 34.9 E

B. Topographical features

Erosion status	Topography	Soil depth
Moderate	Flat dune	Deep (>100 cm)

C. Vegetation status

Tree species with IVI

M. emarginata (162), *C. decidua* (53), *A. tortilis* (49), *P. cineraria* (36)

Shrub species with IVI

L. pyrotechnica (300)

ENS Tree : 2.97 Shrub: 1.00

Herb and Grass

B. diffusa, B. articularis, B. ramose, D. sindicum, E. ciliaris, E. alsinoides, Gisekia pharnacioides, I. cordifolia, I. linnai, J. procumbense, T. purpurea, T. terrestris, P. amarus, T. roxburghii.

D. Wild life

Boselaphus tragocamelus, Lepus nigricolis

E. Soil characteristics

pH	EC (dSm^{-1})	SOC (%)	BD (g/cm^3)	NO_3-N ($mg\ kg^{-1}$)	NH_4-N ($mg\ kg^{-1}$)	PO_4-P ($mg\ kg^{-1}$)	C density ($tons\ ha^{-1}$)
7.06	0.02	0.135	1.55	3.08	6.17	14.24	6.28

F. History and status

It is about 200-250 years old sacred grove. A temple has been established in name of Gogaji, who stopped the horse of the king of Gopalpura at this place according to the local people. The grove is dominated by *Maytenus emarginata* trees.

DATA ON SACRED GROVES OF RAJASTHAN

Sacred Grove No. 36

A. Basic information about the sacred groves

Name of sacred groves	Gosai Ji Ka Oran
Name of village	Berasar
Panchayat	Bambu
Nearest forest Block	Sandawa
Land status	Community land
District	Churu
Area of sacred grove	285 bigha
Altitude (m)	324
Latitude	27° 46' 32.5 N
Longitude	74° 08' 58.2 E

B. Topographical features

Erosion status	Topography	Soil depth
Moderate	Almost plain land	Deep (>100 cm)

C. Vegetation status

Tree species with IVI

S. oleoides (120), *A. tortilis* (84), *P. cineraria* (62), *M. emarginata* (23), *C. decidua* (11).

Shrub species with IVI

Z. nummularia (219), *A. persica* (66), *Cucumis callosus* (15).

ENS Tree : 1.71 Shrub: 2.03

Herb and Grass

Cenchrus species, B. diffusa, B. articularis, B. ramose, D. sindicum, E. ciliaris, E. alsinoides, G. pharnacioides, I. cordifolia, I. linnai, J. procumbense, T. purpurea, T. terrestris, P. amarus, T. roxburghii.

D. Wild-life

Lepus nigricolis, Axis axis, Boselaphus tragocamelus.

E. Soil characteristics

pH	EC (dSm^{-1})	SOC (%)	BD (g/cm^3)	NO_3-N (mg kg^{-1})	NH_4-N (mg kg^{-1})	PO_4-P (mg kg^{-1})	C density (tons ha^{-1})
6.41	0.04	0.195	1.54	1.86	3.28	11.10	9.01

F. History and status

It is about 500 year old grove. Gosaiji is *Kuladevata* of many communities in India and Pakistan. It is considered that God Vishnu put his leg at three places, i.e. Makka Madina, Jhujhala and Berasar, when he took 52nd incarnation (*Avtar*). This temple was built to worship him.

DATA ON SACRED GROVES OF RAJASTHAN

Sacred Grove No. 37

A. Basic information about the sacred groves

Name of sacred groves	Kaludas Baba Ki Dungri
Name of village	Sikra, Baradu, Gijgad
Panchayat	Sikrai
Nearest forest Block	Jagrampura pati
Land status	Community land
District	Dausa
Area of sacred grove	56 bigha
Altitude (m)	374
Latitude	26° 48' 28.0 N
Longitude	76° 37' 53.3 E

B. Topographical features

Erosion status	Topography	Soil depth
Moderate	Hillock with slope	Moderately deep (25-50)

C. Vegetation status

Tree species with IVI

A. pendula (265), *H. integrifolia* (35)

Shrub species with IVI

A. vasica (300)

ENS Tree : 1.15 Shrub: 1.00

Herb and Grass

B. diffusa, B. ramosa, C. tora, E. hirta, J. procumbens, T. procumbens.

D. Wild-life

Lepus nigricolis, *Axia axis*, *Boselaphus tragocamelus*.

E. Soil characteristics

pH	EC (dSm^{-1})	SOC (%)	BD (g/cm^3)	NO_3-N (mg kg^{-1})	NH_4-N (mg kg^{-1})	PO_4-P (mg kg^{-1})	C density (tons ha^{-1})
7.55	0.28	0.465	1.46	3.14	2.45	6.91	7.47

F. History and status

Baba Kaludas ji had meditated here. In his name this sacred groves is maintained and conserved. There is a temple on the hill top, where people come and worshiped. The area is dominated by old crop of *A. pendula* as the most dominant species.

DATA ON SACRED GROVES OF RAJASTHAN

Sacred Grove No. 38

A. Basic information about the sacred groves

Name of sacred groves	Devnarayna Ji Ka Oran
Name of village	Kalota
Panchayat	Dausa
Nearest forest Block	Kalota
Land status	Community land
District	Dausa
Area of sacred grove	20 bigha
Altitude (m)	329
Latitude	26° 58' 29.1 N
Longitude	76° 25' 17.0 E

B. Topographical features

Erosion status	Topography	Soil depth
Moderate	Hill slope	Shallow (<25 cm)

C. Vegetation status

Tree species with IVI

A. nilotica (210), *A. leucophloea* (27), *H. integrifolia* (24), *P. juliflora* (19), *W. tinctoria* (20)

Shrub species with IVI

A. vasica (117), *Euphorbia caducifolia* (91), *Euphorbia neriifolia* (36), *C. procera* (28), *C. sepiaria* (14), *Z. nummularia* (14)

ENS of Tree : 1.83 Shrub: 3.94

Herb and Grass

Blepharis repens, B. diffusa, B. ramosa, C. tora, E. hirta, J. procumbens, T. procumbens.

D. Wild-life

Lepus nigricolis, *Axis axis*

E. Soil characteristics

pH	EC (dSm^{-1})	SOC (%)	BD (g/cm^3)	NO_3-N ($mg\ kg^{-1}$)	NH_4-N ($mg\ kg^{-1}$)	PO_4-P ($mg\ kg^{-1}$)	C density ($tons\ ha^{-1}$)
7.25	0.26	0.326	1.43	3.22	3.26	5.23	7.30

F. History and status

People believe that God Deonarayan- a diety of Gurjar community has taken rebirth in form of a Sandu Mata, where local people use to worship. At present a new temple is under construction.

DATA ON SACRED GROVES OF RAJASTHAN

Sacred Grove No. 39

A. Basic information about the sacred groves

Name of sacred groves	Sawanria Ji Seth Dham
Name of village	Kanawar
Panchayat	Bandikue
Nearest forest Block	Basura
Land status	Community land
District	Dausa
Area	50 bigha
Altitude (m)	336
Latitude	27° 06’ 43.2 N
Longitude	76° 31’ 13.7 E

B. Topographical features

Erosion status	Topography	Soil depth
Moderate	Hill slope	Shallow (<25 cm)

C. Vegetation status

Tree species with IVI

H. integrifolia (230), *A. nilotica* (60), *A. leucophloea* (10)

Shrub species with IVI

A. vasica (300)

ENS Tree : 1.78 Shrub: 1.00

Herb and Grass

Blepharis repens, B. diffusa, B. ramosa, C. tora, E. hirta, J. procumbens, T. purpurea, T. procumbens.

D. Wild-life

Lepus nigricolis, *Canis aureus auveus*, *Sus scrofa*, *Vulpes bengalehsis*, *Psittacula krameri*, *Pavo cristatus*, *Acridotheres tristis* (because of attached area of Sariska sanctuary Tiger use to visit this area also as people says).

E. Soil characteristics

pH	EC (dSm^{-1})	SOC (%)	BD (g/cm^3)	NO_3-N (mg kg^{-1})	NH_4-N (mg kg^{-1})	PO_4-P (mg kg^{-1})	C density (tons ha^{-1})
7.68	0.23	0.360	1.44	4.23	2.82	11.80	9.21

F. History and status

People say that once some body was moving with an idol (murti) of Lord Krishna. Unfortunately the idol fell at the place and could not be taken back again. Later the Murti was established here and worshiped by the locals. The area is presently dominated by *H. integrifolia* tree.

DATA ON SACRED GROVES OF RAJASTHAN

Sacred Grove No. 40

A. Basic information about the sacred groves

Name of sacred groves	Bali Baba Ki Dargah
Name of village	Longpura
Panchayat	-
Nearest forest Block	Longpura
Land status	Community land
District	Dholpur
Area	300 bigha
Altitude (m)	222
Latitude	26° 41' 41.2 N
Longitude	77° 52' 82.1 E

B. Topographical features

Erosion status	Topography	Soil depth
Moderate	Hill slope	Moderately deep (25-50)

C. Vegetation status

Tree species with IVI

P. juliflora (154), *H. integrifolia* (84), *A. pendula* (46), *F. religiosa* (16)

Shrub species with IVI

Z. nummularia (145), *C. procera* (84), *C. sepiaria* (71)

ENS Tree : 1.85 Shrub: 2.95

Herb and Grass

B. ramosa, C. tora, E. hirta, J. procumbens, T. purpurea, T. procumbens.

D. Wild-life

Hyaena hyaena, *Vulpes bengalehsis*, *Lepus nigricolis*, *Varanus griseus*, *Pavo cristatus*.

E. Soil characteristics

pH	EC (dSm^{-1})	SOC (%)	BD (g/cm^3)	NO_3-N ($mg\ kg^{-1}$)	NH_4-N ($mg\ kg^{-1}$)	PO_4-P ($mg\ kg^{-1}$)	C density ($tons\ ha^{-1}$)
7.48	0.19	0.720	1.39	6.15	4.64	12.46	20.28

F. History and status

It is about 700 years old sacred grove situated near Baba Garib Nawaz. The mosque is situated on high altitude.

DATA ON SACRED GROVES OF RAJASTHAN

Sacred Grove No. 41

A. Basic information about the sacred groves

Name of sacred groves	MuchKund
Name of village	Yogpura
Panchayat	-
Nearest forest Block	-
Land status	Community land
District	Dholpur
Area	250 Bigha
Altitude (m)	210
Latitude	26° 40' 69.9 N
Longitude	77° 52' 02.1 E

B. Topographical features

Erosion status	Topography	Soil depth
Moderate	Hill slope	Moderately deep (25-50)

C. Vegetation status

Tree species with IVI

A. pendula (151), *P. juliflora* (69), *A. indica* (38), *A. nilotica* (14), *F. religiosa* (14), *H. integrifolia* (9), *Capparis decidua* (5)

Shrub species with IVI

C. procera (166), *C. sepiaria* (105), *Z. nummularia* (29)

ENS Tree : 3.43 Shrub: 3.18

Herb and Grass

Cassia tora, E. hirta, T. procumbens

D. Wild-life

Hyaena hyaena, Snake, *Varanus griseus*, *Vulpes bengalehsis*, *Pavo cristatus*, *Psittacula krameri*.

E. Soil characteristics

pH	EC (dSm^{-1})	SOC (%)	BD (g/cm^3)	NO_3-N ($mg\ kg^{-1}$)	NH_4-N ($mg\ kg^{-1}$)	PO_4-P ($mg\ kg^{-1}$)	C density ($tons\ ha^{-1}$)
5.20	0.26	0.480	1.47	4.15	1.75	5.00	17.42

F. History and status

It is a historical place and also one of the teerth places of very high belief among the people of nearby places. People take bath in the holy lake here.

DATA ON SACRED GROVES OF RAJASTHAN

Sacred Grove No. 42

A. Basic information about the sacred groves

Name of sacred groves	Mahadeo Ji Ka Oran
Name of village	Sadpura
Panchayat	Seyru
Nearest forest Block	-
Land status	Community land
District	Dholpur
Area	350 bigha
Altitude (m)	172
Latitude	26° 47' 57.0 N
Longitude	77° 44' 54.5 E

B. Topographical features

Erosion status	Topography	Soil depth
Moderate	Hill slope	Moderately deep (25-50)

C. Vegetation status

Tree species with IVI

F. benghalensis (132), *A. indica* (80), *H. integrifolia* (60), *F. religiosa* (17) *Capparis deciduas* (11).

Shrub species with IVI

Z. nummularia (202), *C. sepiaria* (98)

ENS Tree : 3.34 Shrub: 2.99

Herb and Grass

Dctyloctenium sindicum, Cassia tora

D. Wild life

Pavo cristatus, *Psittacula krameri*, Snake.

E. Soil characteristics

pH	EC (dSm^{-1})	SOC (%)	BD (g/cm^3)	NO_3-N ($mg\ kg^{-1}$)	NH_4-N ($mg\ kg^{-1}$)	PO_4-P ($mg\ kg^{-1}$)	C density ($tons\ ha^{-1}$)
8.21	0.32	0.180	1.51	3.15	2.00	6.24	6.35

F. History and status

An ancient temple in the sacred grove in name of Lord Shiva, where people use to come and worship him. The adjoining area is conserved and maintained by the local communities.

DATA ON SACRED GROVES OF RAJASTHAN

Sacred Grove No. 43

A. Basic informations about the sacred groves

Name of sacred groves	Govindguru Chhari Magri
Name of village	Bersa
Panchayat	Simalwara
Nearest forest Block	Dhambola
Land status	Community land
District	Dungarpur
Area of sacred grove	54 bigha
Altitude (m)	229
Latitude	23° 38' 21.5 N
Longitude	73° 49' 59.6 E

B. Topographical features

Erosion status	Topography	Soil depth
Moderate	Flat land	Moderate (50-100)

C. Vegetation status

Tree species with IVI

P. juliflora (264), *D. melanoxylon* (20), *A. indica* (10), *Ehretia laevis* (6).

Shrub species with IVI

Z. nummularia (300)

ENS Tree : 1.20 Shrub: 1.00

Herb and Grass

B. ramosa, B. repens, C. tora, D. sindicum, E. hirta, J. procumbens, I. cordifolia, T. purpurea, T. procumbens.

D. Wild-life

Boselaphus tragocamelus, *Lepus nigricolis*, *Canis aureus auveus*, *Picnonotus cafer*.

E. Soil characteristics

pH	EC (dSm^{-1})	SOC (%)	BD (g/cm^3)	NO_3-N (mg kg^{-1})	NH_4-N (mg kg^{-1})	PO_4-P (mg kg^{-1})	C density (tons ha^{-1})
6.67	0.15	0.780	1.42	2.30	3.20	9.56	17.79

F. History and status

About 1000 years old and is considered sacred by the local people. A temple is situated in the grove where people worship the diety. The grove is presently dominated by an invasive species *P. juliflora*.

DATA ON SACRED GROVES OF RAJASTHAN

Sacred Grove No. 44

A. Basic information about the sacred groves

Name of sacred groves	Bhuneshwar Mahadev Ka Oran
Name of village	Karauli
Panchayat	Karauli
Nearest forest Block	-
Land status	Community land
District	Dungarpur
Area of sacred grove	119 bigha
Altitude (m)	235
Latitude	23° 49' 50.9 N
Longitude	79° 35' 43.3 E

B. Topographical features

Erosion status	Topography	Soil depth
Moderate	Hill top with slope	Shallow (<25 cm)

C. Vegetation status

Tree species with IVI

A. pendula (148), *A. leucophloea* (55), *A. catechu* (33), *A. indica* (23), *M. parviflora* (16), *Madhuca indica* (7), *T. grandis* (6), *D. melanoxylon* (6), *H. integrifolia* (3), *M. emarginata* (3).

Shrub species with IVI

D. cinerea (118), *L. camara* (93), *S. virosa* (29), *Aristolochia bracteata* (29), *Z. nummularia* (16), *C. sepiaria* (15)

ENS Tree : 3.13 Shrub: 5.54

Herb and Grass

B. ramosa, B. repens, D. sindicum, E. hirta, J. procumbens, I. cordifolia, T. procumbens.

D. Wild-life

Boselaphus tragocamelus, Lepus nigricolis, Pavo cristatus.

E. Soil characteristics

pH	EC (dSm^{-1})	SOC (%)	BD (g/cm^3)	NO_3-N ($mg\ kg^{-1}$)	NH_4-N ($mg\ kg^{-1}$)	PO_4-P ($mg\ kg^{-1}$)	C density ($tons\ ha^{-1}$)
7.30	0.27	1.862	1.38	1.09	2.90	7.12	32.47

F. History and status

There are 6 siblings in this oran. It is believed that all siblings are self originated and worshiped by the local people. This grove is dominated by *A. pendula* trees.

DATA ON SACRED GROVES OF RAJASTHAN

Sacred Grove No. 45

A. Basic information about the sacred groves

Name of sacred groves	Neelkanth Mahadev
Name of village	Bhiluda
Panchayat	Bhiluda
Nearest forest Block	-
Land status	Community land
District	Dungarpur
Area of sacred grove	71 bigha
Altitude (m)	200
Latitude	23° 36' 39.5 N
Longitude	74° 05' 13.1 E

B. Topographical features

Erosion status	Topography	Soil depth
Moderate	Flat land	Moderate (50-100)

C. Vegetation status

Tree species with IVI

Prosopis juliflora (182), *A. indica* (55), *F. benghalensis* (24), *A. nilotica* (15), *A. excelsa* (13), *C. decidua* (11)

Shrub species with IVI

Jatropha curcas (151), *C. auriculata* (104), *L. camara* (45)

ENS Tree : 2.02 Shrub: 2.80

Herb and Grass

B. ramosa, B. repens, C. tora, E. hirta, J. procumbens, I. cordifolia, T. purpurea, T. procumbens.

D. Wild-life

Boselaphus tragocamelus, *Lepus nigricolis*, *Canis aureus auveus*, *Pavo cristatus*.

E. Soil characteristics

pH	EC (dSm^{-1})	SOC (%)	BD (g/cm^3)	NO_3-N (mg kg^{-1})	NH_4-N (mg kg^{-1})	PO_4-P (mg kg^{-1})	C density (tons ha^{-1})
7.50	0.18	0.630	1.47	1.80	4.45	12.86	12.01

7. History and status

This is an ancient grove with an old temple situated along a river, where people come and worship lord Shiva. The land is in the name of temple only. However, *P. juliflora* is dominating in the grove area.

DATA ON SACRED GROVES OF RAJASTHAN

Sacred Grove No. 46

A. Basic information about the sacred groves

Name of sacred groves	Bhadra Kali Mata Ji Ka Oran
Name of village	Amarpura
Panchayat	Hanumangarh
Nearest forest Block	Kolu farum
Land status	Community land
District	Hanumangarh
Area of sacred grove	125 bigha
Altitude (m)	192
Latitude	29° 36' 00.6 N
Longitude	74° 23' 02.3 E

B. Topographical features

Erosion status	Topography	Soil depth
Slight	Almost flat land	Moderate (50-100)

C. Vegetation status

Tree species with IVI

P. juliflora (105), *A. nilotica* (48), *Tamarix aphylla* (45), *P. cineraria* (42), *S. oleoides* (29), *C. decidua* (23), *F. religiosa* (8)

Shrub species with IVI

No shrub (0)

ENS Tree : 4.42 Shrub: 1.00

Herb and Grass

Fagonia cretica, Peganum harmala, E. hirta, I. cordifolia, T. procumbens.

D. Wild-life

Passer domesticus, Acridotheres tristis

E. Soil characteristics

pH	EC (dSm^{-1})	SOC (%)	BD (g/cm^3)	NO_3-N ($mg\ kg^{-1}$)	NH_4-N ($mg\ kg^{-1}$)	PO_4-P ($mg\ kg^{-1}$)	C density ($tons\ ha^{-1}$)
7.67	0.24	1.290	1.42	2.56	2.58	5.12	43.58

F. History and status

During the period of Maharaja Ganga Singh, this temple was constructed. People are worshipping here since then. Animals like goat are sacrificed and wine distribution is a custom here.

DATA ON SACRED GROVES OF RAJASTHAN

Sacred Grove No. 47

A. Basic Information about the sacred groves

Name of sacred groves	Chisti Peer ki Dargah
Name of village	Chhistiya
Panchayat	Hanumangarh
Nearest forest Block	Kolu farum
Land status	Community land
District	Hanumangarh
Area of sacred grove	13 bigha
Altitude (m)	180
Latitude	29° 39' 31.6 N
Longitude	74° 14' 23.9 E

B. Topographical features

Erosion status	Topography	Soil depth
Slight	Undulating flatland	Moderate (50-100)

C. Vegetation status

Tree species with IVI

C. decidua (232), *A. nilotica* (20), *S. oleoides* (19), *D. sissoo* (15), *M. emarginata* (14).

Shrub species with IVI

Z. nummularia (244), *Opuntia dillenii* (40), *A. persica* (16).

ENS Tree : 1.50 Shrub: 1.72

Herb and Grass

F. cretica, T. procumbens.

D. Wild-life

Passer domesticus, *Lepus nigricolis*

E. Soil characteristics

pH	EC (dSm^{-1})	SOC (%)	BD (g/cm^3)	NO_3-N ($mg\ kg^{-1}$)	NH_4-N ($mg\ kg^{-1}$)	PO_4-P ($mg\ kg^{-1}$)	C density ($tons\ ha^{-1}$)
8.14	0.29	0.390	1.55	4.80	4.50	10.24	18.14

F. History and status

This grove is conserved and protected in name of Chisti Peer Ki Dargah, where people (mainly Muslim community) come and worship him. The area is dominated by *Capparis decidua* trees.

DATA ON SACRED GROVES OF RAJASTHAN

Sacred Grove No. 48

A. Basic information about the sacred groves

Name of sacred groves	Guru Gorakhnath Ji Ka Oran
Name of village	Gogamed
Panchayat	Nohar
Nearest forest Block	Gogamedi
Land status	Community land
District	Hanumangarh
Area	91 bigha
Altitude (m)	196
Latitude	29° 10' 41.7 N
Longitude	75° 01' 57.1 E

B. Topographical features

Erosion status	Topography	Soil depth
Slight	Flat land	Deep (>100 cm)

C. Vegetation status

Tree species with IVI

S. oleoides (167), *A. indica* (45), *P. cineraria* (25), *D. sissoo* (17), *A. tortilis* (17), *F. benghalensis* (17), *A. nilotica* (12).

Shrub species with IVI

Z. nummularia (300)

ENS Tree : 4.27 Shrub: 1.00

Herb and Grass

B. ramosa, F. cretica, Glinus lotoides, T. procumbens, Solanum nigrum.

D. Wild-life

Boselaphus tragocamelus, *Lepus nigricolis*, *Pavo cristatus*

E. Soil characteristics

pH	EC (dSm^{-1})	SOC (%)	BD (g/cm^3)	NO_3-N (mg kg^{-1})	NH_4-N (mg kg^{-1})	PO_4-P (mg kg^{-1})	C density (tons ha^{-1})
7.67	0.32	0.450	1.52	2.70	2.38	6.52	8.65

F. History and status

People believe that Guru Gorakhnath has mediated here. Since then this place is considered a sacred place and the area is conserved and protected in his name. The grove is dominated by *Salvadora oleoides* trees.

DATA ON SACRED GROVES OF RAJASTHAN

Sacred Grove No. 49

A. Basic information about the sacred groves

Name of sacred groves	Bheru ji Ki Bani
Name of village	Khonagorian
Panchayat	Sangner
Nearest forest Block	Jhalana
Land status	Community land
District	Jaipur
Area of sacred grove	9 bigha
Altitude (m)	406
Latitude	26° 51' 30.1 N
Longitude	75° 50' 32.3 E

B. Topographical features

Erosion status	Topography	Soil depth
Moderate	Undulating plain	Deep (>100 cm)

C. Vegetation status

Tree species with IVI

A. tortilis (129), *P. cineraria* (111), *A. marmelos* (17), *A. indica* (15), *E. genita* (14), *A. digitata* (14)

Shrub species with IVI

C. sepiaria (300)

ENS Tree : 3.23 Shrub: 1.00

Herb and Grass

A. aspera, D. aegyptium, D. sindicum, Ipomoea pes-tigridis, O. canum, P. minima, P. paniculata, Perotis indica, S. cordifolia, S. indica, Xanthium strumarium.

D. Wild-life

Boselaphus tragocamelus, *Passer domesticus*

E. Soil characteristics

pH	EC (dSm^{-1})	SOC (%)	BD (g/cm^3)	NO_3-N (mg kg^{-1})	NH_4-N (mg kg^{-1})	PO_4-P (mg kg^{-1})	C density (tons ha^{-1})
7.50	0.18	0.225	1.55	1.72	6.85	10.35	9.94

F. History and status

A quite old grove, which is protected and conserved by the local communities in name of Bheru ji- the local diety. However, dominant tree is *Acacia tortilis*, appears to be due to plantation activity in the area.

DATA ON SACRED GROVES OF RAJASTHAN

Sacred Grove No. 50

A. Basic information about the sacred groves

Name of sacred groves	Kapil Muni Ashram
Name of village	Goner
Panchayat	Sanganer
Nearest forest Block	Gonu beed
Land status	Community land
District	Jaipur
Area of sacred grove	7 bigha
Altitude (m)	352
Latitude	26° 46' 11.4 N
Longitude	75° 54' 47.5 E

B. Topographical features

Erosion status	Topography	Soil depth
Slight	Flat land	Deep (>100 cm)

C. Vegetation status

Tree species with IVI

A. indica (95), *P. cineraria* (85), *F. religiosa* (24), *Polyalthia longifolia* (20), *A. leucophloea* (17), *Leucaena leucocephala* (16), *S. cummini* (14), *F. glomerata* (12), *C. decidua* (9), *Thespesia populnea* (8).

Shrub species with IVI

C. procera (189), *Z. nummularia* (96) *Nerium odoratum* (15)

ENS Tree : 5.57 Shrub: 2.80

Herb and Grass

B. diffusa, B. ramosa, C. benghalensis, E. acaulis, E. aspera, P. amarus, P. hysterophorus, P. paniculata, Rhynchosia minima, T. roxburghii, V. cinerea.

D. Wild-life

Boselaphus tragocamelus, *Canis lupus*

E. Soil characteristics

pH	EC (dSm^{-1})	SOC (%)	BD (g/cm^3)	NO_3-N (mg kg^{-1})	NH_4-N (mg kg^{-1})	PO_4-P (mg kg^{-1})	C density (tons ha^{-1})
8.14	0.22	1.125	1.49	5.30	2.86	9.08	41.36

F. History and status

This is an ancient grove conserved and protected by local communities in name of Kapil Muni- a Hindu sage old period. However, plantations of many species have been done in this grove.

DATA ON SACRED GROVES OF RAJASTHAN

Sacred Grove No. 51

A. Basic information about the sacred groves

Name of sacred groves	Shiv Dungri
Name of village	Manpur Dungri
Panchayat	Chaksa/Kathawada
Nearest forest Block	Paharia
Land status	Community land
District	Jaipur
Area of sacred grove	12 bigha
Altitude (m)	374
Latitude	26° 37' 42.3 N
Longitude	75° 55' 38.9 E

B. Topographical features

Erosion status	Topography	Soil depth
Slight	Hill top with pediment	Shallow (<25 cm)

C. Vegetation status

Tree species with IVI

P. juliflora (178), *B. monosperma* (81), *P. cineraria* (28), *A. indica* (13)

Shrub species with IVI

Z. nummularia (300)

ENS Tree : 2.02 Shrub: 1.00

Herb and Grass

A. funiculata, A. vaginalis, B. acanthoides, B. articularis, B. eracta, C. barbata, C. ciliaris, C. pumila, D. aegyptium, D. annulatum, D. ciliaris, E. alsinoides, I. cordifolia, I. linnai, J. procumbense, L. trinervis, M. jacqmontii, P. hysterophorus, P. lappacea, P. paniculata, T. purpurea, T. roxburghii, T. tennalus, T. terrestris, T. villosa.

D. Wild-life

Boselaphus tragocamelus, Lepus nigricolis, Pavo cristatus

E. Soil characteristics

pH	EC (dSm^{-1})	SOC (%)	BD (g/cm^3)	NO_3-N (mg kg^{-1})	NH_4-N (mg kg^{-1})	PO_4-P (mg kg^{-1})	C density (tons ha^{-1})
7.60	0.19	0.150	1.57	2.42	4.51	10.35	6.53

F. History and status

This is an old grove conserved and protected by the local community in name of Lord Shiva. However, the grove is now infested by invasive species *P. juliflora*, affecting the water quality of adjoining water body.

DATA ON SACRED GROVES OF RAJASTHAN

Sacred Grove No. 52

A. Basic information about the sacred groves

Name of sacred groves	Ramdevji ka Oran
Name of village	Pokharan
Panchayat	Pokharan
Nearest forest Block	-
Land status	Community land
District	Jaisalmer
Area of sacred grove	12 bigha
Altitude (m)	245
Latitude	26° 59' 37.9 N
Longitude	71° 55' 46.2 E

B. Topographical features

Erosion status	Topography	Soil depth
Moderate	Flat land	Deep (>100 cm)

C. Vegetation status

Tree species with IVI

S. oleoides (134), *A. tortilis* (85), *Mangifera indica* (41), *T. undulata* (40)

Shrub species with IVI

C. burhia (170), *Z. nummularia* (130)

ENS Tree : 3.08 Shrub: 1.70

Herb and Grass

L. sindicus, B. articularis, E. alsinoides, I. cordifolia, I. linnai, M. jacqmontii, T. roxburghii, T. tennalus, T. terrestris.

D. Wild-life

Boselaphus tragocamelus, Lepus nigricolis.

E. Soil characteristics

pH	EC (dSm^{-1})	SOC (%)	BD (g/cm^3)	NO_3-N ($mg\ kg^{-1}$)	NH_4-N ($mg\ kg^{-1}$)	PO_4-P ($mg\ kg^{-1}$)	C density ($tons\ ha^{-1}$)
8.04	0.35	0.235	1.65	1.23	2.33	4.56	11.36

F. History and status

A quite old sacred grove and a temple in name of Ramdev Peer Maharajji – the local diety of almost all communities is available here. Many devotees come across to offer their prayers and gives Baba ji gives his Blessings to all his devottees and fulfill their dreams, which become true also. It is dominated mostly by *S. oleoides* trees.

DATA ON SACRED GROVES OF RAJASTHAN

Sacred Grove No. 53

A. Basic Information about the sacred groves

Name of sacred groves	Bhadariya ji Ka Oran (old)
Name of village	Bhadariya
Panchayat	Pokharan
Nearest forest Block	-
Land status	Community land
District	Jaisalmer
Area of sacred grove	165000 bigha
Altitude (m)	212
Latitude	27° 03' 33.4 N
Longitude	71° 31' 38.9 E

B. Topographical features

Erosion status	Topography	Soil depth
Moderate	Flatland	Deep (>100 cm)

C. Vegetation status

Tree species with IVI

Z. rotundifolia (265), *P. cineraria* (35)

Shrub species with IVI

No shrub

ENS Tree : 1.17 Shrub: 1.00

Herb and Grass

Dactyloctenium sindicum, Cenchrus biflorus, Tephrosia purpurea

D. Wild life

Boselaphus tragocamelus, Oryctolagus cuniculus, Rattus norvegicus

E. Soil characteristics

pH	EC (dSm^{-1})	SOC (%)	BD (g/cm^3)	NO_3-N ($mg\ kg^{-1}$)	NH_4-N ($mg\ kg^{-1}$)	PO_4-P ($mg\ kg^{-1}$)	C density ($tons\ ha^{-1}$)
8.12	0.4	0.325	1.61	1.25	2.56	6.55	14.65

F. History and status

This grove is an ancient one in name of Bhadariya Mataji. Maharawal Gajsinghji of Jaisalmer wanted to celebrate his victory over Bikaner and therefore constructed this temple. This temple belongs to the Shakta sect, but all Hindus worship the temple. The pooja ceremony is punctually performed everyday. According to the practices of Shakta cult, the Goddess is taken to be symbol of power and worship as Amba.

DATA ON SACRED GROVES OF RAJASTHAN

Sacred Grove No. 54

A. Basic Information about the sacred groves

Name of sacred groves	Malani Mataji Oran
Name of village	Dhaysar
Panchayat	Jaisalmer
Forest Block	-
Land status	Community land
District	Jaisalmer
Area	450 bigha
Altitude (m)	276
Latitude	26° 27' 39.4 N
Longitude	71° 09' 54.1 E

B. Topographical features

Erosion status	Topography	Soil depth
Severe	Undulating plain land	Deep (>100 cm)

C. Vegetation status

Tree species with IVI

C. decidua (300)

Shrub species with IVI

Calotropis procera (210), *A. persica* (47), *L. pyrotechnica* (19), *C. burhia* (13), *Z. nummularia* (11)

ENS Tree : 1.00 Shrub: 2.40

Herb and Grass

Convolvulus spp. E. alsinoides, Farsetia hamiltonii, I. cordifolia, T. terrestris, C. biflorus, C. Burhia

D. Wild-life

Varanus bengalensis, *Tetera indica*, *Uromastix hardwickii*, *Axis axis*, *Eryx johnii*, *Francolinus pondicerianus*, *Boselaphus tragocamelus*.

C. Soil characteristics

pH	EC (dSm^{-1})	SOC (%)	BD (g/cm^3)	NO_3-N (mg kg^{-1})	NH_4-N (mg kg^{-1})	PO_4-P (mg kg^{-1})	C density (tons ha^{-1})
8.04	0.29	0.215	1.67	2.05	1.56	4.55	10.58

F. History and status

This is an ancient sacred grove conserved and protected in name of Mataji, the local diety. Local people worship her for blessing to the communities residing in nearby areas.

DATA ON SACRED GROVES OF RAJASTHAN

Sacred Grove No. 55

A. Basic Information about the sacred groves

Name of sacred groves	Bhelani Mataji Oran
Name of village	Bhelai
Panchayat	Jaisalmer
Nearest forest Block	-
Land status	Community land
District	Jaisalmer
Area	430 bigha
Altitude (m)	288
Latitude	26° 35' 37.8 N
Longitude	71° 11' 41.2 E

B. Topographical features

Erosion status	Topography	Soil depth
Moderate	Flat land	Deep (>100 cm)

C. Vegetation status

Tree species with IVI

C. decidua (202), *A. senegal* (98)

Shrub species with IVI

Z. nummularia (196), *C. procera* (104)

ENS Tree : 1.82 Shrub: 1.95

Herb and Grass

Convolvulus species, F. hamiltonii, I. cordifolia, T. terrestris.

D. Wild-life

Varanus bengalensis, *Tetera indica*, *Uromastix hardwickii*, *Axis axis*, *Eryx johnii*, *Francolinus pondicerianus*, *Boselaphus tragocamelus*.

C. Soil characteristics

pH	EC (dSm^{-1})	SOC (%)	BD (g/cm^3)	NO_3-N (mg kg^{-1})	NH_4-N (mg kg^{-1})	PO_4-P (mg kg^{-1})	C density (tons ha^{-1})
7.89	0.34	0.341	1.59	1.89	2.13	7.23	16.06

F. History and status

An ancient sacred grove in name of Mata ji, where local people use to gather and worship. The grove is conserved and protected by the local community. The grove is dominated by *Capparis decidua* and *Acacia senegal*.

DATA ON SACRED GROVES OF RAJASTHAN

Sacred Grove No. 56

A. Basic information about the sacred groves

Name of sacred groves	Dharneswar Mahadeo ka Oran
Name of village	Borwada
Panchayat	Jalore
Nearest forest Block	-
Land status	Community land
District	Jalore
Area	115 bigha
Altitude (m)	123
Latitude	25° 15' 36.6 N
Longitude	72° 16' 30.0 E

B. Topographical features

Erosion status	Topography	Soil depth
Slight	Almost flat land	Deep (>100 cm)

C. Vegetation status

Tree species with IVI

Salvadora oleoides (193), *Prosopis cineraria* (53), *Capparis decidua* (32), *Prosopis juliflora* (12), *Maytenus emarginata* (10).

Shrub species with IVI

Acacia jacquemontii (151), *Cassia auriculata* (111), *Leptadenia pyrotechnica* (38).

ENS Tree : 1.63 Shrub: 1.60

Herb and Grass

A. aspera, A. funiculata, B. ramosa, H. marifolium, J. procumbens, I. cordifolia, M. cerviana, T. procumbens, T. purpurea.

D. Wild life

Varanus griseus, *Lepus nigricolis*, *Pavo cristatus*, *Psittacula krameri*, *Columba livia*, *Francolinus pondicerianus*, *Canis aureus auveus*, *Hyaena hyaena*, *Boselaphus tragocamelus*, *Eryx johnii*, *Eudynamys scolopaceus*.

F. Soil characteristics

pH	EC (dSm^{-1})	SOC (%)	BD (g/cm^3)	NO_3-N ($mg\ kg^{-1}$)	NH_4-N ($mg\ kg^{-1}$)	PO_4-P ($mg\ kg^{-1}$)	C density ($tons\ ha^{-1}$)
7.85	0.18	0.135	1.56	4.38	6.30	14.24	5.46

F. History and status

It is an ancient grove conserved and protected by local communities in name of Lord Shiva. There is a temple in the grove presently taken care of by Mandir trust. The grove is dominated by *Salvadora oleoides*.

DATA ON SACRED GROVES OF RAJASTHAN

Sacred Grove No. 57

A. Basic information about the sacred groves

Name of sacred groves	Dudeshwar Mahadeo Ka Oran
Name of village	Noon
Panchayat	Jalore
Forest Block	Sayla
Land status	Community land
District	Jalore
Area	224 bigha
Altitude (m)	165
Latitude	25° 10' 78.1 N
Longitude	72° 29' 83.0 E

B. Topographical features

Erosion status	Topography	Soil depth
Moderate	Undulating flat land	Deep (>100 cm)

C. Vegetation status

Tree species with IVI

Tecomella undulata (233), *Prosopis cineraria* (21), *Prosopis juliflora* (16), *Acacia leucophloea* (20), *Acacia indica* (10)

Shrub species with IVI

Acacia jacquemontii (264), *Cassia auriculata* (36)

ENS Tree : 2.48 Shrub: 1.14

Herb and Grass

Ipomoea pes-tigridis, H. marifolium, T. purpurea.

D. Wild-life

Varanus griseus, *Lepus nigricolis*, *Pavo cristatus*, *Psittacula krameri*, *Columba livia*, *Francolinus pondicerianus*, *Canis aureus auveus*, *Hyaena hyaena*, *Boselaphus tragocamelus*, Snake, *Uromastix hardwickii*.

F. Soil characteristics

pH	EC (dSm^{-1})	SOC (%)	BD (g/cm^3)	NO_3-N (mg kg^{-1})	NH_4-N (mg kg^{-1})	PO_4-P (mg kg^{-1})	C density (tons ha^{-1})
7.85	0.28	0.090	1.61	3.75	5.12	12.46	4.32

F. History and status

An ancient grove in name of lord Shiva is protected and conserved by the local communities. An ancient temple, presently taken cares of by the local panchayat and is under good condition.

DATA ON SACRED GROVES OF RAJASTHAN

Sacred Grove No. 58

A. Basic information about the sacred groves

Name of sacred groves	Jhootana Mama ji Ka Oran
Name of village	Bhorda
Panchayat	Aahore
Forest Block	-
Land status	Community land
District	Jalore
Area	100 bigha
Altitude (m)	174
Latitude	25° 40’ 18.6 N
Longitude	72° 47’ 34.0 E

B. Topographical features

Erosion status	Topography	Soil depth
Slight	Plane	Deep (>100 cm)

C. Vegetation status

Tree species with IVI

Salvadora oleoides (198), *Prosopis juliflora* (58), *Capparis decidua* (44).

Shrub species with IVI

Cassia auriculata (300)

ENS Tree : 2.25 Shrub: 1.16

Herb and Grass Species

A. aspera, Cleome viscosa, Corchorus tridens, H. marifolium, T. purpurea.

D. Wild-life

Varanus griseus, *Lepus nigricolis*, *Boselaphus tragocamelus*, *Passer domesticus*, *Francolinus pondicerianus*.

E. Soil characteristics

pH	EC (dSm^{-1})	SOC (%)	BD (g/cm^3)	NO_3-N ($mg\ kg^{-1}$)	NH_4-N ($mg\ kg^{-1}$)	PO_4-P ($mg\ kg^{-1}$)	C density ($tons\ ha^{-1}$)
7.36	0.21	0.135	1.55	3.05	3.18	10.24	6.12

F. History and status

It is an ancientsacred grove in name of Jhootana Mataji, the local diety. There is an ancient temple in the grove, which is under care of Mandir trust. The grove is conserved and protected by the local community and dominated by *Salvadora oleoides* as the tree species.

DATA ON SACRED GROVES OF RAJASTHAN

Sacred Grove No. 59

A. Basic information about the sacred groves

Name of sacred groves	Kalla ji Ka Oran
Name of village	Akawad Khurd
Panchayat	Khabpur
Forest Block	-
Land status	Community land
District	Jhalawar
Area	50 bigha
Altitude (m)	326
Latitude	24° 37' 33.0 N
Longitude	76° 29' 34.8 E

B. Topographical features

Erosion status	Topography	Soil depth
Moderate	Almost flat land	Moderately deep (25-50)

C. Vegetation status

Tree species with IVI

A. leucophloea (155), *D. melanoxylon* (80), *P. juliflora* (31), *H. integrifolia* (14), *P. sylvestris* (11), *A. nilotica* (9).

Shrub species with IVI

No shrub

ENS Tree : 3.68 Shrub: 0

Herb and Grass Species

Cynodon dactylon, *H. marifolium*

D. Wild-life

Boselaphus tragocamelus, *Rusa unicolor*, *Sus scrofa*, *Hyaena hyaena*

E. Soil characteristics

pH	EC (dSm^{-1})	SOC (%)	BD (g/cm^3)	NO_3-N ($mg\ kg^{-1}$)	NH_4-N ($mg\ kg^{-1}$)	PO_4-P ($mg\ kg^{-1}$)	C density ($tons\ ha^{-1}$)
7.97	0.12	0.990	1.36	3.22	4.40	10.45	25.53

F. History and status

An ancient grove, which is conserved in name of local legendary. A temple has been established in the grove where people use to come for the cure of snake bite. The grove is dominated by *Acacia leucophloea* trees.

DATA ON SACRED GROVES OF RAJASTHAN

Sacred Grove No. 60

A. Basic information about the sacred groves

Name of sacred groves	Bapulala Smirti Van
Name of village	Taandi Sohanpura
Panchayat	Jhalrapaatan
Forest Block	Bapulala Smirti Van
Land status	Community land
District	Jhalawar
Area	14 bigha
Altitude (m)	371
Latitude	24° 30' 04.2 N
Longitude	76° 23' 09.7 E

B. Topographical features

Erosion status	Topography	Soil depth
Moderate	Hill slope	Shallow (<25 cm)

C. Vegetation status

Tree species with IVI

A. pendula (193), *D. melanoxylon* (95), *M. parviflora* (13)

Shrub species with IVI

No shrub

ENS Tree : 2.10 Shrub: 1.00

Herb and Grass Species

B. eriantha, B. pussilla, E. alsinoides, I. cordifolia, J. procumbense, T. tennala, T. procumbens.

D. Wild-life

Lepus nigricolis, *Pavo cristatus*.

E. Soil characteristics

pH	EC (dSm^{-1})	SOC (%)	BD (g/cm^3)	NO_3-N ($mg\ kg^{-1}$)	NH_4-N ($mg\ kg^{-1}$)	PO_4-P ($mg\ kg^{-1}$)	C density ($tons\ ha^{-1}$)
7.75	0.23	0.585	1.52	6.54	6.50	8.30	15.64

F. History and status

This grove is in name of a nature lover. A temple of Mahadev ji is also here, where people use to worship lord Shiva. In the memory of nature lover Babulal ji, a large fair is organised here on every Shivratri.

DATA ON SACRED GROVES OF RAJASTHAN

Sacred Grove No. 61

A. Basic information about the sacred groves

Name of sacred groves	Devnarayan Ji Ka Asthan
Name of village	Savalpura, Kheria
Panchayat	Bakani
Forest Block	-
Land status	Community land
District	Jhalawar
Area	100 bigha
Altitude (m)	366
Latitude	24° 21' 21.9 N
Longitude	76° 18' 33.7 E

B. Topographical features

Erosion status	Topography	Soil depth
Moderate	Plain, Normal undulating	Moderate (50-100)

C. Vegetation status

Tree species with IVI

A. leucophloea (148), *P. sylvestris* (92), *A. indica* (18), *P. juliflora* (17), *F. religiosa* (10), *H. integrifolia* (8), *Z. mauritiana* (7).

Shrub species with IVI

C. procera (300)

ENS Tree : 3.16 Shrub: 1.00

Herb and Grass Species

C. tora, E. alsinoides, E. hirta, I. cordifolia, T. procumbens.

D. Wild-life

Herpestes edwardsi, Lepus nigricolis, Pavo cristatus.

E. Soil characteristics

pH	EC (dSm^{-1})	SOC (%)	BD (g/cm^3)	NO_3-N ($mg\ kg^{-1}$)	NH_4-N ($mg\ kg^{-1}$)	PO_4-P ($mg\ kg^{-1}$)	C density ($tons\ ha^{-1}$)
7.04	0.14	0.450	1.59	7.23	3.00	5.12	19.21

F. History and status

The grove is in name of God Devnarayan – Kul Devata of Gurjar community. There is small temple in the gove, where use to come and worship the diety. The dominant tree species in the area is *Acacia leucophloea*.

DATA ON SACRED GROVES OF RAJASTHAN

Sacred Grove No. 62

A. Basic information about the sacred groves

Name of sacred groves	Shiv Mandir, Kelkheura
Name of village	Dudharkhedi, Chanderia, Kelkheura
Panchayat	Ghatoli
Forest Block	-
Land status	Community land
District	Jhalawar
Area	400 bigha
Altitude (m)	345
Latitude	24° 18' 53.3 N
Longitude	76° 90' 40.5 E

B. Topographical feature

Erosion status	Topography	Soil depth
Moderate	Hill slope	Shallow (<25 cm)

C. Vegetation status

Tree species with IVI

D. melanoxylon (161), *T. bellerica* (59), *Madhuca indica* (42), *A. pendula* (38)

Shrub species with IVI

No shrub

ENS Tree : 2.30 Shrub: 0

Herb and Grass Species

B. pussilla, B. repens, E. alsinoides, E. hirta, I. cordifolia, J. procumbens, T. procumbens.

D. Wild-life

Panthera pardus, Boselaphus tragocamelus, Cuon alpinus, Rusa unicolor, Sus scrofa, Hyaena hyaena, Lepus nigricolis

E. Soil characteristics

pH	EC (dSm^{-1})	SOC (%)	BD (g/cm^3)	NO_3-N ($mg\ kg^{-1}$)	NH_4-N ($mg\ kg^{-1}$)	PO_4-P ($mg\ kg^{-1}$)	C density ($tons\ ha^{-1}$)
8.43	0.11	0.285	1.63	3.65	3.20	6.50	5.78

F. History and status

This is about 400 year old grove, where there is Samadhi of Guru Makhandas ji Maharaj, at village Kelkherua, Ghatoli. Natural water reservoir is present here. The grove is dominated by *D. melanoxylon* tree.

DATA ON SACRED GROVES OF RAJASTHAN

Sacred Grove No. 63

A. Basic Information about the sacred groves

Name of sacred groves	Bhairav Ji Ka Bag
Name of village	Ladania, Kheda
Panchayat	Khanpur
Forest Block	-
Land status	Community land
District	Jhalawar
Area	15 bigha
Altitude (m)	295
Latitude	24° 39' 03.9 N
Longitude	76° 08' 01.4 E

B. Topographical features

Erosion status	Topography	Soil depth
Moderate	Plain	Deep (>100 cm)

C. Vegetation status

Tree species with IVI

B. monosperma (100), *P. sylvestris* (59), *T. indica* (39), *F. religiosa* (26), *A. leucophloea* (21), *H. integrifolia* (18), *A. nilotica* (17), *M. parviflora* (10), *D. melanoxylon* (5), *T. bellerica* (5)

Shrub species with IVI

C. sepiaria (150), *Cryptolepis buchananii* (112), *F. indica* (38)

ENS Tree : 5.03 Shrub: 1.40

Herb and Grass Species

C. tora, E. alsinoides, I. cordifolia, J. procumbense, T. procumbens.

D. Wild-life

Lepus nigricolis, *Boselaphus tragocamelus*.

E. Soil characteristics

pH	EC (dSm^{-1})	SOC (%)	BD (g/cm^3)	NO_3-N (mg kg^{-1})	NH_4-N (mg kg^{-1})	PO_4-P (mg kg^{-1})	C density (tons ha^{-1})
7.14	0.09	0.375	1.60	2.59	5.00	8.30	16.40

F. History and status

An ancient sacred grove, which is conserved and maintained in name of Bairavji. The grove is dominated by *Butea monosperma* as the tree. A small place of worship is there, where local people use to worship Bairav ji.

DATA ON SACRED GROVES OF RAJASTHAN

Sacred Grove No. 64

A. Basic information about the sacred groves

Name of sacred groves	Bihari ji Ka Mandir
Name of village	Brindavan, Bhadinda Kala
Panchayat	Bhadunda kala
Forest Block	Chirawa
Land status	Community land
District	Jhunjunnu
Area	60 bigha
Altitude (m)	356
Latitude	28° 05' 37.0 N
Longitude	75° 32' 59.9 E

B. Topographical features

Erosion status	Topography	Soil depth
Moderate	Flat land	Deep (>100 cm)

C. Vegetation status

Tree species with IVI

S. oleoides (160), *P. cineraria* (117), *A. nilotica* (23)

Shrub species with IVI

No shrub (0)

ENS Tree : 2.16 Shrub: 1.00

Herb and Grass

Amaranthus species, C. dactylon, C. tora, O. compressa, T. terrestris.

D. Wild-life

Lepus nigricolis, Boselaphus tragocamelus.

E. Soil characteristics

pH	EC (dSm^{-1})	SOC (%)	BD (g/cm^3)	NO_3-N (mg kg^{-1})	NH_4-N (mg kg^{-1})	PO_4-P (mg kg^{-1})	C density (tons ha^{-1})
6.90	0.22	0.165	1.65	1.55	6.91	11.20	8.17

F. History and status

An ancient grove in name of Bihari ji, where a temple has been established 500 years before. This temple is dedicated to Bihariji and is visited by many devotees everyday. The temple is mainly known for its antique paintings and murals. The grove is dominated by *Salvadora oleoides* and cutting of trees is total prohibited here.

DATA ON SACRED GROVES OF RAJASTHAN

Sacred Grove No. 65

A. Basic Information about the sacred groves

Name of sacred groves	Prithviraj Ki Bani
Name of village	Bhachunda Khurd
Panchayat	Bhadunda khurd
Forest Block	Chidawa
Land status	Community land
District	Jhunjunnu
Area	250 bigha
Altitude (m)	332
Latitude	28° 08' 29.0 N
Longitude	75° 30' 36.2 E

B. Topographical features

Erosion status	Topography	Soil depth
Moderate	Almost flat land	Deep (>100 cm)

C. Vegetation status

Tree species with IVI

P. cineraria (218), *C. decidua* (68), *S. oleoides* (15)

Shrub species with IVI

C. procera (300)

ENS Tree : 1.88 Shrub: 1.00

Herb and Grass Species

Amaranthus species, C. dactylon, C. tora, D. sindicum, E. hirta, H. marifolium, I. cordifolia, T. purpurea, T. terrestris, V. cinerea, Verbesina encelioides.

D. Wild-life

Lepus nigricolis, Boselaphus tragocamelus.

E. Soil characteristics

pH	EC (dSm^{-1})	SOC (%)	BD (g/cm^3)	NO_3-N ($mg\ kg^{-1}$)	NH_4-N ($mg\ kg^{-1}$)	PO_4-P ($mg\ kg^{-1}$)	C density ($tons\ ha^{-1}$)
7.46	0.12	0.390	1.62	2.02	1.16	5.20	18.95

F. History and status

This sacred grove is about 200-250 years old dominated by *Prosopis cineraria* trees. This grove is conserved and maintained by local communities in name of Prithiviraj-Kuldev of Jhakhadia Vansh and a temple was established in the grove in his memory.

DATA ON SACRED GROVES OF RAJASTHAN

Sacred Grove No. 66

A. Basic information about the sacred groves

Name of sacred groves	Panchmir Pir Ki Bani
Name of village	Mahawadi
Panchayat	Nawalgarh
Forest Block	Lohagal
Land status	Community land
District	Jhunjunnu
Area	100 bigha
Altitude (m)	450
Latitude	27° 44' 15.3 N
Longitude	75° 19' 30.7 E

B. Topographical features

Erosion status	Topography	Soil depth
Moderate	Almost flat land	Deep (>100 cm)

C. Vegetation status

Tree species with IVI

A. tortilis (125), *P. cineraria* (116), *A. nilotica* (30),*M. emarginata* (16), *C. decidua* (13).

Shrub species with IVI

Z. nummularia (237), *C. procera* (63)

ENS Tree : 2.43 Shrub: 1.81

Herb and Grass

Amaranthus species, C. dactylon, C. tora, D. sindicum, H. marifolium, I. cordifolia, T. purpurea, T. terrestris, V. cinerea.

D. Wild-life

Lepus nigricolis, Passer domesticus.

E. Soil characteristics

pH	EC (dSm^{-1})	SOC (%)	BD (g/cm^3)	NO_3-N (mg kg^{-1})	NH_4-N (mg kg^{-1})	PO_4-P (mg kg^{-1})	C density (tons ha^{-1})
6.18	0.14	0.180	1.63	4.04	4.60	10.40	8.80

F. History and status

About 300 years old grove is maintained and conserved by the locals in name of Panchmir Pir. Peopleuse to come and offer worship to the Pir Baba. *Acacia tortilis* is dominating in this grove and appeared to be due to plantation activity of this species.

DATA ON SACRED GROVES OF RAJASTHAN

Sacred Grove No. 67

A. Basic Information about the sacred groves

Name of sacred groves	Kolu Pabuji Ka Oran
Name of village	Kolu Pabuji
Panchayat	Baap
Forest Block	-
Land status	Community land
District	Jodhpur
Area	20000 bigha
Altitude (m)	262
Latitude	26° 55' 03.1 N
Longitude	72° 18' 44.6 E

B. Topographical features

Erosion status	Topography	Soil depth
Slight	Flat land	Moderate (50-100)

C. Vegetation status

Tree species with IVI

C. decidua (240), *P. cineraria* (60)

Shrub species with IVI

No shrub

ENS Tree : 1.04 Shrub: 1.00

Herb and Grass Species

D. sindicum, Aerva pseudotomentosa

D. Wild-life

Boselaphus tragocamelus, Gazella gazella, Naja naja

E. Soil characteristics

pH	EC (dSm^{-1})	SOC (%)	BD (g/cm^3)	NO_3-N ($mg\ kg^{-1}$)	NH_4-N ($mg\ kg^{-1}$)	PO_4-P ($mg\ kg^{-1}$)	C density ($tons\ ha^{-1}$)
7.23	0.23	0.220	1.62	2.22	5.26	5.66	10.22

F. History and status

It is one of the largest groves in the region in name of Pabuji- a folk-deity of Rajasthan in India. He lived in the 14th century in Rajasthan. He is now widely worshipped as a deity by Rabari herdsmen and others. There is about 500 old temple of Pabuji, also known as the god of Camel. People use to come here to worship Pabuji. This grove is mostly dominated by *P. cineraria*, *Capparis decidua* and *Z. nummularia* trees.

DATA ON SACRED GROVES OF RAJASTHAN

Sacred Grove No. 68

A. Basic information about the sacred groves

Name of sacred groves	Karni Mata Ji Oran
Name of village	Khari Khurd
Panchayat	Luni
Forest Block	Kharikurd
Land status	Community land
District	Jodhpur
Area	175 bigha
Altitude (m)	230
Latitude	26° 10' 26.9 N
Longitude	73° 21' 10.5 E

B. Topographical features

Erosion status	Topography	Soil depth
Moderate	Almost flatland	Deep (>100 cm)

C. Vegetation status

Tree species with IVI

C. decidua (139), *P. juliflora* (125), *P. cineraria* (20), *S. oleoides* (16).

Shrub species with IVI

Z. nummularia (82), *G. tenax* (80), *C. procera* (78), *C. burhia* (30), *A. persica* (30).

ENS Tree : 2.47 Shrub: 4.71

Herb and Grass Species

C. tora, D. sindicum, E. ciliaris, H. marifolium, I. cordifolia, T. purpurea, T. terrestris, V. cinerea.

D. Wild-life

Cuon alpinus Funambulus palmarum.

E. Soil characteristics

pH	EC (dSm^{-1})	SOC (%)	BD (g/cm^3)	NO_3-N ($mg\ kg^{-1}$)	NH_4-N ($mg\ kg^{-1}$)	PO_4-P ($mg\ kg^{-1}$)	C density ($tons\ ha^{-1}$)
8.18	0.38	0.210	1.63	2.96	6.43	6.44	10.17

F. History and status

It is an ancient grove in name of Karni Mata, whose main temple is in Deshnok. There is an old temple in the grove, where local people come and worship Mataji. The grove is dominated mostly by *Capparis decidua* and *P. juliflora*.

DATA ON SACRED GROVES OF RAJASTHAN

Sacred Grove No. 69

A. Basic information about the sacred groves

Name of sacred groves	Ramdeo Ji ka Oran
Name of village	Nandiya, Birahi
Panchayat	Bavdi
Forest Block	Bilara
Land status	Community land
District	Jodhpur
Area	345 bigha
Altitude (m)	273
Latitude	26° 31' 05.6 N
Longitude	73° 17' 48.5 E

B. Topographical features

Erosion status	Topography	Soil depth
Moderate	Undulating flat land	Deep (>100 cm)

C. Vegetation status

Tree species with IVI

P. cineraria (235), *P. juliflora* (65)

Shrub species with IVI

C. procera (300)

ENS Tree : 1.74 Shrub: 1.00

Herb and Grass Species

C. tora, D. sindicum, E. hirta, E. ciliaris, H. marifolium, I. cordifolia, T. purpurea, T. terrestris, V. cinerea.

D. Wild-Life

Cuon alpinus, *Funambulus palmarum*, *Passer domesticus.*

E. Soil characteristics

pH	EC (dSm^{-1})	SOC (%)	BD (g/cm^3)	NO_3-N (mg kg^{-1})	NH_4-N (mg kg^{-1})	PO_4-P (mg kg^{-1})	C density (tons ha^{-1})
8.24	0.24	0.300	1.59	4.46	5.32	4.90	11.61

F. History and status

An old Oran has been kept a side in name of Baba Ramdeo, the local diety in western Rajsthan. Local people use to worship Baba Ramdeo. The dominant tree species is *P. cineraria*, though *P. juliflora* has infested this grove.

DATA ON SACRED GROVES OF RAJASTHAN

Sacred Grove No. 70

A. Basic information about the sacred groves

Name of sacred groves	Pabuji Ka Oran
Name of village	Nandiya Prabhavati
Panchayat	Bhopalgarh
Forest Block	Nandiya
Land status	Community land
District	Jodhpur
Area	40 bigha
Altitude (m)	349
Latitude	26° 33' 14.0 N
Longitude	73° 20' 56.8 E

B. Topographical features

Erosion status	Topography	Soil depth
Moderate	Hill slope	Shallow (<25 cm)

C. Vegetation status

Tree species with IVI

P. juliflora (300)

Shrub species with IVI

No shrub (0)

ENS Tree > 30: 1.00 Shrub: 0

Herb and Grass Species

B. diffusa, B. pussilla, B. ramosa, B. repens, C. tora, D. sindicum, E. hirta, E. ciliaris, H. marifolium, I. cordifolia, T. purpurea, T. terrestris, V. cinerea

D. Wild-Life

Funambulus palmarum, *Passer domesticus*, *Vulpes bengalehsis*

E. Soil characteristics

pH	EC (dSm^{-1})	SOC (%)	BD (g/cm^3)	NO_3-N ($mg\ kg^{-1}$)	NH_4-N ($mg\ kg^{-1}$)	PO_4-P ($mg\ kg^{-1}$)	C density ($tons\ ha^{-1}$)
7.82	0.28	0.030	1.71	2.30	7.14	6.24	1.52

F. History and status

This grove has also been put a side by the local people in name of Pabuji- a folk-deity of Rajasthan, worshipped particularly by Rabari herdsmen and others. There is a temple on the hill top, where people worship the diety. The grove has been infested solely by *P. juliflora*.

DATA ON SACRED GROVES OF RAJASTHAN

Sacred Grove No. 71

A. Basic information about the sacred groves

Name of sacred groves	Joleyali Oran
Name of village	Joleyali
Panchayat	Mandor
Forest Block	
Land status	Community land
District	Jodhpur
Area	260 bigha
Altitude (m)	195
Latitude	26° 16' 21.4 N
Longitude	72° 44' 59.9 E

B. Topographical features

Erosion status	Topography	Soil depth
Gullied	Undulating plain	Deep (>100 cm)

C. Vegetation status

Tree species with IVI

P. cineraria (267), *A. sericea* (33)

Shrub species with IVI

No shrub (0)

ENS Tree : 1.26 Shrub: 0

Herb and Grass Species

B. diffusa, B. ramosa, C. tora, D. sindicum, E. hirta, E. ciliaris, H. marifolium, I. cordifolia, T. purpurea, T. terrestris.

D. Wild-Life

Naja naja, Passer domesticus.

E. Soil characteristics

pH	EC (dSm^{-1})	SOC (%)	BD (g/cm^3)	NO_3-N ($mg\ kg^{-1}$)	NH_4-N ($mg\ kg^{-1}$)	PO_4-P ($mg\ kg^{-1}$)	C density ($tons\ ha^{-1}$)
7.67	0.34	0.300	1.66	2.18	6.16	4.50	5.00

F. History and status

An ancient grove along with the water body and has been put aside in name of local diety. Though the grove is dominated by *P. cineraria*, but quite old trees of *Anogeissus sericea*- an endangered species are also present in this grove.

DATA ON SACRED GROVES OF RAJASTHAN

Sacred Grove No. 72

A. Basic information about the sacred groves

Name of sacred groves	Jardabala Ji Ka Oran
Name of village	Langra
Panchayat	Sapotra
Forest Block	Langra
Land status	Community land
District	Karauli
Area	2 bigha
Altitude (m)	304
Latitude	26° 24' 46.3 N
Longitude	77° 07' 30.9 E

B. Topographical features

Erosion status	Topography	Soil depth
Moderate	Valley (riverbed)	Moderately deep (25-50)

C. Vegetation status

Tree species with IVI

T. indica (71), *Mangifera indica* (50), *H. integrifolia* (31), *F. glomerata* (30), *F. benghalensis* (28), *A. leucophloea* (27), *C. nurvala* (21), *A. procera* (12), *P. pinnata* (12), *Syzygium cummini* (11), *M. hortensis* (7).

Shrub species with IVI

C. procera (150), *Z. nummularia* (100), *S. virosa* (27), *D. cinerea* (24)

ENS Tree : 8.75 Shrub: 2.85

Herb and Grass Species

B. pussilla, B. ramosa, B. repens, C. tora, E. hirta, E. ciliaris, H. marifolium, I. cordifolia, L. trinervis, M. jacquemontii, T. purpurea, V. cinerea.

D. Wild-Life

Psittacula krameri, Funambulus palmarum, Pavo cristatus.

E. Soil characteristics

pH	EC (dSm^{-1})	SOC (%)	BD (g/cm^3)	NO_3-N ($mg\ kg^{-1}$)	NH_4-N ($mg\ kg^{-1}$)	PO_4-P ($mg\ kg^{-1}$)	C density ($tons\ ha^{-1}$)
6.38	0.26	0.855	1.32	3.45	1.82	6.40	29.61

F. History and status

An old grove is being conserved and protected by local communities in name of local diety. Dominant tree species in the grove is *Tamarindus indica*. People use to gather here for different social function in the grove.

DATA ON SACRED GROVES OF RAJASTHAN

Sacred Grove No. 73

A. Basic information about the sacred groves

Name of sacred groves	Barbasni Mata Ji Ka Oran
Name of village	Dudherai
Panchayat	Amarwa
Forest Block	Barbasni
Land status	Community land
District	Karauli
Area	252 bigha
Altitude (m)	279
Latitude	26° 17' 00.3 N
Longitude	76° 48' 40.4 E

B. Topographical features

Erosion status	Topography	Soil depth
Moderate	Hill slope	Moderately deep (25-50)

C. Vegetation status

Tree species with IVI

A. pendula (215), *W. tinctoria* (46), *D. melanoxylon* (24), *L. coromandelica* (15)

Shrub species with IVI

No shrub (0)

ENS Tree : 1.79 Shrub: 0

Herb and Grass:

B. pussilla, B. ramosa, B. repens, C. tora, E. hirta, I. cordifolia, L. trinervis, M. Jacquemontii, V. cinerea.

D. Wild-Life

Boselaphus tragocamelus, Rusa unicolor

E. Soil characteristics

pH	EC (dSm^{-1})	SOC (%)	BD (g/cm^3)	NO_3-N ($mg\ kg^{-1}$)	NH_4-N ($mg\ kg^{-1}$)	PO_4-P ($mg\ kg^{-1}$)	C density ($tons\ ha^{-1}$)
6.97	0.31	0.390	1.52	3.35	7.75	13.00	13.25

F. History and status

This grove is maintained and conserved since quite ancient time. There is an old temple of Barbasni Mata Ji – the local Goddess, to whom the people worship. The grove is dominated by old trees of *A.pendula*.

DATA ON SACRED GROVES OF RAJASTHAN

Sacred Grove No. 74

A. Basic information about the sacred groves

Name of sacred groves	Devnarayan Ji Ki Banni
Name of village	Shyampur
Panchayat	Sapotra
Forest Block	Langra
Land status	Community land
District	Karauli
Area	21 bigha
Altitude (m)	375
Latitude	26° 22' 09.9 N
Longitude	77° 09' 04.6 E

B. Topographical features

Erosion status	Topography	Soil depth
Moderate	Hill slope	Shallow (<25 cm)

C. Vegetation status

Tree species with IVI

A. pendula (300)

Shrub species with IVI

D. cinerea (265), *Z. nummularia* (35)

ENS Tree : 1.00 Shrub: 1.33

Herb and Grass Species:

B. pussilla, B. ramosa, B. repens, Corchorus tridens, C. pumila, E. granularis, E. hirta, H. granularis, I. cordifolia, L. trinervis, M. jacquemontii, V. cinerea.

D. Wild-Life

Boselaphus tragocamelus, *Cuon alpinus*, *Passer domesticus*.

E. Soil characteristics

pH	EC (dSm^{-1})	SOC (%)	BD (g/cm^3)	NO_3-N ($mg\ kg^{-1}$)	NH_4-N ($mg\ kg^{-1}$)	PO_4-P ($mg\ kg^{-1}$)	C density ($tons\ ha^{-1}$)
7.85	0.18	0.435	1.45	7.09	3.90	9.00	14.74

F. History and status

This temple was built up 25 years ago. People of about 24 villages of Gunjan community were forced by Sh. Swami Ganeshsanand Baba to build this temple up. All families use to provide about 250 g ghee for temple's Akhand Jyoti.

DATA ON SACRED GROVES OF RAJASTHAN

Sacred Grove No. 75

A. Basic information about the sacred groves

Name of sacred groves	Dardevi Mata ji Ka Oran
Name of village	Prahladpura, Umedganj
Panchayat	Laadpura
Forest Block	Dardevi
Land status	Community land
District	Kota
Area	80 bigha
Altitude (m)	289
Latitude	25° 04' 38.1 N
Longitude	75° 55' 08.6 E

B. Topographical features

Erosion status	Topography	Soil depth
Slight	Flat land	Moderate (50-100)

C. Vegetation status

Tree species with IVI

T. arjuna (95), *M. parviflora* (41), *A. pendula* (37), *P. juliflora* (20), *P. dulce* (16), *L. coromandelica* (15), *P. sylvestris* (14), *B. monosperma* (12), *A. leucophloea* (10), *E. camaldulensis* (8), *A. indica* (6), *Sterculia urens* (5), *W. tinctoria* (5), *H. integrifolia* (3), *C. fistula* (3).

Shrub species with IVI

Z. nummularia (196), *F. indica* (68), *C. carrundas* (36)

ENS Tree : 8.84 Shrub: 2.28

Herb and Grass Species

B. ramosa, B. repens, C. tora, C. pumila, E. granularis, E. hirta, H. granularis, I. cordifolia, L. trinervis, M. Jacquemontii, T. procumbens, V. cinerea.

D. Wild-Life

Axis axis, *Boselaphus tragocamelus*, *Canis aureus auveus*.

E. Soil characteristics

pH	EC (dSm^{-1})	SOC (%)	BD (g/cm^3)	NO_3-N ($mg\ kg^{-1}$)	NH_4-N ($mg\ kg^{-1}$)	PO_4-P ($mg\ kg^{-1}$)	C density ($tons\ ha^{-1}$)
7.71	0.57	0.495	1.44	2.35	4.46	6.52	18.94

F. History and status

In ancient times King of Kota used to come here to worship the Goddess, The Dar Devi Mataji. People of the region use to continue this tradition. Goats are offered and sacrificed here as a custom. Before some years, *Pendanus odoratissinus* was found here but has disappear now and grove is dominated by *Terminalia arjuna* and quite rich in floral diversity.

DATA ON SACRED GROVES OF RAJASTHAN

Sacred Grove No. 76

A. Basic information about the sacred groves

Name of sacred groves	Devnarayan Bhagwan Ka Oran
Name of village	Keechalheda
Panchayat	Laadpura
Forest Block	-
Land status	Community land
District	Kota
Area	50 bigha
Altitude (m)	278
Latitude	25° 01' 14.7 N
Longitude	76° 03' 38.7 E

B. Topographical features

Erosion status	Topography	Soil depth
Moderate	Almost flat land	Deep (> 100 cm)

C. Vegetation status

Tree species with IVI

A. catechu (120), *A. nilotica* (48), *B. monosperma* (44), *M. parviflora* (24), *Z. mauritiana* (17), *A. leucophloea* (17), *Randia dumetorum* (9), *D. falcata* (8), *Ehretia aspera* (7), *C. decidua* (6)

Shrub species with IVI

D. cinerea (300)

ENS Tree : 4.50 Shrub: 1.00

Herb and Grass Species:

B. ramosa, B. repens, C. tora, E. alsinoides, E. granularis, E. hirta, I. cordifolia, T. procumbens, V. cinerea.

D. Wild-Life

Axis axis, *Boselaphus tragocamelus*, *Lepus nigricolis*.

E. Soil characteristics

pH	EC (dSm^{-1})	SOC (%)	BD (g/cm^3)	NO_3-N (mg kg^{-1})	NH_4-N (mg kg^{-1})	PO_4-P (mg kg^{-1})	C density (tons ha^{-1})
7.73	0.45	0.945	1.33	2.36	3.78	4.98	20.61

F. History and status

This grove is about 200 year old, where a temple of lord Devanarayan has been built in those days and worshiped by the local communities. The grove is dominated by *Acacia catechu* and indicator of this type of forests in the region.

DATA ON SACRED GROVES OF RAJASTHAN

Sacred Grove No. 77

A. Basic information about the sacred groves

Name of sacred groves	Devnarayan Ji Ka Oran
Name of village	Khajuri
Panchayat	Sangoda
Forest Block	-
Land status	Forest land
District	Kota
Area	600 bigha
Altitude (m)	268
Latitude	25° 03' 34.1 N
Longitude	76° 05'43.8 E

B. Topographical features

Erosion status	Topography	Soil depth
Slight	Almost flat land	Deep (>100 cm)

C. Vegetation status

Tree species with IVI

E. camaldulensis (208), *P. sylvestris* (60), *P. juliflora* (32)

Shrub species with IVI

No shrub (0)

ENS Tree : 1.89 Shrub: 1.00

Herb and Grass Species:

B. ramosa, C. tora, E. alsinoides, E. hirta, I. cordifolia, T. procumbens, V. cinerea.

D. Wild-Life

Axis axis, *Boselaphus tragocamelus*, *Lepus nigricolis*, *Sus scrofa*, *Bos taurus*.

E. Soil characteristics

pH	EC (dSm^{-1})	SOC (%)	BD (g/cm^3)	NO_3-N (mg kg^{-1})	NH_4-N (mg kg^{-1})	PO_4-P (mg kg^{-1})	C density (tons ha^{-1})
8.33	0.35	0.720	1.68	1.53	7.12	2.32	22.85

F. History and status

This is a quite old grove. In ancient time trees of Khajur (*Phoenix sylvestris*) were found in this sacred grove in much abundance, hence the name of the village is Khajuriya. At present plantation of *Eucalyptus camaldulensis* has been done.

DATA ON SACRED GROVES OF RAJASTHAN

Sacred Grove No. 78

A. Basic Information about the sacred groves

Name of sacred groves	Devnarayan Ji Ka Oran
Name of village	Chatarpura
Panchayat	Sangoda
Forest Block	-
Land status	Community land
District	Kota
Area	25 bigha
Altitude (m)	269
Latitude	24° 52' 10.0 N
Longitude	76° 20' 18.0 E

B. Topographical features

Erosion status	Topography	Soil depth
Moderate	Almost flat land	Deep (>100 cm)

C. Vegetation status

Tree species with IVI

A. nilotica (92), *A. leucophloea* (92), *P. sylvestris* (33), *P. juliflora* (31), *C. decidua* (27), *A. indica* (10), *D. falcata* (9), *Pithecellobium dulce* (6)

Shrub species with IVI

D. cinerea (177), *C. sepiaria* (81), *C. procera* (42)

ENS Tree : 4.91 Shrub: 2.86

Herb and Grass Species:

B. ramosa, C. tora, E. alsinoides, E. hirta, I. cordifolia, T. procumbens.

D. Wild-Life

Axis axis, *Boselaphus tragocamelus.*

E. Soil characteristics

pH	EC (dSm^{-1})	SOC (%)	BD (g/cm^3)	NO_3-N (mg kg^{-1})	NH_4-N (mg kg^{-1})	PO_4-P (mg kg^{-1})	C density (tons ha^{-1})
7.76	0.45	1.035	1.40	1.26	9.00	17.00	24.83

F. History and status

An ancient grove and is dominated by *Acacia nilotica*. The grove is conserved and protected by local communities. There is an old temple, where people come and worship Lord Devnarayan in this place.

DATA ON SACRED GROVES OF RAJASTHAN

Sacred Grove No. 79

A. Basic information about the sacred groves

Name of sacred groves	Dhokad Wali Mata ji Ka Oran
Name of village	Daarupura
Panchayat	Kherabad
Forest Block	-
Land status	Community land
District	Kota
Area	7 bigha
Altitude (m)	347
Latitude	24° 45' 51.0 N
Longitude	75° 55' 53.9 E

B. Topographical features

Erosion status	Topography	Soil depth
Slight	Flat land	Deep (>100 cm)

C. Vegetation status

Tree species with IVI

A. pendula (211), *A. leucophloea* (59), *A. nilotica* (16), *D. falcata* (14).

Shrub species with IVI

C. sepiaria (300)

ENS Tree : 1.77 Shrub: 1.00

Herb and Grass Species:

E. alsinoides

D. Wild-Life

Lepus nigricolis, *Canis aureus auveus*.

E. Soil characteristics

pH	EC (dSm^{-1})	SOC (%)	BD (g/cm^3)	NO_3-N ($mg\ kg^{-1}$)	NH_4-N ($mg\ kg^{-1}$)	PO_4-P ($mg\ kg^{-1}$)	C density ($tons\ ha^{-1}$)
7.99	0.31	0.210	1.62	1.14	5.20	9.00	7.21

F. History and status

A very old grove is being protected and conserved in name of Dhokad Wali Mata ji. An old temple is also situated in the grove. As told by the local people, the whole area was covered with dense forest in relatively older times.

DATA ON SACRED GROVES OF RAJASTHAN

Sacred Grove No. 80

A. Basic Information about the sacred groves

Name of sacred groves	Devnarayan Ji Ka Oran
Name of village	Minyakhedi
Panchayat	Kherabad
Forest Block	-
Land status	Community land
District	Kota
Area	23 bigha
Altitude (m)	336
Latitude	24° 48' 00.5 N
Longitude	75° 56' 54.0 E

B. Topographical features

Erosion status	Topography	Soil depth
Moderate	Undulating plain	Deep (>100 cm)

C. Vegetation status

Tree species with IVI

A. catechu (163), *C. decidua* (38), *A. pendula* (25), *M. parviflora* (24), *B. monosperma* (18), *B. aegyptiaca* (15), *A. leucophloea* (10), *P. juliflora* (7)

Shrub species with IVI

D. cinerea (127), *C. sepiaria* (97), *Z. nummularia* (52), *C. procera* (24)

ENS Tree : 2.97 Shrub: 3.60

Herb and Grass Species:

B. diffusa, B. ramosa, C. tora, E. alsinoides, E. hirta, E. ciliaris, I. cordifolia, T. procumbens, V. cinerea.

D. Wild-Life

Lepus nigricolis, Canis aureus auveus.

E. Soil characteristics

pH	EC (dSm^{-1})	SOC (%)	BD (g/cm^3)	NO_3-N (mg kg^{-1})	NH_4-N (mg kg^{-1})	PO_4-P (mg kg^{-1})	C density (tons ha^{-1})
7.90	0.33	0.660	1.40	2.33	7.20	13.00	17.46

F. History and status

This is a quite old sacred grove dominated by *Acacia catechu*. The grove is conserved and protected by the local people in name of Lord Devanarayan, whose temple is there in the grove.

DATA ON SACRED GROVES OF RAJASTHAN

Sacred Grove No. 81

A. Basic information about the sacred groves

Name of sacred groves	Hanuman Ji ka Oran
Name of village	Kushalipura
Panchayat	Kurad
Forest Block	-
Land status	Community land
District	Kota
Area	17 bigha
Altitude (m)	268
Latitude	25° 02' 49.7 N
Longitude	76° 07' 31.3 E

B. Topographical features

Erosion status	Topography	Soil depth
Moderate	Rocky flat land	Moderately deep (25-50)

C. Vegetation status

Tree species with IVI

B. monosperma (271), *C. nurvala* (29)

Shrub species with IVI

Z. nummularia (202), *D. cinerea* (69), *S. virosa* (29)

ENS Tree : 1.06 Shrub: 2.08

Herb and Grass Species:

B. eriantha, B. diffusa, B. ramosa, C. tora, E. alsinoides, E. hirta, E. ciliaris, I. cordifolia, T. procumbens, Urginea indica, V. cinerea.

D. Wild-Life

Axis axis, Boselaphus tragocamelus, Lepus nigricolis.

E. Soil characteristics

pH	EC (dSm^{-1})	SOC (%)	BD (g/cm^3)	NO_3-N (mg kg^{-1})	NH_4-N (mg kg^{-1})	PO_4-P (mg kg^{-1})	C density (tons ha^{-1})
7.66	0.24	0.945	1.32	1.18	9.60	13.00	24.15

F. History and status

This sacred grove is about 260 years old. It is conserved and protected by local communities in name of Lord Hanuman ji. People believe that diseases are cured by coming to this land. The grove is dominated by *Butea monosperma* trees.

DATA ON SACRED GROVES OF RAJASTHAN

Sacred Grove No. **82**

A. Basic information about the sacred groves

Name of sacred groves	Devnarayan Ji ka Oran
Name of village	Bambali
Panchayat	Sultanpur
Forest Block	-
Land status	Community land
District	Kota
Area	125 bigha
Altitude (m)	276
Latitude	25° 05' 06.1 N
Longitude	76° 05' 22.8 E

B. Topographical features

Erosion status	Topography	Soil depth
Moderate	Almost flat land	Deep (>100 cm)

C. Vegetation status

Tree species with IVI

P. juliflora (102), *M. parviflora* (81), *C. decidua* (40), *B. aegyptiaca* (27), *A. catechu* (27), *A. indica* (12), *H. integrifolia* (11)

Shrub species with IVI

C. sepiaria (300)

ENS Tree : 4.35 Shrub: 1.00

Herb and Grass Species:

B. diffusa, B. ramosa, C. tora, E. alsinoides, E. hirta, I. cordifolia, T. procumbens, V. cinerea.

D. Wild-Life

Axis axis, Boselaphus tragocamelus, Lepus nigricolis.

E. Soil characteristics

pH	EC (dSm^{-1})	SOC (%)	BD (g/cm^3)	NO_3-N ($mg\ kg^{-1}$)	NH_4-N ($mg\ kg^{-1}$)	PO_4-P ($mg\ kg^{-1}$)	C density ($tons\ ha^{-1}$)
8.05	0.29	0.780	1.44	1.17	5.20	9.00	13.05

F. History and status

An old sacred grove, where people use to come and worship Lord Devnarayanji in the temple built longback in the grove. Quite important species of relatively moist climate, but are presently dominated by *P. juliflora*.

DATA ON SACRED GROVES OF RAJASTHAN

Sacred Grove No. 83

A. Basic information about the sacred groves

Name of sacred groves	Kalaji-Gora Ji Oran
Name of village	Mohanpura
Panchayat	Sangod
Forest Block	-
Land status	Community land
District	Kota
Area	24 bigha
Altitude (m)	268
Latitude	24° 56' 16.8 N
Longitude	76° 03' 00.3 E

B. Topographical features

Erosion status	Topography	Soil depth
Moderate	Rocky pediment plain	Moderate (50-100)

C. Vegetation status

Tree species with IVI

A. pendula (158), *P. sylvestris* (42), *B. monosperma* (32), *P. juliflora* (11), *T. arjuna* (11), *D. regia* (10), *D. falcata* (10), *A. nilotica* (9), *Z. mauritiana* (9), *C. siamia* (8)

Shrub species with IVI

Z. nummularia (165), *F. indica* (84), *C. sepiaria* (51)

ENS Tree : 3.19 Shrub: 2.34

Herb and Grass Species:

B. eriantha, B. diffusa, B. ramosa, C. tora, E. alsinoides, E. hirta, I. cordifolia, J. procumbens, T. procumbens, V. cinerea.

D. Wild-Life

Axis axis, Boselaphus tragocamelus, Lepus nigricolis, Sus scrofa.

E. Soil characteristics

pH	EC (dSm^{-1})	SOC (%)	BD (g/cm^3)	NO_3-N ($mg\ kg^{-1}$)	NH_4-N ($mg\ kg^{-1}$)	PO_4-P ($mg\ kg^{-1}$)	C density ($tons\ ha^{-1}$)
7.34	0.36	0.645	1.48	1.09	7.00	13.00	23.85

F. History and status

An ancient sacred grove and is conserved and protected by the local communities in name of Kalaji- Goraji, deities in Hindu culture. The main deity at Kala- Gora Bhairav is Bhairavji who is usually kept at the entrance of the Hindu temples across the country as per custom. Two Bhairav brothers namely Kala Bhairav and Gaura Bhairav are worshipped in general. The grove is dominated by *Anogeissus pendula* trees.

DATA ON SACRED GROVES OF RAJASTHAN

Sacred Grove No. 84

A. Basic information about the sacred groves

Name of sacred groves: Bheruji Ka Oran
Name of village: Theakariyawas
Panchayat: Deedwana
Forest Block: -
Land status: Community land
District: Nagaur
Area: 23 bigha
Altitude (m): 333
Latitude: 27° 26' 33.2 N
Longitude: 74° 26' 41.8 E

B. Topographical features

Erosion status	Topography	Soil depth
Severe	Almost flat land	Deep (>100 cm)

C. Vegetation status

Tree species with IVI

P. cineraria (111), *M. emarginata* (110), *A. senegal* (40), *F. religiosa* (27), *C. decidua* (12).

Shrub species with IVI

A. jacquemontii (133), *C. burhia* (55), *C. procera* (37), *L. pyrotechnica* (34), *Z. nummularia* (16), *L. barbarum* (13), *A. pseudotomentosa* (12).

ENS Tree : 3.57 Shrub: 5.77

Herb and Grass Species:

A. funiculata, B. articularis, B. eracta, C. ciliaris, D. aegyptium, D. annulatum, D. ciliaris, E. alsinoides, I. cordifolia, I. linnai, J. procumbense, M. jacqmontii, P. paniculata, T. purpurea, T. roxburghii, T. tennalus, T. terrestris. V. encelioides, V. cinerea.

D. Wild-Life

Axis axis, *Boselaphus tragocamelus*, *Lepus nigricolis*

E. Soil characteristics

pH	EC (dSm^{-1})	SOC (%)	BD (g/cm^3)	NO_3-N ($mg\ kg^{-1}$)	NH_4-N ($mg\ kg^{-1}$)	PO_4-P ($mg\ kg^{-1}$)	C density ($tons\ ha^{-1}$)
8.37	0.27	0.525	1.55	4.80	2.28	7.32	15.72

F. History and status

A quite old temple of Bheruji- a local diety is there in this sacred grove, where people use to come and worship the diety. The grove is dominated by *P. cineraria* the important tree of Indian Desert.

DATA ON SACRED GROVES OF RAJASTHAN

Sacred Grove No. 85

A. Basic information about the sacred groves

Name of sacred groves	Bheruji Ka Oran
Name of village	Dujar
Panchayat	Laadnu
Forest Block	Bhanglaw
Land status	Community land
District	Nagaur
Area	14 bigha
Altitude (m)	338
Latitude	27° 35' 41.5 N
Longitude	74° 24' 35.9 E

B. Topographical features

Erosion status	Topography	Soil depth
Slight	Undulating plain	Deep (>100 cm)

C. Vegetation status

Tree species with IVI

P. juliflora (133), *A. senegal* (85), *C. decidua* (61), *M. emarginata* (12), *A. tortilis* (8)

Shrub species with IVI

L. barbarum (202), *C. procera* (81), *L. pyrotechnica* (17)

ENS Tree : 2.98 Shrub: 1.90

Herb and Grass Species:

A. funiculata, B. articularis, B. eracta, Corchorus depressus, C. ciliaris, D. aegyptium, D. annulatum, D. ciliaris, E. alsinoides, I. cordifolia, I. linnai, J. procumbense, M. jacqmontii, P. erioptera, P. paniculata, T. purpurea, T. roxburghii, T. tennalus, T. terrestris, V. cinerea, V. encelioides.

D. Wild-Life

Axis axis, *Boselaphus tragocamelus*, *Lepus nigricolis*, *Psittacula krameri*.

E. Soil characteristics

pH	EC (dSm^{-1})	SOC (%)	BD (g/cm^3)	NO_3-N ($mg\ kg^{-1}$)	NH_4-N ($mg\ kg^{-1}$)	PO_4-P ($mg\ kg^{-1}$)	C density ($tons\ ha^{-1}$)
7.25	0.22	0.075	1.68	3.50	4.46	8.30	3.69

F. History and status

This sacred grove is also in name of God Bheruji, the local diety in the region. A temple has also been established in the grove, where people come and worship him. The grove is dominated presently by *P. juliflora*-an invasive species.

DATA ON SACRED GROVES OF RAJASTHAN

Sacred Grove No. 86

A. Basic Information about the sacred groves

Name of sacred groves	Gosai Ji Ka Oran
Name of village	Junjala
Panchayat	Jaayal
Forest Block	-
Land status	Community land
District	Nagaur
Area	420 bigha
Altitude (m)	317
Latitude	27° 01' 44.5 N
Longitude	73° 55' 58.4 E

B. Topographical features

Erosion status	Topography	Soil depth
Moderate	Sloppy flat land	Deep (>100 cm)

C. Vegetation status

Tree species with IVI

A. sericea subsp *nummularia* (114), *C. decidua* (96), *P. cineraria* (90)

Shrub species with IVI

Z. nummularia (300)

ENS Tree : 2.70 Shrub: 1.00

Herb and Grass Species:

A. funiculata, B. articularis, B. diffusa, C. depressus, C. ciliaris, D. aegyptium, D. ciliaris, E. ciliaris, E. alsinoides, I. cordifolia, I. linnai, J. procumbense, M. jacqmontii, P. erioptera, T. purpurea, T. roxburghii, T. tennalus, T. terrestris, V. cinerea.

D. Wild-Life

Axis axis, *Boselaphus tragocamelus*, *Lepus nigricolis*, *Psittacula krameri*, *Pavo cristatus*, *Vulpes bengalehsis*.

E. Soil characteristics

pH	EC (dSm^{-1})	SOC (%)	BD (g/cm^3)	NO_3-N ($mg\ kg^{-1}$)	NH_4-N ($mg\ kg^{-1}$)	PO_4-P ($mg\ kg^{-1}$)	C density ($tons\ ha^{-1}$)
8.23	0.29	0.315	1.48	1.50	2.36	10.52	13.73

F. History and status

This is an old sacred grove in Name of Gosaiji, the local diety. There is a temple in the grove, where people use to gather and offer prayer. This grove is dominated by old trees of an endangered species *Anogeissus sericea*

DATA ON SACRED GROVES OF RAJASTHAN

Sacred Grove No. 87

A. Basic information about the sacred groves

Name of sacred groves	Jogmaya Ji Ka Oran
Name of village	Sathika Kala
Panchayat	Nagaur
Forest Block	-
Land status	Community land
District	Nagaur
Area	3000 bigha
Altitude (m)	318
Latitude	27° 01' 11.4 N
Longitude	73° 11' 30.1 E

B. Topographical features

Erosion status	Topography	Soil depth
Severe	Undulating flat land	Deep (>100 cm)

C. Vegetation status

Tree species with IVI

Capparis decidua (194), *M. emarginata* (77), *P. cineraria* (29)

Shrub species with IVI

Z. nummularia (168), *A. pseudotomentosa* (60), *L. pyrotechnica* (36),*C. procera* (24), *A. persica* (12).

ENS Tree : 1.85 Shrub: 3.94

Herb and Grass Species:

A. funiculata, B. diffusa, C. ciliaris, Cassia italica, D. aegyptium, D. ciliaris, E. ciliaris, E. alsinoides, I. cordifolia, I. linnai, J. procumbense, M. jacqmontii, P. erioptera, T. purpurea, T. roxburghii, T. tennalus, T. terrestris, V. cinerea.

D. Wild-Life

Axis axis, Boselaphus tragocamelus, Lepus nigricolis, Psittacula krameri, Pavo cristatus.

E. Soil characteristics

pH	EC (dSm^{-1})	SOC (%)	BD (g/cm^3)	NO_3-N ($mg\ kg^{-1}$)	NH_4-N ($mg\ kg^{-1}$)	PO_4-P ($mg\ kg^{-1}$)	C density ($tons\ ha^{-1}$)
8.52	0.31	0.255	1.49	2.80	4.10	6.33	10.57

F. History and status

This is a quite old sacred grove conserved and protected by local people in name of Jog Maya, Kuldevi of Rajpurohit. Local people use to come and worship her here. The grove is dominated by *Capparis decidua* trees.

DATA ON SACRED GROVES OF RAJASTHAN

Sacred Grove No. 88

A. Basic information about the sacred groves

Name of sacred groves	Laba ji Maharaj Ka Oran
Name of village	Khariya Soda
Panchayat	Sojat
Forest Block	-
Land status	Community land
District	Pali
Area	130 bigha
Altitude (m)	288
Latitude	25° 49' 28.3 N
Longitude	73° 38' 20.4 E

B. Topographical features

Erosion status	Topography	Soil depth
Slight	Almost flatland	Deep (>100 cm)

C. Vegetation status

Tree species with IVI

A. sericea (279), *A. indica* (21)

Shrub species with IVI

No shrub (0)

ENS Tree : 1.19 Shrub: 0

Herb and Grass Species

B. diffusa, E. ciliaris, E. alsinoides, I. cordifolia, I. linnai, J. procumbense, M. jacqmontii, P. erioptera, T. tennalus, T. terrestris, V. cinerea.

D. Wild-Life

Boselaphus tragocamelus, *Passer domesticus*

E. Soil characteristics

pH	EC (dSm^{-1})	SOC (%)	BD (g/cm^3)	NO_3-N ($mg\ kg^{-1}$)	NH_4-N ($mg\ kg^{-1}$)	PO_4-P ($mg\ kg^{-1}$)	C density ($tons\ ha^{-1}$)
7.08	0.17	0.345	1.58	3.12	3.23	5.66	16.03

F. History and status

A very old sacred grove is being conserved and protected by the local people. It is dominated by the old trees of *Anogiesus sericea*-an endangered species. A temple of Mamaji is there in the grove, where local people people come and worship Mamaji.

DATA ON SACRED GROVES OF RAJASTHAN

Sacred Grove No. 89

A. Basic information about the sacred groves

Name of sacred groves	Mama ji Ka Oran
Name of village	Nimbuda (Busi)
Panchayat	Pali
Forest Block	Desuri
Land status	Community land
District	Pali
Area	450 bigha
Altitude (m)	277
Latitude	26° 33' 06.3 N
Longitude	73° 27' 33.0 E

B. Topographical features

Erosion status	Topography	Soil depth
Moderate	Almost flatland	Deep (>100 cm)

C. Vegetation status

Tree species with IVI

P. juliflora (126), *S. oleoides* (114), *P. cineraria* (38), *C. decidua* (22)

Shrub species with IVI

C. auriculata (263), *G. tenax* (37)

ENS Tree : 2.09 Shrub: 1.14

Herb and Grass Species

B. diffusa, C. tora, C. italic, E. ciliaris, E. alsinoides, I. cordifolia, I. linnai, J. procumbense, M. jacqmontii, P. erioptera, T. tennalus, T. terrestris, V. cinerea.

D. Wild-Life

Boselaphus tragocamelus, *Passer domesticus*, *Francolinus pondicerianus*.

E. Soil characteristics

pH	EC (dSm^{-1})	SOC (%)	BD (g/cm^3)	NO_3-N ($mg\ kg^{-1}$)	NH_4-N ($mg\ kg^{-1}$)	PO_4-P ($mg\ kg^{-1}$)	C density ($tons\ ha^{-1}$)
7.11	0.15	0.360	1.55	2.56	2.56	6.23	14.84

F. History and status

This is a very old sacred grove conserved and protected by the local people in name of Mamaji, the local diety in the region. It is remnant of *Cassia auriculata*-type forest, but presently invaded by invasive-*Prosopis juliflora*.

DATA ON SACRED GROVES OF RAJASTHAN

Sacred Grove No. 90

A. Basic information about the sacred groves

Name of sacred groves	Devnarayanji Ka Oran
Name of village	Digai (Gundoj)
Panchayat	Pali
Forest Block	Gundoj
Land status	Community land
District	Pali
Area	30 bigha
Altitude (m)	246
Latitude	25° 34' 28.4 N
Longitude	73° 17' 21.6 E

B. Topographical features

Erosion status	Topography	Soil depth
Moderate	Almost flat land	Deep (>100 cm)

C. Vegetation status

Tree species with IVI

S. oleoides (196), *C. decidua* (79), *P. juliflora* (24)

Shrub species with IVI

C. auriculata (300)

ENS Tree : 2.59 Shrub: 1.00

Herb and Grass Species:

B. diffusa, C. tora, C. italica, E. ciliaris, E. alsinoides, I. cordifolia, I. linnai, J. procumbense, T. procumbens, T. terrestris.

D. Wild-Life

Boselaphus tragocamelus, *Passer domesticus*, Snake.

E. Soil characteristics

pH	EC (dSm^{-1})	SOC (%)	BD (g/cm^3)	NO_3-N ($mg\ kg^{-1}$)	NH_4-N ($mg\ kg^{-1}$)	PO_4-P ($mg\ kg^{-1}$)	C density ($tons\ ha^{-1}$)
7.15	0.12	0.360	1.56	1.56	2.15	4.58	14.36

F. History and status

Old Oran is kept intact in name of Lord Devnarayanji- Kuldevata of Gurjar community. The grove is dominated by *Salvadora oleoides* trees. However, it is a remnant of *Cassia auriculata* type of forest according to Cahmpon and Seth (1968).

DATA ON SACRED GROVES OF RAJASTHAN

Sacred Grove No. 91

A. Basic information about the sacred groves

Name of sacred groves	Ashapura Mata Ji Ka Oran
Name of village	Guda Keshor Singh
Panchayat	Nadol
Forest Block	-
Land status	Community land
District	Pali
Area	80
Altitude (m)	306
Latitude	25° 22' 51.1 N
Longitude	73° 28' 02.9 E

B. Topographical features

Erosion status	Topography	Soil depth
Moderate	Almost flat land	Deep (>100 cm)

C. Vegetation status

Tree species with IVI

S. oleoides (199), *P. juliflora* (45), *A. nilotica var. cupressiformis* (25), *A. nilotica* (20), *A. leucophloea* (11).

Shrub species with IVI

C. sepiaria (228), *C. auriculata* (72)

ENS Tree : 2.69 Shrub: 1.67

Herb and Grass Species:

B. diffusa, B. ramosa, C. tora, C. italica, D. sindicum, E. ciliaris, E. alsinoides, I. cordifolia, I. linnai, J. procumbense, T. procumbens, T. terrestris, V. cinerea.

D. Wild- life

Boselaphus tragocamelus, Hyaena hyaena, Lepus nigricolis, Pavo cristatus, Psittacula krameri, Acridotheres tristis.

E. Soil characteristics

pH	EC (dSm^{-1})	SOC (%)	BD (g/cm^3)	NO_3-N ($mg\ kg^{-1}$)	NH_4-N ($mg\ kg^{-1}$)	PO_4-P ($mg\ kg^{-1}$)	C density ($tons\ ha^{-1}$)
7.11	0.11	0.300	1.55	2.12	2.56	6.58	10.54

F. History and status

The history of this place is related to the great Hindu King Prithviraj Chauhan. Before going to any battle Prithivi raj used to come here for blessing of mata Ashapurna. She is also "Kul devi" of Chauhan Vansh. The grove is dominated by *Salvadora oleoides* trees.

DATA ON SACRED GROVES OF RAJASTHAN

Sacred Grove No. 92

A. Basic information about the sacred groves

Name of sacred groves	Mata ji Ka Oran
Name of village	Shree Mangleshwar Mahadev Dham
Panchayat	Desuri
Forest Block	Mada
Land status	Community land
District	Pali
Area	7 bigha
Altitude (m)	349
Latitude	25° 12' 53.5 N
Longitude	73° 26' 41.6 E

B. Topographical features

Erosion status	Topography	Soil depth
Slight	Almost flat land	Deep (>100 cm)

C. Vegetation status

Tree species with IVI

A. leucophloea (149), *P. cineraria* (43), *C. decidua* (34), *B. aegyptiaca* (32), *P. juliflora* (26), *A. nilotica* (16).

Shrub species with IVI

C. auriculata (210), *Z. nummularia* (51), *D. cinerea* (28), *S. virosa* (11).

ENS Tree : 4.36 Shrub: 1.43

Herb and Grass Species:

B. diffusa, B. ramose, C. tora, D. sindicum, E. ciliaris, E. alsinoides, I. cordifolia, I. linnai, J. procumbense, T. procumbens, T. purpurea, T. terrestris, V. cinerea.

D. Wild- life

Boselaphus tragocamelus, *Hyaena hyaena*, *Lepus nigricolis*, *Pavo cristatus*, *Psittacula krameri*, *Acridotheres tristis*.

E. Soil characteristics

pH	EC (dSm^{-1})	SOC (%)	BD (g/cm^3)	NO_3-N ($mg\ kg^{-1}$)	NH_4-N ($mg\ kg^{-1}$)	PO_4-P ($mg\ kg^{-1}$)	C density ($tons\ ha^{-1}$)
7.25	0.23	0.330	1.52	2.22	2.22	7.12	10.87

F. History and status

An old sacred grove is being conserved and maintained by the local communities in name of Mataji, to whom the people use to worship here. A type of forest, i.e. *Acacia leucophloea* and *Cassia auriculata* is represented by the grove.

DATA ON SACRED GROVES OF RAJASTHAN

Sacred Grove No. 93

A. Basic information about the sacred groves

Name of sacred groves	Dhundhlimata Ki Banni
Name of village	Banshi
Panchayat	Badi Sadri
Forest Block	-
Land status	Community land
District	Pratapgarh
Area	500 bigha
Altitude (m)	599
Latitude	24° 18' 1.4 N
Longitude	72° 23' 49.9 E

B. Topographical features

Erosion status	Topography	Soil depth
Moderate	Hill slope	Shallow (<25 cm)

C. Vegetation status

Tree species with IVI

T. grandis (132), *W. tinctoria* (35), *D. melanoxylon* (22), *H. integrifolia* (19), *H. antidysenterica* (17), *M. tomentosa* (15), *Soymida febrifuga* (11), *A. marmelos* (10), *B. monosperma* (9), *L. coromandelica* (6), *G. tiliifolia* (6), *A. lebbeck* (4), *Sapindus mukorossi* (4), *E. laevis* (4), *Z. mauritiana* (3), *B. racemosa* (3).

Shrub species with IVI

F. indica (208), *Z. nummularia* (92).

ENS Tree : 4.76 Shrub: 1.82

Herb and Grass Species:

A. vaginalis, B. diffusa, B. ramose, B. repens, E. acaulis, E. hirta, C. tora, D. sindicum, E. alsinoides, I. cordifolia, J. procumbense, T. procumbens.

D. Wild- life

Boselaphus tragocamelus, *Semnopithecus entellus, Hyaena hyaena*, *Lepus nigricolis*, *Pavo cristatus*, *Psittacula krameri*, *Acridotheres tristis*.

E. Soil characteristics

pH	EC (dSm^{-1})	SOC (%)	BD (g/cm^3)	NO_3-N ($mg\ kg^{-1}$)	NH_4-N ($mg\ kg^{-1}$)	PO_4-P ($mg\ kg^{-1}$)	C density ($tons\ ha^{-1}$)
6.42	0.22	0.630	1.40	5.78	2.41	6.16	19.55

F. History and status

A saint Sh Dhundhari Mal had taken Samadhi here 20 years ago. A temple is build here and local people worship Dhoondhali Mata- a local diety. Grove is dominated by *T. grandis* trees.

DATA ON SACRED GROVES OF RAJASTHAN

Sacred Grove No. 94

A. Basic information about the sacred groves

Name of sacred groves	Bijasau Mata Ji Bani
Name of village	Vijania
Panchayat	Dhariavad
Forest Block	-
Land status	Community land
District	Pratapgarh
Area	35 bigha
Altitude (m)	263
Latitude	24° 04' 34.1 N
Longitude	74° 27' 41.8 E

B. Topographical features

Erosion status	Topography	Soil depth
Moderate	Hill slope	Shallow (<25 cm)

C. Vegetation status

Tree species with IVI

T. grandis (133), *H. integrifolia* (82), *D. melanoxylon* (40), *B. monosperma* (18), *A. leucophloea* (15), *A. marmelos* (12)

Shrub species with IVI

L. camara (300)

ENS Tree : 3.68 Shrub: 1.00

Herb and Grass Species:

A. vaginalis, B. diffusa, B. ramosa, B. repens, C. pumila, E. granularis, E. acaulis, E. hirta, C. tora, D. sindicum, E. alsinoides, I. cordifolia, J. procumbense, T. procumbens.

D. Wild- life

Boselaphus tragocamelus, Semnopithecus entellus, Hyaena hyaena, Lepus nigricolis, Pavo cristatus, Psittacula krameri, Acridotheres tristis.

E. Soil characteristics

pH	EC (dSm^{-1})	SOC (%)	BD (g/cm^3)	NO_3-N ($mg\ kg^{-1}$)	NH_4-N ($mg\ kg^{-1}$)	PO_4-P ($mg\ kg^{-1}$)	C density ($tons\ ha^{-1}$)
7.60	0.17	1.515	1.35	3.90	8.56	9.12	14.03

F. History and status

This grove was created long back in name of Bisau mata ji. There is a temple where people use to come and worship Mataji. This grove is dominated by *Tectona grandis*, but is now invaded by *Lantana camara*-an invasive species.

DATA ON SACRED GROVES OF RAJASTHAN

Sacred Grove No. 95

A. Basic information about the sacred groves

Name of sacred groves	Chandimata ji Ka Oran
Name of village	Barol
Panchayat	Choti sadri
Forest Block	-
Land status	Community land
District	Pratapgarh
Area	40 bigha
Altitude (m)	464
Latitude	24° 18' 09.4 N
Longitude	74° 43' 37.8 E

B. Topographical features

Erosion status	Topography	Soil depth
Moderate	Degraded hill slope	Shallow (<25 cm)

C. Vegetation status

Tree species with IVI

T. arjuna (105), *B. monosperma* (100), *P. sylvestris* (43), *M. parviflora* (22), *H. integrifolia* (18), *D. cordifolia* (12)

Shrub species with IVI

L. camara (300)

ENS Tree: 4.11 Shrub: 1.00

Herb and Grass Species:

C. tora, D. sindicum, E. acaulis, E. hirta, E. alsinoides, I. cordifolia, J. procumbense, T. procumbens.

D. Wild- life

Panthera pardus, Melursus ursinus, Boselaphus tragocamelus, Canis lupus, Semnopithecus entellus, Hyaena hyaena, Lepus nigricolis, Pavo cristatus, Psittacula krameri, Acridotheres tristis.

E. Soil characteristics

pH	EC (dSm^{-1})	SOC (%)	BD (g/cm^3)	NO_3-N (mg kg^{-1})	NH_4-N (mg kg^{-1})	PO_4-P (mg kg^{-1})	C density (tons ha^{-1})
6.05	0.08	0.570	1.42	3.94	2.01	9.23	17.46

F. History and status

An ancient sacred grove, where an old temple of Chandimata is situated. People come and worship Mataji. This grove appears to be a remnant of relatively moist forest as the grove is dominated by *Terminalia arjuna*.

DATA ON SACRED GROVES OF RAJASTHAN

Sacred Grove No. 96

A. Basic information about the sacred groves

Name of sacred groves	Devnarayan ji Ki Bani
Name of village	Chhotidati
Panchayat	Bara barda
Forest Block	-
Land status	Community land
District	Pratapgarh
Area	35 bigha
Altitude (m)	412
Latitude	24° 12' 05.2 N
Longitude	74° 42' 25.6 E

B. Topographical features

Erosion status	Topography	Soil depth
Moderate	Degraded hill slope	Shallow (<25 cm)

C. Vegetation status

Tree species with IVI

T. grandis (227), *B. monosperma* (73)

Shrub species with IVI

L. camara (250), *C. sepiaria* (50)

ENS Tree : 1.54 Shrub: 1.46

Herb and Grass Species:

A. vaginalis, B. diffusa, B. ramosa, B. repens, E. acaulis, E. hirta, C. tora, D. sindicum, E. alsinoides, I. cordifolia, J. procumbense, T. procumbens.

D. Wild- life

Boselaphus tragocamelus, *Semnopithecus entellus, Hyaena hyaena*, *Lepus nigricolis*, *Pavo cristatus*, *Psittacula krameri*, *Acridotheres tristis*.

E. Soil characteristics

pH	EC (dSm^{-1})	SOC (%)	BD (g/cm^3)	NO_3-N (mg kg^{-1})	NH_4-N (mg kg^{-1})	PO_4-P (mg kg^{-1})	C density (tons ha^{-1})
6.55	0.19	0.512	1.44	2.66	3.66	6.33	13.71

F. History and status

This grove is maintained by the local communities in name of Lord Devnarayanji. There is an old temple in the grove, where people use to come and worship Lord Devanarayan. This is an example of dry teak forest.

DATA ON SACRED GROVES OF RAJASTHAN

Sacred Grove No. 97

A. Basic information about the sacred groves

Name of sacred groves	Devnarayan ji ki Bani
Name of village	Hira Ki Bassi
Panchayat	Devgarh
Forest Block	Bagna
Land status	Community land
District	Rajsamand
Area	40 bigha
Altitude (m)	657
Latitude	25° 32' 01 N
Longitude	73° 52' 04.8 E

B. Topographical features

Erosion status	Topography	Soil depth
Moderate	Hill slope	Shallow (<25 cm)

C. Vegetation status

Tree species with IVI

A. pendula (253), *A. indica* (28), *P. cineraria* (19)

Shrub species with IVI

D. cinerea (194), *G. tenax* (106)

ENS Tree: 1.17 Shrub: 1.75

Herb and Grass Species

A. vaginalis, B. diffusa, B. pussilla, B. ramosa, B. repens, C. tora, D. sindicum, E. acaulis, E. hirta, E. alsinoides, I. cordifolia, J. procumbense, T. procumbens, T. rhomboidea, V. cinerea, Zornia gibbosa.

D. Wild- life

Boselaphus tragocamelus, Canis lupus, Semnopithecus entellus, Hyaena hyaena, Lepus nigricolis, Pavo cristatus, Psittacula krameri, Acridotheres tristis.

E. Soil characteristics

pH	EC (dSm^{-1})	SOC (%)	BD (g/cm^3)	NO_3-N (mg kg^{-1})	NH_4-N (mg kg^{-1})	PO_4-P (mg kg^{-1})	C density (tons ha^{-1})
6.58	0.12	0.675	1.46	2.12	6.52	8.30	20.92

F. History and status

This oran is conserved and protected by the people of nearby villagers for the sake of Lord Devnarayan- The 'Kuldevata' of Gurjar community in Rajasthan and the adjoining states. It is mostly dominated by old trees of *A. pendula*.

DATA ON SACRED GROVES OF RAJASTHAN

Sacred Grove No. 98

A. Basic information about the sacred groves

Name of sacred groves	Bheru Nath ji Ki Vani
Name of village	Swadari
Panchayat	Devgarh
Forest Block	Devgarh Beed
Land status	Community land
District	Rajsamand
Area	75 bigha
Altitude (m)	652
Latitude	25° 29' 51.1 N
Longitude	73° 54' 02.4 E

B. Topographical features

Erosion status	Topography	Soil depth
Moderate	Hill slope and valley	Shallow (<25 cm)

C. Vegetation status

Tree species with IVI

A. pendula (130), *A. leucophloea* (60), *W. tinctoria* (40), *M. emarginata* (31), *F. benghalensis* (11), *H. integrifolia* (8), *P. juliflora* (8), *B. monosperma* (7), *A. nilotica* (5).

Shrub species with IVI

G. tenax (197), *Euphorbia caducifolia* (103)

ENS Tree : 4.24 Shrub: 1.75

Herb and Grass Species:

A. vaginalis, B. diffusa, B. pussilla, B. ramosa, B. repens, C. tora, D. sindicum, E. acaulis, E. hirta, E. alsinoides, G. bosvallia, I. cordifolia, J. procumbense, S. cordifolia, T. rhomboidea, T. procumbens, T. tennalus, V. cinerea, Z. gibbosa.

D. Wild- life

Panther, *Boselaphus tragocamelus, Semnopithecus entellus, Hyaena hyaena, Lepus nigricolis, Pavo cristatus, Psittacula krameri, Acridotheres tristis.*

E. Soil characteristics

pH	EC (dSm^{-1})	SOC (%)	BD (g/cm^3)	NO_3-N ($mg\ kg^{-1}$)	NH_4-N ($mg\ kg^{-1}$)	PO_4-P ($mg\ kg^{-1}$)	C density ($tons\ ha^{-1}$)
7.12	0.15	0.525	1.49	1.59	8.30	12.46	15.19

F. History and status

This oran is kept in name of Bheruji, the local diety in the region. People use to worship the diety and maintain and conserved the grove. This grove is dominated by *Anogeissus pndula* trees and rich in floral diversity.

DATA ON SACRED GROVES OF RAJASTHAN

Sacred Grove No. 99

A. Basic information about the sacred groves

Name of sacred groves	Chamunda Mata Ji Ki Bani
Name of village	Chamunda Khoot
Panchayat	Amet
Forest Block	-
Land status	Community land
District	Rajsamand
Area	36 bigha
Altitude (m)	602
Latitude	25° 23' 59.3 N
Longitude	74° 00' 51.3 E

B. Topographical features

Erosion status	Topography	Soil depth
Moderate	Hill slope	Moderately deep (25-50)

C. Vegetation status

Tree species with IVI

A. pendula (216), *M. emarginata* (84)

Shrub species with IVI

R. mysurensis (300)

ENS Tree: 1.75 Shrub: 1.00

Herb and Grass Species:

A. vaginalis, B. diffusa, B. pussilla, B. ramosa, B. repens, C. tora, D. sindicum, E. acaulis, E. hirta, E. alsinoides, G. bosvallia, I. cordifolia, J. procumbense, L. trinervis, P. erioptera, S. cordifolia, T. rhomboidea, T. procumbens, T. tennalus, V. cinerea, Z. gibbosa.

D. Wild- life

Boselaphus tragocamelus, Semnopithecus entellus, Hyaena hyaena, Lepus nigricolis, Pavo cristatus, Psittacula krameri, Acridotheres tristis, Picnonotus cafer.

E. Soil characteristics

pH	EC (dSm^{-1})	SOC (%)	BD (g/cm^3)	NO_3-N ($mg\ kg^{-1}$)	NH_4-N ($mg\ kg^{-1}$)	PO_4-P ($mg\ kg^{-1}$)	C density ($tons\ ha^{-1}$)
7.25	0.18	0.600	1.52	1.26	4.52	10.46	18.24

F. History and status

Being an ancient sacred grove, this area is conserved and protected by the communities of the nearby villages in name of Goddess Chamunda Mata ji, whose temple is there in the grove. It is dominated by old trees of *A. pendula*.

DATA ON SACRED GROVES OF RAJASTHAN

Sacred Grove No. 100

A. Basic information about the sacred groves

Name of sacred groves	Haumanji Ki Dungri
Name of village	Neenadra
Panchayat	Sawaimadhopur
Forest Block	Dunda
District	Sawaimadhopur
Area	8 bigha
Altitude (m)	267
Latitude	26° 08' 01.0 N
Longitude	76° 20' 17.5 E

B. Topographical features

Erosion status	Topography	Soil depth
Moderate	Hillock and hill slope	Shallow (<25 cm)

C. Vegetation status

Tree species with IVI

P. juliflora (141), *A. pendula* (139), *A. nilotica* (20)

Shrub species with IVI

C. procera (300)

ENS Tree : 2.17 Shrub: 1.00

Herb and Grass Species:

A. hispidum, Alysicarpus monilifer, A. vaginalis, B. diffusa, B. pussilla, B. ramosa, B. repens, Cyanotis cristata, C. tora, D. sindicum, E. acaulis, E. hirta, E. alsinoides, G. bosvallia, I. cordifolia, J. procumbense, L. trinervis, P. erioptera, S. cordifolia, T. rhomboidea, T. procumbens, T. tenellus, V. cinerea, Z. gibbosa.

D. Wild- life

Boselaphus tragocamelus, Cuon alpinus, Semnopithecus entellus, Hyaena hyaena, Lepus nigricolis, Pavo cristatus, Psittacula krameri, Acridotheres tristis.

E. Soil characteristics

pH	EC (dSm^{-1})	SOC (%)	BD (g/cm^3)	NO_3-N ($mg\ kg^{-1}$)	NH_4-N ($mg\ kg^{-1}$)	PO_4-P ($mg\ kg^{-1}$)	C density ($tons\ ha^{-1}$)
7.48	0.27	0.630	1.50	0.18	6.50	16.44	9.07

F. History and status

The area around God Hanuman ji temple has been conserved and protected by the local communities. Local people come and worship the god. The grove was formerly dominated by *A. pendula* but at present it is dominated by *P. juliflora*.

DATA ON SACRED GROVES OF RAJASTHAN

Sacred Grove No. 101

A. Basic information about the sacred groves

Name of sacred groves: Jasram Baba Ji Ki Bani

Name of village: Naroda, Meesakhedi

Panchayat: khardar

Forest Block: -

Land status: Community land

District: Sawaimadhopur

Area: 12 bigha

Altitude (m): 207

Latitude: 25° 52' 30.0 N   Longitude: 76° 35' 00.4 E

B. Topographical features

Erosion status	Topography	Soil depth
Gullied	Undulating (Ravine)	Deep (>100 cm)

C. Vegetation status

Tree species with IVI

M. parviflora (87), *P. juliflora* (55), *C. dichotoma* (43), *F. infectoria* (34), *A. indica* (31), *B. aegyptiaca* (24), *A. nilotica* (16), *A. pendula* (10).

Shrub species with IVI

C. phlomidis (221), *C. sepiera* (48), *C. carrundus* (31).

ENS Tree : 6.12 Shrub: 2.14

Herb and Grass Species:

A. hispidum, A. monolifer, B. diffusa, B. pusilla, B. ramosa, B. repens, E. acaulis, E. hirta, E. alsinoides, I. cordifolia, J. procumbense, L. trinervis, P. erioptera, S. cordifolia, T. rhomboidea, T. procumbens, T. tenellus, V. cinerea.

D. Wild- life

Crocodile, *Boselaphus tragocamelus*, *Cuon alpinus*, *Semnopithecus entellus*, *Sus scrofa*, *Hyaena hyaena*, *Lepus nigricolis*, *Pavo cristatus*, *Psittacula krameri*.

E. Soil characteristics

pH	EC (dSm^{-1})	SOC (%)	BD (g/cm^3)	NO_3-N ($mg\ kg^{-1}$)	NH_4-N ($mg\ kg^{-1}$)	PO_4-P ($mg\ kg^{-1}$)	C density ($tons\ ha^{-1}$)
7.58	0.26	0.660	1.49	0.13	6.52	14.36	9.96

F. History and status

Jasram baba had taken "Jal-samadhi" here in Chambal river so people come here to worship him. Temple has been build along the Chambal river and surrounding area has been conserved by temple priest. The grove is dominated by *Mitragyna parviflora* (Kadamb) trees.

DATA ON NAME OF SACRED GROVES OF RAJASTHAN

Sacred Grove No. 102

A. Basic information about the sacred groves

Name of sacred groves	Peer Baba Ki Dargah
Name of village	Garadwas
Panchayat	Sawai Madhopur
Forest Block	Chauth ka barbada
Land status	Community land
District	Sawai Madhopur
Area	40 bigha
Altitude (m)	329
Latitude	26° 00' 09.6 N
Longitude	76° 07' 51.3 E

B. Topographical features

Erosion status	Topography	Soil depth
Slight	Hill slope and top	Shallow (<25 cm)

C. Vegetation status

Tree species with IVI

A. pendula (212), *P. juliflora* (35), *C. decidua* (19), *A. indica* (14), *C. nurvala* (11), *A. nilotica* (9).

Shrub species with IVI

C. sepiaria (300)

ENS Tree: 2.09 Shrub: 1.00

Herb and Grass Species:

A. monolifer, A. vaginalis, B. diffusa, B. pusilla, B. ramosa, B. repens, Cyanotis cristata, C. tora, D. sindicum, E. acaulis, E. hirta, E. alsinoides, G. bosvallea, H. granularis, O. thomaeum, I. cordifolia, J. procumbense, L. trinervis, P. erioptera, S. cordifolia, T. rhomboidea, T. procumbens, T. tennalus, V. cinerea, Z. gibbosa.

D. Wild- life

Panthera pardus, Boselaphus tragocamelus, Cuon alpinus, Axis axis, Sus scrofa, Hyaena hyaena, Lepus nigricolis, Pavo cristatus, Psittacula krameri.

E. Soil characteristics

pH	EC (dSm^{-1})	SOC (%)	BD (g/cm^3)	NO_3-N (mg kg^{-1})	NH_4-N (mg kg^{-1})	PO_4-P (mg kg^{-1})	C density (tons ha^{-1})
7.66	0.23	0.562	1.55	0.23	5.23	14.12	9.03

F. History and status

This place is conserved and protected by the villagers because of Peer Baba Mosque is at the top of the hill. There is a penalty of Rs. 1100, if anyone cut the trees. The grove is dominated by old trees of *A. pendula*, though *P. juliflora* has invaded this grove.

DATA ON NAME OF SACRED GROVES OF RAJASTHAN

Sacred Grove No. 103

A. Basic information about the sacred groves

Name of sacred groves	Bheruji Ki Bani
Name of village	Tadi
Panchayat	Pipradi
Forest Block	Sangarwa
Land status	Community land
District	Sikar
Area	50 bigha
Altitude (m)	480
Latitude	27° 33' 98.3 N
Longitude	75° 17' 05.5 E

B. Topographical features

Erosion status	Topography	Soil depth
Slight	Undulating flat land	Moderate (50-100)

C. Vegetation status

Tree species with IVI

P. cineraria (125), *A. tortilis* (93), *A. nilotica* (67), *M. emarginata* (16)

Shrub species with IVI

C. procera (191), *L. pyrotechnica* (90), *C. callosus* (19)

ENS Tree: 3.43 Shrub: 2.20

Herb and Grass Species:

Aristida adscensionis, A. funiculata, B. diffusa, C. ciliaris, D. aegyptium, D. ciliaris, E. ciliaris, E. alsinoides, I. cordifolia, I. linnai, J. procumbense, P. erioptera, T. purpurea, T. roxburghii, T. terrestris, V. cinerea.

D. Wild- life

Lepus nigricolis, *Boselaphus tragocamelus*, *Axis axis*, *Herpestes edwardsi*, *Pavo cristatus*, *Francolinus pondicerianus*.

E. Soil characteristics

pH	EC (dSm^{-1})	SOC (%)	BD (g/cm^3)	NO_3-N (mg kg^{-1})	NH_4-N (mg kg^{-1})	PO_4-P (mg kg^{-1})	C density (tons ha^{-1})
6.20	0.12	0.210	1.62	2.16	5.30	7.32	10.21

F. History and status

This is about 2000 years old sacred grove in name of Bheruji, the local deity. The grove is conserved and protected by the people of the nearby villages. It is dominated by *P. cineraria* trees though plantation of *Acacia tortilis* is there.

DATA ON NAME OF SACRED GROVES OF RAJASTHAN

Sacred Grove No. 104

A. Basic information about the sacred groves

Name of sacred groves	Kheri Walai Balaji ka Oran
Name of village	Raipura
Panchayat	Pipradi
Forest Block	Jeen mata
Land status	Community land
District	Sikar
Area	257 bigha
Altitude (m)	482
Latitude	27° 28' 27.2 N
Longitude	73° 17' 45.2 E

B. Topographical features

Erosion status	Topography	Soil depth
Slight	Flat land	Moderate (50-100)

C. Vegetation status

Tree species with IVI

C. decidua (104), *M. emarginata* (67), *P. cineraria* (57), *A. tortilis* (45), *A. catechu* (27).

Shrub species with IVI

Z. nummularia (171), *C. procera* (129)

ENS Tree: 4.44 Shrub: 1.96

Herb and Grass Species:

A. adscensionis, B. diffusa, C. ciliaris, D. aegyptium, D. ciliaris, E. ciliaris, E. alsinoides, I. cordifolia, I. linnai, J. procumbense, P. erioptera, T. purpurea, T. roxburghii, T. terrestris.

D. Wild- life

Lepus nigricolis, Boselaphus tragocamelus, Psittacula krameri.

E. Soil characteristics

pH	EC (dSm^{-1})	SOC (%)	BD (g/cm^3)	NO_3-N ($mg\ kg^{-1}$)	NH_4-N ($mg\ kg^{-1}$)	PO_4-P ($mg\ kg^{-1}$)	C density ($tons\ ha^{-1}$)
7.06	0.09	0.390	1.60	2.38	6.09	11.32	18.72

F. History and status

This is about 1500 years old sacred grove along with a temple of Balaji –Lord Hanuman. People use to come and worship here in the temple. The grove is protected since ancient times. It is dominated by *Capparis decidua* trees, plantation of *Acacia tortilis* has been done long back.

DATA ON NAME OF SACRED GROVES OF RAJASTHAN

Sacred Grove No. 105

A. Basic Information about the sacred groves

Name of sacred groves	Satimata Ji Ka Mandir
Name of village	Moklawas
Panchayat	Danta Ramgarh
Forest Block	-
Land status	Community land
District	Sikar
Area	500 bigha
Altitude (m)	431
Latitude	27° 17' 26.6 N
Longitude	75° 14' 08.3 E

B. Topographical features

Erosion status	Topography	Soil depth
Moderate	Almost flat land	Deep (>100 cm)

C. Vegetation status

Tree species with IVI

P. cineraria (184), *C. decidua* (116)

Shrub species with IVI

C. procera (300)

ENS Tree : 1.99 Shrub: 1.00

Herb and Grass Species:

A. adscensionis, B. diffusa, C. tora, C. ciliaris, D. aegyptium, D. ciliaris, E. ciliaris, E. alsinoides, I. cordifolia, I. linnai, J. procumbense, H. marifolium, O. compressa, P. amarus, P. erioptera, T. purpurea, T. roxburghii, T. terrestris.

D. Wild- life

Boselaphus tragocamelus, *Axis axis*, *Columba livia*, *Pavo cristatus*, *Psittacula krameri*, *Eudynamys scolopaceus*, *Lepus nigricolis*.

E. Soil characteristics

pH	EC (dSm^{-1})	SOC (%)	BD (g/cm^3)	NO_3-N (mg kg^{-1})	NH_4-N (mg kg^{-1})	PO_4-P (mg kg^{-1})	C density (tons ha^{-1})
6.70	0.34	0.225	1.59	3.68	6.11	14.30	10.73

F. History and status

This grove was established about 70 years before, as Smt. Krishna kumari w/o Shri Satidhan Singh had become sati here with her husband. A temple has been established and people come and worship here. The grove is dominated by *P. cineraria* trees.

DATA ON NAME OF SACRED GROVES OF RAJASTHAN

Sacred Grove No. 106

A. Basic information about the sacred groves

Name of sacred groves	Sidheswar Mahadeo Mandir
Name of village	Dhanta, Makrod, Khera
Panchayat	Sirohi
Forest Block	Meerpur
Land status	Community land
District	Sirohi
Area	850 bigha
Altitude (m)	338
Latitude	24° 48' 13.5 N
Longitude	72° 50' 16.9 E

B. Topographical features

Erosion status	Topography	Soil depth
Moderate	Hill slope undulating	Shallow (<25 cm)

C. Vegetation status

Tree species with IVI

Acacia senegal (134), *Anogeissus pendula* (86), *Prosopis juliflora* (44), *Balanites aegyptiaca* (28), *Prosopis cineraria* (8).

Shrub species with IVI

Ziziphus nummularia (183), *Dichrostachys cinerea* (117)

ENS Tree : 2.82 Shrub: 1.70

Herb and Grass Species:

A. aspera, Borreria stricta, B. ramosa, D. ciliaris, E. alsinoides, I. cordifolia, P. amarus, Rhynchosia minima, T. purpurea, T. tenellus.

D. Wild-life

Varanus bengalensis, *Lepus nigricolis*, *Boselaphus tragocamelus*, *Axis axis*, *Herpestes edwardsi*, *Columba livia*, *Pavo cristatus*, *Psittacula krameri*, *Eudynamys scolopaceus*.

E. Soil characteristics

pH	EC (dSm^{-1})	SOC (%)	BD (g/cm^3)	NO_3-N ($mg\ kg^{-1}$)	NH_4-N ($mg\ kg^{-1}$)	PO_4-P ($mg\ kg^{-1}$)	C density ($tons\ ha^{-1}$)
7.82	0.16	0.225	1.50	4.03	10.52	22	9.59

F. History and status

This is an ancient sacred grove, and a temple of Lord Shiva (Shidheswar Mahadeo) has been established since then. The grove is dominated by *Acacia senegal* trees. Presently under care of gram panchayat and appears to be in good condition.

DATA ON NAME OF SACRED GROVES OF RAJASTHAN

Sacred Grove No. 107

A. Basic information about the sacred groves

Name of sacred groves	Varada Hanuman ji Ka Oran
Name of village	Varada
Panchayat	Sirohi
Forest Block	-
Land status	Community land
District	Sirohi
Area	2380 bigha
Altitude (m)	222
Latitude	25° 02' 31.0 N
Longitude	72° 40' 59.7 E

B. Topographical features

Erosion status	Topography	Soil depth
Moderate	Almost flat land	Deep (>100 cm)

C. Vegetation status

Tree species with IVI

S. oleoides (197), *P. juliflora* (79), *C. decidua* (24)

Shrub species with IVI

C. auriculata (300)

ENS Tree : 2.15 Shrub: 1.00

Herb and Grass Species:

A. funiculata, B. diffusa, B. ramosa, C. viscosa, D. ciliaris, E. alsinoides, I. cordifolia, P. amarus, T. purpurea, T. tennalus.

D. Wild- life

Varanus bengalensis, *Rattus norvegicus*, *Lepus nigricolis*, *Boselaphus tragocamelus*, *Axis axis*, *Herpestes edwardsi*, *Columba livia*, *Pavo cristatus*.

E. Soil characteristics

pH	EC (dSm^{-1})	SOC (%)	BD (g/cm^3)	NO_3-N ($mg\ kg^{-1}$)	NH_4-N ($mg\ kg^{-1}$)	PO_4-P ($mg\ kg^{-1}$)	C density ($tons\ ha^{-1}$)
7.82	0.16	0.225	1.55	3.87	6.52	12.46	9.53

F. History and status

An ancient grove is being conserved and protected by the local community in name of Lord Hanumanji. A temple has also been established here, that is a very famous temple of world according to the local residence and second largest Hanumanji Temple in Rajasthan. Dominant tree species of the grove is *Salvadora oleoides*.

DATA ON NAME OF SACRED GROVES OF RAJASTHAN

Sacred Grove No. 108

A. Basic information about the sacred groves

Name of sacred groves	Magriwali Mata ji Ka Oran
Name of village	Pesua
Panchayat	Pindawara
Forest Block	Rampura
Land status	Community land
District	Sirohi
Area	25 bigha
Altitude (m)	409
Latitude	24° 43' 45.3 N
Longitude	72° 58' 43.3 E

B. Topographical features

Erosion status	Topography	Soil depth
Slight	Hill top	Shallow (<25 cm)

C. Vegetation status

Tree species with IVI

Anogeissus pendula (155), *Holoptelea integrifolia* (37), *Capparis decidua* (30), *Acacia leucophloea* (28), *Butea monosperma* (25), *Prosopis juliflora* (15), *Ficus benghalensis* (10)

Shrub species with IVI

Grewia tenax (300)

ENS Tree: 2.88 Shrub: 1.00

Herb and Grass Species:

A. aspera, B. diffusa, B. ramosa, C. italica, D. ciliaris, E. acaulis, E. alsinoides, I. cordifolia, P. amarus, T. procumbens, T. tennalus.

D. Wild- life

Lepus nigricolis, *Boselaphus tragocamelus*, *Axis axis*, *Herpestes edwardsi*, *Columba livia*, *Pavo cristatus*, *Francolinus pondicerianus*.

E. Soil characteristics

pH	EC (dSm^{-1})	SOC (%)	BD (g/cm^3)	NO_3-N (mg kg^{-1})	NH_4-N (mg kg^{-1})	PO_4-P (mg kg^{-1})	C density (tons ha^{-1})
6.99	0.19	0.300	1.41	3.06	9.32	20.00	2.07

F. History and status

An ancient temple is situated in the grove, which is being protected and conserved by the local people. Presently under care of local panchayat and under good condition. The grove is dominated by *A. pendula* trees.

DATA ON NAME OF SACRED GROVES OF RAJASTHAN

Sacred Grove No. 109

A. Basic information about the sacred groves

Name of sacred groves	Gosai Ji Ka Jod
Name of village	Daidaspura
Panchayat	Suratgarh
Forest Block	-
Land status	Community land
District	Sri Ganganagar
Area	216 bigha
Altitude (m)	205
Latitude	29° 00' 50.5 N
Longitude	73° 58' 45.5 E

B. Topographical features

Erosion status	Topography	Soil depth
Severe	Almost flat land	Deep (>100 cm)

C. Vegetation status

Tree species with IVI

C. decidua (199), *P. cineraria* (56), *Z. mauritiana* (28), *A. indica* (17)

Shrub species with IVI

Z. nummularia (188), *A. persica* (112)

ENS Tree : 1.71 Shrub: 1.96

Herb and Grass Species:

B. diffusa, B. ramosa, G. pharnacioides, E. alsinoides, I. cordifolia, T. procumbens.

D. Wild- life

Varanus griseus, *Rattus norvegicus*, *Lepus nigricolis*, *Boselaphus tragocamelus*, *Axis axis*, *Herpestes edwardsi*, *Columba livia*, *Pavo cristatus*, *Francolinus pondicerianus*, *Corvus macrorhynchos*, *Passer domesticus*.

E. Soil characteristics

pH	EC (dSm^{-1})	SOC (%)	BD (g/cm^3)	NO_3-N (mg kg^{-1})	NH_4-N (mg kg^{-1})	PO_4-P (mg kg^{-1})	C density (tons ha^{-1})
6.99	0.19	0.300	1.62	3.15	4.32	8.24	11.65

F. History and status

Because of a temple of God Gosai ji- a local ditey in the region, people use to come here to worship him. The area is conserved by Panchayat of the nearby villages. The dominant tree species in the grve is *Capparis decidua*.

DATA ON NAME OF SACRED GROVES OF RAJASTHAN

Sacred Grove No. 110

A. Basic Information about the sacred groves

Name of sacred groves	Baba Gadanpur Ki Dargah
Name of village	Hindo, Swaroopsar
Panchayat	Anoopgarh
Forest Block	-
Land status	Community land
District	Sri Ganganagar
Area	8 bigha
Altitude (m)	167
Latitude	29° 21' 37.9 N
Longitude	73° 42' 37.0 E

B. Topographical features

Erosion status	Topography	Soil depth
Slight	Almost flat land	Deep (>100 cm)

C. Vegetation status

Tree species with IVI

C. decidua (99), *A. nilotica* (98), *S. oleoides* (73), *P. juliflora* (30)

Shrub species with IVI

No shrub

ENS Tree : 3.18 Shrub: 0

Herb and Grass Species

Amaranthus species, B. diffusa, B. ramosa, Cadaba fruticosa, Cistanche tubulosa, G. pharnacioides, Heliotropium subulatum, E. alsinoides, I. cordifolia, T. procumbens.

D. Wild- life

Boselaphus tragocamelus, Pavo cristatus, Psittacula krameri, Eudynamys scolopaceus, Acridotheres tristis, Lepus nigricolis.

E. Soil characteristics

pH	EC (dSm^{-1})	SOC (%)	BD (g/cm^3)	NO_3-N ($mg\ kg^{-1}$)	NH_4-N ($mg\ kg^{-1}$)	PO_4-P ($mg\ kg^{-1}$)	C density ($tons\ ha^{-1}$)
7.78	0.26	0.465	1.60	3.50	4.92	10.19	16.66

F. History and status

An ancient mosqe of Baba Gadanpur is situated here in this grove. People use to come and worshiped here. The grove is protected and conserved by the local community. The dominant tree species of the grove is *Capparis decidua*.

DATA ON NAME OF SACRED GROVES OF RAJASTHAN

Sacred Grove No. 111

A. Basic Information about the sacred groves

Name of sacred groves	Miyan Mama Peer Baba
Name of village	Sardargarh
Panchayat	-
Forest Block	-
Land status	Community land
District	Sri Ganganagar
Area	50 bigha
Altitude (m)	192
Latitude	29° 27' 23.5 N
Longitude	73° 44' 24.5 E

B. Topographical features

Erosion status	Topography	Soil depth
Slight	Almost flat land	Deep (>100 cm)

C. Vegetation status

Tree species with IVI

A. nilotica (172), *C. decidua* (72), *A. tortilis* (45), *P. juliflora* (11)

Shrub species with IVI

C. procera (170), *Haloxylon salicornicum* (130)

ENS Tree: 2.41 Shrub: 2.00

Herb and Grass Species

Amaranthus species, B. diffusa, B. ramosa, Cistanche tubulosa, G. pharnacioides, Heliotropium subulatum, E. alsinoides, I. cordifolia, T. procumbens.

D. Wild- life

Boselaphus tragocamelus, *Canis aureus auveus*, *Pavo cristatus*, *Psittacula krameri*, *Eudynamys scolopaceus*, *Acridotheres tristis*, *Lepus nigricolis*.

E. Soil characteristics

pH	EC (dSm^{-1})	SOC (%)	BD (g/cm^3)	NO_3-N ($mg\ kg^{-1}$)	NH_4-N ($mg\ kg^{-1}$)	PO_4-P ($mg\ kg^{-1}$)	C density ($tons\ ha^{-1}$)
7.53	0.18	0.855	1.44	1.80	2.38	6.10	31.77

F. History and status

It is a cremation place so nobody can enter in this grove. Miyan Mama Peer mosque is present here. This grove is conserved by the villagers. Rs 500/ is use to impose as penalty if anybody is seen cutting trees from this place. Dominant tree species in the grove in *Acacia nilotica*.

DATA ON NAME OF SACRED GROVES OF RAJASTHAN

Sacred Grove No. 112

A. Basic Information about the sacred groves

Name of sacred groves	Baba Rahupeer Dargah
Name of village	2MSR
Panchayat	Anoopgarh
Forest Block	5/7 PGM
Land status	Community land
District	Sri Ganganagar
Area	210 bigha
Altitude (m)	157
Latitude	29° 11' 58.2 N
Longitude	73° 12' 49.0 E

B. Topographical features

Erosion status	Topography	Soil depth
Slight	Almost flat land	Deep (>100 cm)

C. Vegetation status

Tree species with IVI

S. oleoides (180), *A. nilotica* (46), *C. decidua* (37), *P. juliflora* (27), *P. cineraria* (10).

Shrub species with IVI

Z. nummularia (300)

ENS Tree: 3.22 Shrub: 1.00

Herb and Grass Species:

B. diffusa, B. ramosa, G. pharnacioides, E. alsinoides, I. cordifolia, T. procumbens.

D. Wild- life

Axis axis, *Boselaphus tragocamelus*, *Pavo cristatus*, *Psittacula krameri*, *Eudynamys scolopaceus*, *Acridotheres tristis*, *Lepus nigricolis*,

E. Soil characteristics

pH	EC (dSm^{-1})	SOC (%)	BD (g/cm^3)	NO_3-N ($mg\ kg^{-1}$)	NH_4-N ($mg\ kg^{-1}$)	PO_4-P ($mg\ kg^{-1}$)	C density ($tons\ ha^{-1}$)
7.70	0.18	0.450	1.58	3.50	4.52	8.30	15.38

F. History and status

More than 65 years old mosque is here in the grove. People of all community come here for fulfillment of their aspiration. It is being conserved by local Muslims in name of diety Rahupeer Baba. The grove is dominated by *Salvadora oleoides* trees followed by *A. nilotica*.

DATA ON NAME OF SACRED GROVES OF RAJASTHAN

Sacred Grove No. 113

A. Basic Information about the sacred groves

Name of sacred groves	Mata Ji Ka Darra
Name of village	Beldi
Panchayat	Uniara
Forest Block	-
Land status	Community land
District	Tonk
Area	4 bigha
Altitude (m)	345
Latitude	25° 54' 40.5 N
Longitude	75° 45' 29.4 E

B. Topographical features

Erosion status	Topography	Soil depth
Moderate	Almost flat land	Moderate (50-100)

C. Vegetation status

Tree species with IVI

Prosopis juliflora (152), *Anogeissus pendula* (148).

Shrub species with IVI

Capparis sepiaria (300)

ENS Tree: 1.89 Shrub: 1.00

Herb and Grass Species:

B. diffusa, B. ramosa, J. procumbens, E. alsinoides, I. cordifolia, T. procumbens.

D. Wild- life

Melursus ursinus, *Canis lupus*, *Hyaena hyaena*, *Boselaphus tragocamelus*, *Picnonotus cafer*, *Eudynamys scolopaceus*, *Pavo cristatus*, *Psittacula krameri*, *Lepus nigricolis.*

E. Soil characteristics

pH	EC (dSm^{-1})	SOC (%)	BD (g/cm^3)	NO_3-N (mg kg^{-1})	NH_4-N (mg kg^{-1})	PO_4-P (mg kg^{-1})	C density (tons ha^{-1})
7.45	0.24	0.075	1.68	3.15	5.03	6.49	1.67

F. History and status

An ancient sacred grove is being protected and conserved by local people in name of Mataji. There is a temple, where people use to come and worship. Formerly it was dominated by *A. pendula*, but presently dominated by *P. juliflora*.

DATA ON NAME OF SACRED GROVES OF RAJASTHAN

Sacred Grove No. 114

A. Basic Information about the sacred groves

Name of sacred groves	Devnarayan Ka Asthan
Name of village	Devgaon, Mandavar, Sitarampur
Panchayat	Marelawar
Forest Block	Soyala
District	Tonk
Area	13 bigha
Altitude (m)	301
Latitude	26° 09' 22.3 N
Longitude	75° 59' 17.4 E

B. Topographical features

Erosion status	Topography	Soil depth
Moderate	Hill slope and pediment	Moderately deep (25-50)

C. Vegetation status

Tree species with IVI

A. pendula (207), *W. tinctoria* (29), *A. nilotica* (20), *C. decidua* (16), *P. cineraria* (11), *C. religiosa* (9), *C. nurvala* (8)

Shrub species with IVI

C. sepiaria (300)

ENS Tree : 1.64 Shrub: 1.00

Herb and Grass Species:

B. diffusa, B. ramosa, J. procumbens, E. alsinoides, I. cordifolia, T. procumbens, V. cinerea, Z. gibbosa.

D. Wild- life

Boselaphus tragocamelus, Pavo cristatus, Psittacula krameri, Sus scrofa, Hyaena hyaena, Lepus nigricolis.

E. Soil characteristics

pH	EC (dSm^{-1})	SOC (%)	BD (g/cm^3)	NO_3-N ($mg\ kg^{-1}$)	NH_4-N ($mg\ kg^{-1}$)	PO_4-P ($mg\ kg^{-1}$)	C density ($tons\ ha^{-1}$)
7.24	0.22	0.180	1.65	3.12	6.23	7.82	5.44

F. History and status

An ancient sacred grove is being conserved and protected by the local people in name of Lord Devnarayan, whose temple has also been built at that time. People use to come and worship here. The grove is dominated by *A. pendula* trees.

DATA ON NAME OF SACRED GROVES OF RAJASTHAN

Sacred Grove No. 115

A. Basic Information about the sacred groves

Name of sacred groves	Devmaharaj Ka Asthan
Name of village	Devpura, Bhanvarthal
Panchayat	Sakhar
Forest Block	Garota
Land status	Community land
District	Tonk
Area	15 bigha
Altitude (m)	332
Latitude	25° 57' 04.7 N
Longitude	75° 44' 45.3 E

B. Topographical features

Erosion status	Topography	Soil depth
Moderate	Hillock with hillslope	Shallow (<25 cm)

C. Vegetation status

Tree species with IVI

A. pendula (223), *S. oleoides* (35), *C. decidua* (16), *A. indica* (10), *A. leucophloea* (9), *P. juliflora* (7).

Shrub species with IVI

No shrub

ENS Tree : 1.35 Shrub: 0

Herb and Grass Species:

B. diffusa, B. ramosa, B. repens, C. Tora, J. procumbens, E. alsinoides, I. cordifolia, T. procumbens, V. cinerea, Z. gibbosa.

D. Wild- life

Boselaphus tragocamelus, Pavo cristatus, Psittacula krameri, Sus scrofa, Hyaena hyaena, Lepus nigricolis.

E. Soil characteristics

pH	EC (dSm^{-1})	SOC (%)	BD (g/cm^3)	NO_3-N (mg kg^{-1})	NH_4-N (mg kg^{-1})	PO_4-P (mg kg^{-1})	C density (tons ha^{-1})
7.33	0.26	0.235	1.55	2.56	9.26	8.22	7.13

F. History and status

An ancient sacred grove has been put a side in name of local diety Shri Devmaharaj. There is a temple on the hillock, where local people use to worship. The grove is dominated by the old trees of *A. pendua*.

DATA ON NAME OF SACRED GROVES OF RAJASTHAN

Sacred Grove No. 116

A. Basic Information about the sacred groves

Name of sacred groves	Bhairon ji Ka Oran, Pava Bab ji
Name of village	Pava
Panchayat	Girva
Forest Block	Alsigarh marvad
Land status	Community land
District	Udaipur
Area	50 bigha
Altitude (m)	865
Latitude	24° 25' 25.3 N
Longitude	23° 32' 55.5 E

B. Topographical features

Erosion status	Topography	Soil depth
Moderate	Hill slope	Modrate deep (50-100cm)

C. Vegetation status

Tree species with IVI

W. tinctoria (112), *M. tomentosa* (40), *T. bellerica* (37), *L. coromandelica* (21), *B. retusa* (14), *A. latifolia* (13), *C. fistula* (11), *M. parviflora* (11), *H. integrifolia* (9), *E. officinalis* (7), *Ougeinia oojeinensis* (5), *D. melanoxylon* (4), *B. racemosa* (4), *P. sylvestris* (4), *B. monosperma* (4), *C. dichotoma* (4).

Shrub species with IVI

Grewia villosa (132), *L. camara* (87), *G. flavescence* (81)

ENS Tree: 6.20 Shrub: 2.66

Herb and Grass Species:

B. diffusa, B. ramosa, B. repens, C. tora, J. procumbens, E. alsinoides, I. cordifolia, T. rhomboidea, T. procumbens, V. cinerea, Z. gibbosa.

D. Wild- life

Panthera pardus, *Melursus ursinus*, *Boselaphus tragocamelus*, *Cuon alpinus*, *Pavo cristatus*, *Psittacula krameri*, *Sus scrofa*, *Hyaena hyaena*, *Lepus nigricolis*.

E. Soil characteristics

pH	EC (dSm^{-1})	SOC (%)	BD (g/cm^3)	NO_3-N (mg kg^{-1})	NH_4-N (mg kg^{-1})	PO_4-P (mg kg^{-1})	C density (tons ha^{-1})
6.78	0.11	0.465	1.44	1.73	3.30	13.34	10.85

F. History and status

This area is conserved naturally since long back as people believe that the area is protected by miraculous power of God Bhairon Ji. So no one dare to disturb the area. The grove is dominated by the tree *Wrightia tinctoria* and rich in floral diversity.

DATA ON NAME OF SACRED GROVES OF RAJASTHAN

Sacred Grove No. **117**

A. Basic Information about the sacred groves

Name of sacred groves	Hanuman Ji Ka Oran
Name of village	Madri
Panchayat	-
Forest Block	-
Land status	Community land
District	Udaipur
Area	8 bigha
Altitude (m)	704
Latitude	24°14' 49.8 N
Longitude	73° 29' 15.9 E

B. Topographical features

Erosion status	Topography	Soil depth
Moderate	Flat land	Moderate (50-100)

C. Vegetation status

Tree species with IVI

F. benghalensis (277), *P. sylvestris* (23)

Shrub species with IVI

No shrub (0)

ENS Tree : 1.23 Shrub: 1.00

Herb and Grass Species:

B. diffusa, B. ramosa, C. tora, J. procumbens, E. alsinoides, I. cordifolia, T. rhomboidea, T. procumbens.

D. Wild- life

Vulpes bengalehsis, Lepus nigricolis, Pavo cristatus, Psittacula krameri, Spilopelia chinensis.

E. Soil characteristics

pH	EC (dSm^{-1})	SOC (%)	BD (g/cm^3)	NO_3-N ($mg\ kg^{-1}$)	NH_4-N ($mg\ kg^{-1}$)	PO_4-P ($mg\ kg^{-1}$)	C density ($tons\ ha^{-1}$)
6.27	0.14	0.480	1.51	3.57	3.58	12.06	13.10

F. History and status

This is a small but ancient grove alongwith a temple of Lord Hanuman. The area is protected and conserved by the local people in name of God Hanuman ji.The grove is dominated by *Ficus bengalensis*.

DATA ON NAME OF SACRED GROVES OF RAJASTHAN

Sacred Grove No. 118

A. Basic Information about the sacred groves

Name of sacred groves	Shringi Rishi Ji Ka Oran
Name of village	Bhikarni/ Chirva
Panchayat	Udaipur
Forest Block	Shringi Rishi
District	Udaipur
Area	200 bigha
Altitude (m)	732
Latitude	73° 45' 52.3 N
Longitude	24° 42' 06.2 E

B. Topographical features

Erosion status	Topography	Soil depth
Moderate	Hill slope	Moderately deep (25-50)

C. Vegetation status

Tree species with IVI

A. senegal (106), *W. tinctoria* (54), *A. leucophloea* (45), *H. integrifolia* (42), *A. leucophloea* (24), *E. camaldulensis* (11), *Z. mauritiana* (10), *Diospyros cordifolia* (8).

Shrub species with IVI

G. tenax (119), *Dyerophytum indicum* (65), *D. cinerea* (43), *Annona squamosa* (42), C. *sepiaria* (31)

ENS Tree : 5.22 Shrub: 4.17

Herb and Grass Species:

B. diffusa, B. ramosa, C. tora, J. procumbens, E. alsinoides, I. cordifolia, Rungia pectinata, T. rhomboidea, T. procumbens.

D. Wild- life

Panthera pardus, *Melursus ursinus*, *Vulpes bengalehsis*, *Acridotheres tristis*, *Herpestes edwardsi*, *Boselaphus tragocamelus*, *Cuon alpinus*, *Pavo cristatus*, *Psittacula krameri*, *Sus scrofa*, *Hyaena hyaena*, *Lepus nigricolis.*

E. Soil characteristics

pH	EC (dSm^{-1})	SOC (%)	BD (g/cm^3)	NO_3-N ($mg\ kg^{-1}$)	NH_4-N ($mg\ kg^{-1}$)	PO_4-P ($mg\ kg^{-1}$)	C density ($tons\ ha^{-1}$)
8.16	0.18	1.845	1.28	4.56	4.23	11.13	30.47

F. History and status

In ancient time an ascetic saint Shree Shringi Rishi used to worship Lord Shiva here. From that time people use to come here and worship. The grove is maintained since that time. It is dominated by *Acacia senegal* trees.

DATA ON NAME OF SACRED GROVES OF RAJASTHAN

Sacred Grove No. 119

A. Basic Information about the sacred groves

Name of sacred groves	Bhanvar Mata ji Ki Bani
Name of village	Sakroda
Panchayat	Girva
Forest Block	-
Land status	Community land
District	Udaipur
Area	625 bigha
Altitude (m)	642
Latitude	24° 34' 07.6 N
Longitude	73° 51' 11.5 E

B. Topographical features

Erosion status	Topography	Soil depth
Moderate	Hill top and pediment	Shallow (<25 cm)

C. Vegetation status

Tree species with IVI

A. pendula (188), *M. parviflora* (34), *L. coromandelica* (19), *A. indica* (15), *P. cineraria* (12), *A. senegal* (11), *C. decidua* (11), *A. leucophloea* (10).

Shrub species with IVI

Euphorbia caducifolia (163), *G. tenax* (82), *Commiphora wightii* (23), *D. cinerea* (17), *Calotropis procera* (15).

ENS Tree : 2.17 Shrub: 2.88

Herb and Grass Species:

B. diffusa, B. pussilla, B. ramosa, C. tora, J. procumbens, E. acaulis, E. granularis, E. alsinoides, I. cordifolia, S. cordata, T. rhomboidea, T. procumbens.

D. Wild- life

Lepus nigricolis, Boselaphus tragocamelus, Pavo cristatus, Psittacula krameri, Hyaena hyaena, Canis aureus auveus, Semnopithecus entellus.

E. Soil characteristics

pH	EC (dSm^{-1})	SOC (%)	BD (g/cm^3)	NO_3-N ($mg\ kg^{-1}$)	NH_4-N ($mg\ kg^{-1}$)	PO_4-P ($mg\ kg^{-1}$)	C density ($tons\ ha^{-1}$)
6.60	0.32	1.305	1.25	4.58	12.59	17.38	15.58

F. History and status

Because of presence of Local Goddess Bhanvar Mata ji, the whole area is conserved by villagers, who worship here, The mataji. The grove is dominated by *Anogeissus pendula* followed by *Mitragyna parviflora*.

DATA ON NAME OF SACRED GROVES OF RAJASTHAN

Sacred Grove No. 120

A. Basic information about the sacred groves

Name of sacred groves	Kali Pahari Mata ji Ka Oran
Name of village	Gumanpura
Panchayat	Bhinder
Forest Block	Kali Pahari
Land status	Community land
District	Udaipur
Area	10 bigha
Altitude (m)	525
Latitude	24° 41' 53.4 N
Longitude	74° 00' 20.4 E

B. Topographical features

Erosion status	Topography	Soil depth
Slight	Almost flat land	Moderate (50-100)

C. Vegetation status

Tree species with IVI

P. juliflora (180), *A. senegal* (74), *A. catechu* (35), *B. serrata* (11)

Shrub species with IVI

No shrub

ENS Tree : 2.22 Shrub: 0

Herb and Grass Species:

B. diffusa, B. eriantha, B. pussilla, B. ramosa, C. tora, J. procumbens, E. acaulis, E. granularis, E. alsinoides, I. cordifolia, P. paniculata, S. cordata, T. rhomboidea, T. procumbens, V. cinerea.

D. Wild- life

Lepus nigricolis, *Boselaphus tragocamelus*, *Pavo cristatus*, *Hyaena hyaena*, *Canis aureus auveus*, Hanuman Langur.

E. Soil characteristics

pH	EC (dSm^{-1})	SOC (%)	BD (g/cm^3)	NO_3-N ($mg\ kg^{-1}$)	NH_4-N ($mg\ kg^{-1}$)	PO_4-P ($mg\ kg^{-1}$)	C density ($tons\ ha^{-1}$)
8.14	0.22	0.135	1.40	7.95	5.92	4.71	2.29

F. History and status

An ancient sacred grove is being conserved and protected by the local communities residing in the nearby villages. It is conserved in name of Local Goddess- Mata ji. The grove is dominated by *P. juliflora*, which had invaded the grove largely.

DATA ON NAME OF SACRED GROVES OF RAJASTHAN

Sacred Grove No. 121

A. Basic information about the sacred groves

Name of sacred groves	Kundeshwar Mahadevji Oran
Name of village	Basaliya
Panchayat	Iswal
Forest Block	-
Land status	Community land
District	Udaipur
Area	13 bigha
Altitude (m)	708
Latitude	24° 45' 06.3 N
Longitude	73° 39'13.6 E

B. Topographical features

Erosion status	Topography	Soil depth
Moderate	Hillslope with valley	Moderate (50-100)

C. Vegetation status

Tree species with IVI

F. benghalensis (141), *F. infectoria* (33), *Syzygium cummini* (31), *D. cordifolia* (29), *B. retusa* (14), *Santalum album* (9), *M. indica* (7), *Polyalthia longifolia* (7), *C. myxa* (6), *A. indica* (6), *C. fistula* (6), *H. integrifolia* (6), *T. indica* (5).

Shrub species with IVI

C. sepiaria (183), *A. squamosa* (117)

ENS Tree : 5.83 Shrub: 1.89

Herb and Grass Species:

B. diffusa, B. pussilla, B. ramosa, J. procumbens, E. acaulis, E. granularis, E. alsinoides, I. cordifolia, S. cordata, T. rhomboidea, T. procumbens, V. cinerea.

D. Wild- life

Lepus nigricolis, *Boselaphus tragocamelus*, *Pavo cristatus*, *Psittacula krameri*, *Hyaena hyaena*, *Semnopithecus entellus,* Snake etc.

E. Soil characteristics

pH	EC (dSm^{-1})	SOC (%)	BD (g/cm^3)	NO_3-N ($mg\ kg^{-1}$)	NH_4-N ($mg\ kg^{-1}$)	PO_4-P ($mg\ kg^{-1}$)	C density ($tons\ ha^{-1}$)
7.08	0.18	1.620	1.25	8.20	4.34	14.85	19.96

F. History and status

An oldest temple of Lord Shiva since the ancient time is available in this sacred grove, which is relatively smaller in area. People come here to worship and the surroundings area is being protected by the villagers. The grove is dominated by *Ficus benghalensis*.

DATA ON NAME OF SACRED GROVES OF RAJASTHAN

Sacred Grove No. 122

A. Basic information about the sacred groves

Name of sacred groves	Tarkeshwar Mahadev Oran
Name of village	Delwas
Panchayat	Sirada
Forest Block	-
Land status	Community land
District	Udaipur
Area	30 bigha
Altitude (m)	466
Latitude	24° 13' 07.2 N
Longitude	73° 42' 47.8 E

B. Topographical features

Erosion status	Topography	Soil depth
Moderate	Hill slope	Moderately deep (25-50)

C. Vegetation status

Tree species with IVI

Putranjiva roxburghii (53), *H. integrifolia* (52), *M. indica* (37), *T. indica* (24), *F. benghalensis* (22), *S. cummini* (16), *F.limonia* (14), *F. glomerata* (14), *D. melanoxylon* (13), *M. tomentosa* (12), *Cocos nucifera* (11), *C. dichotoma* (10), *Pendanus odoratissinus* (9), *F. infectoria* (6), *C. fistula* (4), *Guazuma ulmifolia* (3)

Shrub species with IVI

A. squamosa (168), *Jatropha curcas* (74), *C. auriculata* (25), *Cordia gharaff* (17), *C. carrandus* (16).

ENS Tree : 10.47 Shrub: 2.81

Herb and Grass Species:

B. diffusa, Blumea mollis, B. pussilla, B. ramosa, Hyptis suaveolens, J. procumbens, L. trinervis, E. acaulis, E. granularis, E. alsinoides, I. cordifolia, S. cordata, T. tennalus, T. rhomboidea, T. procumbens, V. cinerea.

D. Wild- life

Panthera pardus, Boselaphus tragocamelus, Cuon alpinus, Pavo cristatus, Psittacula krameri, Sus scrofa, Hyaena hyaena, Lepus nigricolis.

E. Soil characteristics

pH	EC (dSm^{-1})	SOC (%)	BD (g/cm^3)	NO_3-N ($mg\ kg^{-1}$)	NH_4-N ($mg\ kg^{-1}$)	PO_4-P ($mg\ kg^{-1}$)	C density ($tons\ ha^{-1}$)
7.96	0.16	0.645	1.47	4.46	3.05	14.06	8.73

F. History and status

This sacred grove was established long back, where one of the oldest temple of Lord Shiva is available since ancient time. People use to come and worship Lord Shiva. This grove is highly biodiverse and is being conserved by villagers. It is dominated by an important species of tree, i.e. *Putranjiva roxburghii*.

DATA ON NAME OF SACRED GROVES OF RAJASTHAN

Sacred Grove No. 123

A. Basic information about the sacred groves

Name of sacred groves	Ubeshwar ji Oran
Name of village	Dhar
Panchayat	Badgano
Nearest forest Block	Ubeshwar ji
Land status	Community land
District	Udaipur
Area of sacred grove	700 bigha
Altitude (m)	880
Latitude	24° 37' 13.2 N
Longitude	73° 33' 40.7 E

B. Topographical features

Erosion status	Topography	Soil depth
Moderate	Hill slope	Moderately deep (25-50)

C. Vegetation status

Tree species with IVI

E. camaldulensis (83), *B. monosperma* (70), *D. cordifolia* (35), *M. parviflora* (24), *Acacia auriculiformis* (20), *Syzygium cummini* (17), *A. leucophloea* (15), *Mangifera indica* (9), *P. cineraria* (8), *F. infectoria* (8), *A. latifolia* (7), *R. dumentorum* (4).

Shrub species with IVI

L. camara (136), *A. vasica* (63), *Jatropha curcas* (36), *G. flavescence* (18), *Agave americana* (14), *Z. nummularia* (13), *E. nerrifolium* (10), *C. sepiaria* (10)

ENS Tree : 6.72 Shrub: 3.41

Herb and Grass Species:

B. diffusa, B. pussilla, B. ramosa, J. procumbens, L. trinervis, E. acaulis, E. granularis, E. alsinoides, I. cordifolia, S. cordata, T. tennalus, T. rhomboidea, T. procumbens, V. cinerea.

D. Wild-life

Boselaphus tragocamelus, *Cuon alpinus*, *Sus scrofa*, *Hyaena hyaena*, *Lepus nigricolis*, *Pavo cristatus*, *Psittacula krameri.*

E. Soil characteristics

pH	EC (dSm^{-1})	SOC (%)	BD (g/cm^3)	NO_3-N (mg kg^{-1})	NH_4-N (mg kg^{-1})	PO_4-P (mg kg^{-1})	C density (tons ha^{-1})
7.98	0.17	1.905	1.25	5.44	13.55	10.52	40.80

F. History and status

An old sacred grove, where a temple of Lord Shiva is available since the time of Maharana Kumbha of Mewar. People offer their prayers here and help in conserving the surroundings area rich in biodiversity. However, plantation of *E. camaldulensis* has resulted in its dominance in the grove area.

4

BIOLOGICAL DIVERSITY OF SACRED GROVES

Sacred groves are areas of virgin forest rich in biodiversity, because they have been protected for centuries by the local people for their cultural and religious beliefs and taboos. Though biotic interactions affect regional biodiversity, but also contribute to it sometimes by accelerating adaptation and speciation in the region. The benefits of increased biodiversity, i.e. number of species, are that different species promotes ecosystem functioning during different years, at different places, for different functions and under different environmental change scenarios. Further, the species required to provide one function during multiple years are not the same as those needed to provide multiple functions within one year (Isbell et al., 2011). The interrelationship among the human beings, plants and animals in their environments is much more important. However, rapidly declining biodiversity worldwide, resulting from anthropogenic alterations, decreases ecosystem functioning and services (Cardinale et al., 2011). Though species may appear functionally redundant when one function is considered under one set of environmental conditions, but many species are needed to maintain multiple functions at multiple times and places in a changing world (Isbell et al., 2011). This heightens the need to enhance biological diversity of our ecosystem and develop suitable management strategies that conserve species and maintain ecosystem services. Careful selection of plants through evaluation and assessment in managed landscapes can augment much needed management goals (Bennett and Gratton, 2013). In this chapter diversity of trees categorized into trees of >30 cm girth and 10-30 cm girth, tree saplings (> 3.0 cm to 10 cm girth) and shrub species have been described. It also covers faunal diversity in general, based on the interaction with the local people and sighting.

1. BACKGROUND

Biological diversity of traditional common resources symbolizes a pool of human experiences since ancient time. Ethnic groups residing in remote areas have cultural believes and some feelings in something or somewhere, i.e. sacred groves, sacred sites or sacred places. Because of long and diverse culture sacred groves demonstrates ancient link between human beings and their immediate environment. Sacred groves are stands of trees or patches of forest that local communities conserve primarily due to their religious significance. These groves serve economic, medicinal, social, and cultural functions and are very rich in biological diversity and harbour many endangered plant species including rare herbs and medicinal plants (Tiwari et al., 1998). Likewise, some plant species also provide emergency foods during periods of drought, crop failure and famine. In fact these sacred groves link religious faiths with protected areas and help conserve and promote biological diversity (Kotwal and Banerjee, 2000). Many ethnobotanical studies have been conducted on the sacred groves and highlight the importance of utilization of plants in human well being (Boraiah et al., 2003; Khumbongmayum et al., 2005a, Mala et al., 2013). A study relating tree wealth in the life and economy of the tribal people in Andhra Pradesh indicates that various species are used by the different ethnic groups for various purposes, such as the treatment of some common diseases and disorders (Rani et al., 2003). Likewise, Tolchha-Bhotiya sub-community inhabiting the buffer zone villages of Nanda Devi Biosphere Reserve and many other areas shows strong faith and belief in traditional healthcare system including animals and insect bites through herbal treatment (Maikhuri et al. 1998; Mala et al., 2013; Rajkumar et al., 2014). Holy Basil (*Ocimum* spp.), Asoka (*Saraca asoca*), Banyan tree (*Ficus benghalensis*), Pipal (*Ficus religiosa*), Banana (*Musa paradisica*), Neem (*Azadirachta indica*), Aam (*Mangifera indica*), Bel (*Aegle marmelos*), Khezri (*Prosopis cineraria*) etc. are sacred plant species in India and have many social and economic values (Christopher, 2014; Kour and Sharma, 2014).

2. FLORAL DIVERSITY IN SACRED GROVES

Sacred groves in Rajasthan, normally known as 'Oran', are an important source of natural wealth like fodder, fuel, timber, berries, roots and herbs plants of medicinal values. Orans have been regarded earlier as a symbol of prosperity for the community that owned the grove. Moreover, sacred groves also promote a flourishing livestock based economy and growth of livestock rearing communities, particularly in Western Rajasthan. According to a study, about 41% livestock is dependent on the sacred groves in Barmer district of Rajasthan (Singh and Shishodia, 2003). The groves of Rajasthan harbour many important trees species like *Prosopis cineraria, Tecomella undulata, Anogeussus sericea, Salvadora* spp., *Zizyphus* spp., *Commiphora wightii* etc. Some of these provide famine food whereas some other appear to be endangered in the region (Meena and Singh, 2012; Singh, 2015). Study conducted in 123 sacred groves in Rajasthan, indicates a total number of 92 tree species categorized into >30 cm girth and 10-30 cm girth classes. These tree species belong to 36 families. Likewise 39 shrub species belonging to 23 families have also been identified in these sacred groves (Annexure I). Thus, all these 131 woody species (both trees and shrubs species) are covered under 48 botanical families (Table 4.1). In this highest number of species have been recorded from Fabaceae (27) followed by Euphorbiaceae (7), Combretaceae (6) and Capparaceae (6). Some of them are rare and are important on religious and socioeconomic point of view (Plate 4.1).

Table 4.1. Number of trees and shrubs belonging to various families recorded in different sacred groves of Rajasthan.

SNo.	Family	No. of tree species	No. of shurb species	Total
1	Acanthaceae	-	2	2
2	Agavaceae	-	1	1
3	Amaranthaceae	-	2	2
4	Anacardiaceae	2	1	3
5	Annonaceae	2	1	3
6	Apocynaceae	2	2	4
7	Arecaceae	2	-	2
8	Aristolochiaceae	-	1	1
9	Asclepiadaceae	-	2	2
10	Balanitaceae	1	-	1
11	Bignoniaceae	3	-	3
12	Bombacaceae	2	-	2
13	Boraginaceae	5	-	5
14	Burseraceae	1	1	2
15	Cactaceae	-	1	1
16	Capparaceae	3	3	6
17	Celastraceae	1	1	2
18	Chenopodiaceae	-	1	1
19	Combretaceae	6	-	6
20	Ebenaceae	2	-	2
21	Elaeocarpaceae	1	-	1
22	Euphorbiaceae	2	5	7
23	Fabaceae	23	4	27
24	Flacourtiaceae	-	1	1
25	Lythraceae	1	-	1
26	Malvaceae	1	-	1
27	Meliaceae	2	-	2
28	Moraceae	5	-	5
29	Moringaceae	1	-	1
30	Myrtaceae	2	-	2
31	Pandanaceae	-	1	1
32	Plumbaginaceae	-	1	1
33	Polygonaceae	-	1	1
34	Putranjivaceae	1	-	1
35	Rhamnaceae	2	1	3
36	Rubiaceae	5	-	5
37	Rutaceae	2	-	2
38	Salvadoraceae	2	-	2
39	Santalaceae	1	-	1
40	Sapindaceae	1	-	1

SNo.	Family	No. of tree species	No. of shurb species	Total
41	Sapotaceae	1	-	1
42	Simaroubaceae	1	-	1
43	Solanaceae	-	1	1
44	Sterculiaceae	2	-	2
45	Tamaricaceae	1		1
46	Tiliaceae	1	3	4
47	Ulmaceae	1	-	1
48	Verbenaceae	1	2	3
	Total	**92**	**39**	**131**

2.1 Population and species richness

In general, trees of high girth class are common in the sacred groves because of community protection and belief. Regeneration status of the trees appears to be more in the sacred groves because of sacredness and protection. The study of Khumbongmayum et al. (2006) carried out in four sacred groves based on strength of different age groups in their population indicates good regeneration. However, high occurrence of some additional species in the groves is due to the invasion of other species through dispersal from the other areas. The prevailing favourable microenvironment in the sacred groves contributes to their establishment and growth. But absence of seedlings and saplings of some of the species in the groves appears to be due to their poor seed germination and establishment of seedlings in the forest (Khumbongmayum et al. 2005b; Bharathi and Dev Prasad, 2015), although harsh environmental conditions may also be responsible for poor regeneration of the tree species (Pandey, 2000).

2.1.1 Tree species

Average number of tree (>30 cm girth) species per plot across Rajasthan has been observed to be 5.5 tree species per site (ranging from 1 species in Mata Ji Ka Oran, Barmer, Karni Mata Oran, Bikaner, Malani Oran, Jaisalmer, Dev Narain Ji Ki Banni, Karauli to 16 species in Tarkeshwar Mahadev Ji ka Oran, Udaipur), indicating almost 2.9-fold change in the number of tree species between different groves. Average number of trees per plot is 78.6, ranging from 15 trees per 1.25 ha in Kolu Pabuji Ka Oran, Jodhpur to 272 trees per 1.25 ha in Rupu Ka Wash Oran, Alwar (Annexure II). This indicates significant spatial variation in tree density depending upon edaphic and climatic factors as observed in a study carried out in Tamil Nadu, where tree population density varies from 126 tree ha^{-1} to 162 tree ha^{-1} among the groves (Sundarapandian and Subbiah, 2015). Number of tree species in category of 10-30 cm dbh (diameter at breast height) has been 2.6 tree species per site (ranging almost 1 species in 28 sacred groves including Shringi Rishi Ji ka Oran, Udaipur to 10 species in Pava Babji, Udaipur). There are 3.8 fold variations in number of species in 10-30 cm girth class among the studied sacred groves. Average population of trees in this category across Rajasthan is 25.9 trees per 1.25 ha that ranges from 1 number in 9 sacred groves to 373 trees per 1.25 ha in Dagia Bherogi Ka Oran, Banswara. Unfortunately, 13% sacred groves have no tree under this girth class (Annexure II).

The low value of population density and tree species richness in the present study appears to be due to anthropogenic pressures such as lopping, extraction of minor forest produce (fruits, seeds etc.) and animal grazing. These factors also affect tree regeneration through seedling recruitment and stunted growth due to lopping of the trees. This process leads to small openings in the canopy of sacred groves favouring regeneration of other species available outside the sacred groves like *Prosopis juliflora*, which has been observed invading many sacred groves in the periphery. This inhibits the regeneration of native species due to allelopathic effect (Sundarapandian and Subbiah, 2015).

2.1.2 Shrub species

Twenty one sacred groves do not show shrubs as the ground vegetation, whereas 33 sacred groves have only one shrub species. Average number of shrub species across the study sites is 2.23 species per site, where number of species ranges between 1 and 10 species (highest in Shiv Ji Maharaj ki Devbani, Alwar). Average population of the shrubs is 18.1 individuals per 125 m^2 (sum of five plots of 5 m × 5 m size). Lowest population of the shrubs has been recorded in 7 sacred groves, i.e. 1 individual, whereas highest numbers of shrub (197 individuals) are in Malani Oran, Jaisalmer (Annexure III).

2.1.3 Tree saplings

Climatic, edaphic and anthropogenic factors play an important role in regeneration and recruitment of the tree seedlings. Effects of disturbances on seedling recruitments also vary in the region. For example Laloo et al. (2006) have recorded higher seedling recruitment in four species under disturbed condition of the sacred grove, though per cent conversion of the seedlings into saplings has been observed more in the undisturbed forest, except in the case of *E. kunthianum* species. The gaps generated show facilitative effects in per cent conversion of saplings in majority of the trees species (Laloo et al., 2006). While studying eight sacred groves, Bharathi and Devi Prasad (2015) have recorded regeneration in about 38% tree species, whereas 31% species have not shown regeneration and 31% have considered as reappearing as the mature individuals, whose parents are not available in the sampled area. The average seedling and sapling density among the study sites have been 3.3 individuals per m^2 and 5.2 individuals per m^2, respectively. In a study conducted in Jodhpur district of Rajasthan on the regeneration of different tree species in different land uses including 45 sacred groves, show highest number of seedlings along road side, followed by forestland and sacred groves (Table 4.2). There is strong spatial variation in number of seedlings per hectare, i.e. highest in Shergarh and Bhopalgarh forest ranges and the lowest in Phalodi range (Singh, 2015a).

Out of 123 sacred groves studied in Rajasthan, about 45% sacred groves have no saplings on the ground. Average number of tree species, for which the saplings are recorded is less than 1 (i.e., 0.89 species) across the state. Highest numbers of individuals are for 6 species in Tarkeshwar Mahadev Ji ka Oran in Udaipur, followed by 4 numbers of species in Pava Bab Ji sacred grove, Udaipur. Otherwise most of the sacred groves show regeneration of one or two species. Population of saplings varies from 1 in Ramdevji ka Oran, Jaisalmer to 62 individuals per site in Tarkeshwar Mahadev Ji ka Oran, Udaipur (Annexure III).

Table 4.2. Population (numbers ha^{-1}) of tree saplings influenced by land-use and forest ranges in Jodhpur district of Rajasthan. Values are mean ±SE of five replications. Source: Singh (2015a).

Forest range	Land use					Mean
	Agriculture	Forest	Oran	Pasture	Roadside	
Baap	11.2±4.9	150.0±88.9	94.0±43.5	36.0±8.1	49.6±26.3	68.15±21.3
Balesar	17.8±11.8	98.0±41.6	106.0±49.3	32.0±6.6	96.1±33.3	69.97±15.5
Bhopalgarh	50.2±33.5	92.0±29.2	136.0±35.9	62.0±28.5	72.7±42.5	82.59±15.3
Bilara	17.6±9.3	80.0±34.5	88.0±22.2	116.0±36.7	418.4±144.9	143.99±40.4
Luni	13.6±7.0	44.0±15.0	102.0±7.7	104.0±20.8	463.6±162.4	145.44±35.9
Mandor	45.8±15.7	234.0±82.1	230.0±53.2	125.0±62.5	100.9±32.2	147.14±22.5
Osian	15.8±6.2	198.0±115.1	78.0±11.6	82.0±32.6	82.9±29.6	91.34±25.6
Phalodi	16.6±4.8	66.0±17.8	62.0±22.0	52.0±14.6	62.4±24.8	51.80±8.3
Shergarh	8.0±5.1	106.0±47.1	136.0±49.6	256.0±58.3	57.8±20.4	112.77±24.0
Mean	21.84±4.8	118.67±20.6	114.67±15.6	96.11±14.4	156.04±32.9	

2.2. Diversity index and Effective Number of Species

Species count (i.e., richness), Shannon-Weiner diversity Index (H') and Simpson's index (D) are three important indices describing diversity of an area. Effective number of species described as the number of equally abundant species needed to obtain the same mean proportional species abundance as that observed in the dataset of interest, i.e. where all species may not be equally abundant (Jost, 2006). Species diversity consists of two main components like species richness and species evenness (Jost, 2010). Anthropogenic activities adversely affect species richness as was observed in a study where human activities have decreased species richness significantly in the disturbed sacred grove as compared to the undisturbed ones (Laloo et al., 2006; Bharathi and Devi Prasad, 2015).

2.2.1 Tree species

For trees of >30 cm girth, six sacred groves namely Mata Ji Ka Oran, Barmer, Karni Mata Ji Ka Oran, Bikaner, Malani Oran, Jaisalmer, Pabu ji Ka Oran and Shri Devnarayan Ji Ki Banni, Karauli show only one species and hence almost zero diversity index (H'). For the other sacred groves, H' varies from 1.04 in Kolu Pabu ji Ka Oran, Jodhpur (with ENS value of 1.04) to 2.35 (ENS value of 10.47) in Tarkeshwar Mahadev Ji ka Oran, Udaipur. Average values of H' and ENS across the sites are 0.96 and 2.60 for the tree category of >30 cm girth (Annexure II). For tree category of 10-30 cm girth, about 40 (i.e., 33%) sacred groves including Devnarayan Ji ka Oran and Bhuna Bai Ji Ka Oran, Ajmer, Rupu Ka Vas Oran, Alwar, Prathurinath Ji Ka Oran, Banswara, Devnarayan Ji ka Oran, Baran, Viratra Mata Ka Oran, Mata Ji Ka Oran, Khuwas Mata Ji Ka Oran and Mama Ji Ka Selenadi, Bamer, Adibadri Mahatmya Oran, Bharatpur, Dennaryan Ji Ki Bani and Devnayan Oran, Bhilwara, Goga Ji Ka Oran and karni Mata Oran, Bikaner, Devnarayan Ji Ka Oran,

Chittorgarh and others have zero value of H' with ENS value of 1 indicating almost dominance of one tree species. In the rest of the sacred groves H' ranges between 0.04 in Mataji ka Darra, Tonk and 2.08 in Pawa Babji ka Oran, Udaipur with respective ENS value of 1.05 and 8.02. Average value of H' and ENS across the sites are 0.56 and 1.76, respectively (Annexure II). This indicates relatively low tree diversity in the sacred groves studied in Rajasthan when compared with a general range of 0 to 5.

2.2.2 Shrub species

In case of shrub species, about 42% of the studied sacred groves have zero H' value and 1 ENS, indicating availability of single dominant shrub in these sacred groves (Annexure III). In rest of the sacred groves H' varies from 0.13 in Dhudheshwar Mahadeoji Ka Oran, Jalore and Mamaji Ka Oran, Pali to 1.75 in Bheruji Ka Oran, Nagaur district. The corresponding ENS varies from 1.14 to 5.77 at the respective site. Average values of H' and ENS across the sites are 0.50 and 1.65. This indicates significant variation in these diversity variables between the sacred groves situated throughout Rajasthan.

2.2.3 Tree sapling

For regeneration status in the form of tree saplings, about 78% of the sacred groves show dominance of a single species regeneration indicated by zero H' value and 1 ENS value (Annexure III). In rest of the sacred groves, the diversity index varies from 0.34 in Bhairav Ji Ka Bag, Jhalawar to 1.55 in Tarkeshwar Mahadev Ji ka Oran, Udaipur. The ENS values vary from 1.40 to 4.77 in the respective sacred groves. Average diversity and effective number of species across the studied sites in Rajasthan are 0.16 and 1.17, respectively for the tree saplings.

2.3 Reciprocal of Simpson index (1/D)

The highest value of reciprocal of Simpson dominance of tree >30 is 10.77 in Shri Parbateshwar Mahadeo Ji Ki Bani in Bundi district, whereas lowest value of this variable is 0.55 in Gosaiji Ka Oran in Churu district. Average across the site, 1/D value of 1.81 is observed for the trees of this category. For the trees of 10 to 30 cm girth, 1/D is highest in Pava Bab Ji ka Oran (6.66) in Udaipur district, whereas almost 28 sacred groves studied show value of it as 1. The lowest value of 1 is indicative of dominance of one tree species of this girth category in the particular groves. Its value across the studied sites is 1.82, which is relatively greater as compared to the trees of >30 cm girth class. The highest 1/D value for tree sapling and shrub have been recorded in Tarkeshwar Mahadev Ji ka Oran (4.01), Udaipur and Bhairo Ji Ka Oran (5.11) in Nagaur district, respectively. The lowest values of 1/D in these categories are almost in 41 and 32 sacred grove, respectively. Its average value for sapling and shrub are 2.25 and 1.86, respectively. This indicates that these sacred groves are more diverse in terms of tree sapling diversity as compared to the shrub diversity.

3. SPECIES RICHNESS AND ENS IN DISTRICTS

Species richness and effective number of species (ENS) for the trees of >30 cm girth range from 2.20 in Jodhpur to 10.5 in Bharatpur district and from 1.43 in Ajmer to 6.30 in Bharatapur, respectively. In girth class 10-30 cm, there are no trees in Karauli districts, whereas in rest of the district, these variables range from 0.6 in Barmer to 5.0 in Bharatpur, and from 1.0 in Churu to 4.19 in Bharatpur district, respectively. Relatively greater variations between species richness and ENS in case of trees of >30 cm girth as compared to trees of size 10-30 cm girth indicate large difference in population of different species in >30 cm girth class trees. Species richness for the shrubs range between 0.50 in Bharatpur and 5.0 in Alwar district; but the value of ENS is 1 in Bharatpur and Tonk to the highest value in Ajmer district, i.e. 3.57. Average number of species richness for tree saplings is observed almost zero in Bharatpur, Churu, Dholpur, Jhunjhunu and Sikar. It is 1 in most of the districts, however, the highest value of ENS is observed in Chittorgarh, i.e. 2.28.

Table 4.3. Average values of species richness and effective number of species in the sacred groves of different districts of Rajasthan.

District	Number of species				Effective number of species			
	>30 cm	10-30 cm	Shrubs	Sapling	>30 cm	10-30 cm	Shrubs	Sapling
Ajmer	4.33	1.33	4.00	1.00	1.43	1.33	3.57	1.00
Alwar	7.20	4.60	5.20	0.40	4.38	3.90	3.41	1.00
Banswara	5.33	2.33	2.33	2.00	2.17	1.73	2.38	2.06
Baran	9.00	3.00	3.33	2.67	4.69	2.13	2.78	1.89
Barmer	3.00	0.60	3.60	1.00	2.27	1.07	3.04	1.38
Bharatpur	10.50	5.00	0.50	0.00	6.30	4.19	1.00	0.00
Bhilwara	6.33	1.33	2.00	0.67	4.19	1.82	2.11	1.00
Bikaner	3.00	1.33	2.33	0.33	2.49	1.38	1.69	1.00
Bundi	5.67	4.67	2.33	0.33	4.88	3.18	2.14	1.00
Chittorgarh	4.33	3.00	2.33	2.33	1.42	2.27	1.98	2.28
Churu	4.33	1.00	2.67	0.00	2.13	1.00	2.12	0.00
Dausa	3.33	2.67	2.33	0.67	1.59	2.18	1.98	1.00
Dholpur	5.00	4.67	3.33	0.00	2.87	3.04	3.04	0.00
Dungarpur	6.00	3.00	3.67	1.00	2.12	1.47	3.11	1.00
Hanumangarh	6.33	2.33	1.33	1.00	3.40	2.25	1.24	1.22
Jaipur	7.00	3.33	1.67	0.67	3.61	2.81	1.60	1.00
Jaisalmer	2.25	2.00	2.25	0.50	1.77	1.80	1.76	1.00
Jalore	4.33	2.00	2.33	0.67	2.12	1.53	1.30	1.00
Jhalawar	6.00	2.80	0.60	0.60	3.25	2.38	1.08	1.08
Jhunjunnu	3.67	1.00	1.00	0.00	2.16	1.33	1.27	0.00
Jodhpur	2.20	1.60	1.20	0.80	1.50	1.73	1.74	1.00
Karauli	5.33	0.00	2.00	0.33	3.85	0.00	1.73	1.00
Kota	7.44	3.11	2.11	1.22	3.72	2.31	1.91	1.38
Nagaur	4.00	1.50	4.00	0.25	2.78	1.43	3.15	1.00
Pali	4.00	3.20	1.80	1.60	2.58	1.74	1.25	1.43

District	Number of species				Effective number of species			
	>30 cm	10-30 cm	Shrubs	Sapling	>30 cm	10-30 cm	Shrubs	Sapling
Pratapgarh	7.50	3.75	1.50	0.25	3.52	1.98	1.32	1.00
Rajsamand	4.67	1.33	1.67	1.33	2.39	1.33	1.50	1.33
SawaiMadhopur	5.67	2.67	1.67	0.33	3.46	2.58	1.38	1.00
Sikar	3.67	2.00	2.00	0.00	3.29	1.89	1.72	0.00
Sirohi	5.00	2.33	1.33	1.67	2.62	2.04	1.23	1.64
Sriganganagar	4.50	2.25	1.25	1.00	2.63	1.64	1.49	1.32
Tonk	5.00	2.67	0.67	1.33	1.63	1.60	1.00	1.30
Udaipur	9.88	4.38	3.50	1.63	5.01	3.30	2.48	1.89

4. TREE DOMINANCE AND DIVERSITY

Out of 123 numbers of sacred groves studied, 24 sacred groves are dominated by *A. pendula*, followed by *P. juliflora*, which is dominant in 16 sacred groves. *Salvadora oleoides* and *Capparis decidua* have dominated in 13 and 11 sacred groves, respectively (Annexure IV). Among the studied sacred groves 7 sacred groves are dominated by *P. cineraria* and 6 sacred groves have been observed dominated by *Tectona grandis*. *Acacia leucophloea*, *Acacia nilotica*, *Butea monosperma* and *Zizyphus rotundifolia* have been observed dominant in 5, 4, 4 and 4 sacred groves, respectively (Fig 4.1). Other species are found dominant in relatively less number of sacred groves.

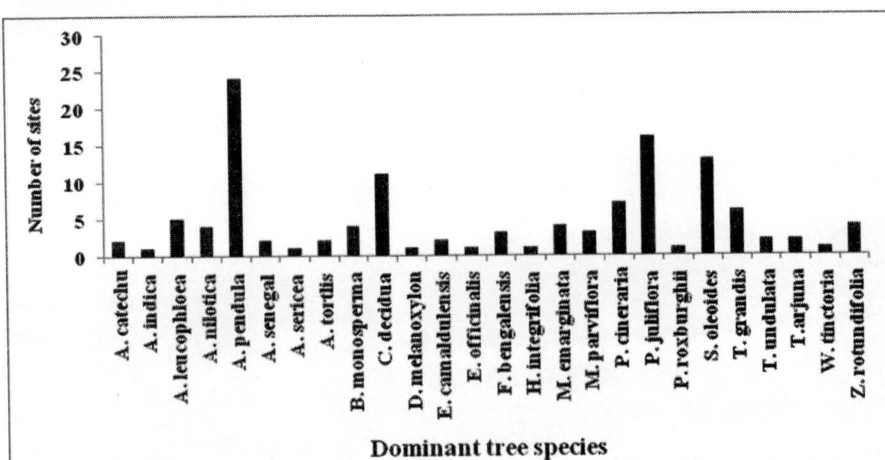

Figure 4.1. Number of sites dominated by different tree species of girth >30 cm.

4.1 Tree diversity

For trees of >30 cm girth, species richness (numbers of species), H' and ENS are highest in the sacred groves dominated by *Putranjiva roxburghii*, whereas 1/D is highest in *E. officinalis* dominated sacred grove. All above mentioned diversity variables including tree population itself show their lowest values in the sacred grove dominated by *Anogeissus sericea* (Table 4.2). Population of the dominant species is highest in sacred groves dominated by *Acacia catechu*. Similar trend in both number of tree species (TNOSP) and ENS values exists (Fig 4.2) but variation in values is due to effectiveness of the species in terms of their population.

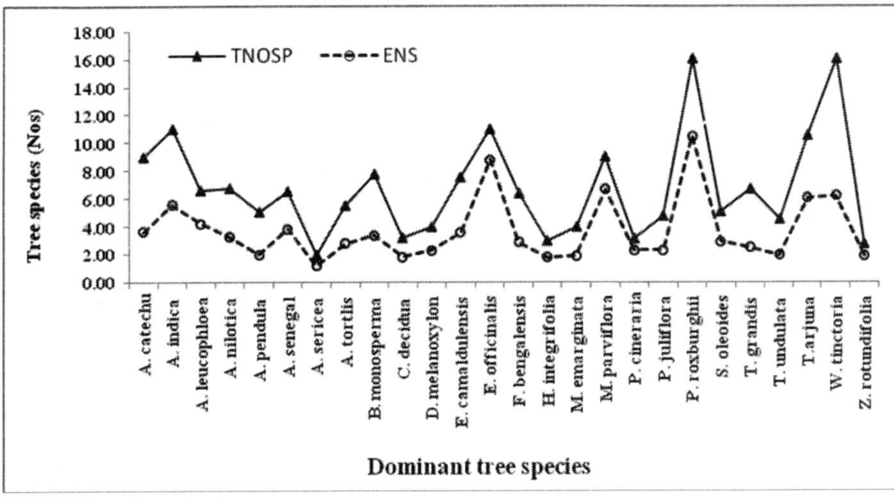

Figure 4.2. Trends tree species richness (nos of species) and ENS for dominant tree species in different sacred groves in Rajasthan.

For trees in categories of 10-30 cm girth, species richness, H' and ENS values are significantly high in the sacred groves dominated by *Wrightia tinctoria*, but the lowest values are observed in the sacred groves dominated by *Z. rotundifolia*. Value of 1/D is found highest in sacred groves dominated by *Holoptelia integrifolia*, whereas the sacred groves dominated by *Tecomella undulata* shows lowest 1/D (Table 4.4). Population of trees of 10-30 cm girth (sub-dominant trees) class is significantly greater and lower in the sacred groves dominated by *Tectona grandis* and *Azadirachta indica*, respectively. Unfortunately there are no trees under 10-30 cm girth class in the sacred groves, where *E. officinalis* is the dominant tree species. Furthermore, average population of trees of girth class 10-30 cm are significantly less as compared to the trees of >30 cm girth class a clear indication of low level of seedling recruitments and its conversion into tree category in most of the sacred groves of Rajasthan.

Table 4.4. Diversity variables (in 1.25 ha) of trees of different girth class in sacred groves dominated by different tree species in Rajasthan. Values are mean±SE of multiple replicates.

SNo.	Dominant tree species	Tree (>30 cm girth)			Tree (10-30 cm girth)		
		H'	ENS	1/D	H'	ENS	1/D
1	*A. catechu*	1.30±0.21	3.74±0.77	2.41±0.40	0.71±0.71	2.57±1.57	2.18±1.18
2	*A. indica*	1.72±0.00	5.57±0.00	3.83±0.00	0.64±0.00	1.89±0.00	1.8±0.00
3	*A. leucophloea*	1.44±0.10	4.30±0.41	3.38±0.38	0.74±0.34	2.64±0.94	2.18±0.94
4	*A. nilotica*	1.18±0.26	3.58±0.85	2.71±0.62	0.95±0.18	2.72±0.51	2.36±0.47
5	*A. pendula*	0.69±0.09	2.18±0.20	1.76±0.15	0.56±0.09	1.95±0.18	1.57±0.17
6	*A. senegal*	1.35±0.31	4.02±1.20	3.08±0.88	0.53±0.53	1.94±0.94	2.00±1.00

SNo.	Dominant tree species	Tree (>30 cm girth)			Tree (10-30 cm girth)		
		H'	ENS	1/D	H'	ENS	1/D
7	*A. sericea*	0.17±0.00	1.19±0.00	1.09±0.00	0.44±0.00	1.55±0.00	1.37±0.00
8	*A. tortilis*	1.03±0.14	2.83±0.40	2.16±0.40	0.87±0.18	2.42±0.42	2.34±0.34
9	*B. monosperma*	1.21±0.39	3.99±1.03	2.95±0.76	0.76±0.32	2.48±0.77	1.91±0.76
10	*C. decidua*	0.61±0.15	2.06±0.32	1.82±0.28	0.34±0.13	1.53±0.21	1.28±0.26
11	*D. melanoxylon*	0.83±0.00	2.30±0.00	1.78±0.00	0.67±0.00	1.95±0.00	1.90±0.00
12	*E. camaldulensis*	1.28±0.64	4.31±2.42	3.08±1.57	0.92±0.66	3.06±1.77	2.44±1.29
13	*E. officinalis*	2.17±0.00	8.75±0.00	7.36±0.00	0.00±0.00	1.00±0.00	0.00+0.00
14	*F. bengalensis*	1.06±0.45	3.47±1.33	2.69±0.79	0.91±0.49	3.06±1.26	3.36±1.69
15	*H. integrifolia*	0.58±0.00	1.78±0.00	1.55±0.00	0.00±0.00	1.00±0.00	1.00±0.00
16	*M. emarginata*	0.65±0.15	1.98±0.33	1.57±0.27	0.17±0.17	1.25±0.25	1.00±0.40
17	*M. parviflora*	1.90±0.11	6.76±0.79	6.80±2.00	1.09±0.55	3.76±1.40	3.82±1.47
18	*P. cineraria*	0.84±0.16	2.49±0.37	2.26±0.34	0.25±0.17	1.41±0.29	0.96±0.42
19	*P. juliflora*	0.84±0.14	2.69±0.41	2.13±0.30	0.56±0.15	2.11±0.35	1.99±0.32
20	*P. roxburghii*	2.35±0.00	10.47±0.00	6.75±0.00	0.90±0.00	2.45±0.00	1.76±0.00
21	*S. oleoides*	1.06±0.10	3.10±0.37	2.05±0.20	0.55±0.18	2.18±0.52	1.93±0.49
22	*T. grandis*	0.92±0.20	2.75±0.53	1.98±0.29	0.61±0.17	1.99±0.36	1.54±0.37
23	*T. undulata*	0.67±0.18	1.99±0.36	1.63±0.36	0.38±0.08	1.47±0.12	1.30±0.11
24	*T.arjuna*	1.80±0.39	6.48±2.37	4.59±1.32	0.60±0.45	2.00±0.83	1.76±0.69
25	*W. tinctoria*	1.82±0.00	6.20±0.00	3.49±0.00	2.08±0.00	8.02±0.00	6.67±0.00
26	*Z. rotundifolia*	0.62±0.32	2.16±0.69	1.64±0.35	0.19±0.19	1.29±0.29	0.95±0.36

4.2 Diversity of tree saplings and shrubs

Sacred groves dominated by *A. indica*, *D. melanoxylon, E. officinalis* and *M. parviflora* trees of >30 cm girth have not shown sapling at the ground surface (i.e., no saplings). Values of sapling population, sapling species richness, diversity index (H'), ENS and 1/D have been observed significantly high in the sacred groves dominated by *P. roxburghii* trees (Table 4.5). Value of H' has been recorded almost zero in the sacred groves dominated by *A. indica, A. nilotica, A. tortilis, D. melnoxylan, E. officinalis, F. bengalensis, H. integrifolia, M. emarginata, M. parviflora, P. cineraria* and *Z. rotundifolia,* where ENS value is 1. This indicates the availability of saplings of only one species. Lowest values of H' and ENS including species evenness are in the sacred groves dominated by *Capparis decidua* (Table 4.5). Sacred groves dominated by *W. tinctoria* show even distribution of sapling species. Lowest numbers of saplings and their species richness are in the sacred groves dominated by *H. integrifolia* and *P. cineraria* tree, respectively. There are no shrubs in the sacred groves dominated by *A. sericea* and *D. melanoxylan* (Table 4.3). In remaining, population of shrubs is significantly high in the sacred grove dominated by *W. tinctoria*, whereas the lowest population of shrubs is recorded in the sacred groves dominated by *T. grandis*. While species richness for the shrubs is highest in the sacred groves dominated by *P. roxburghii*, the lowest species richness for shrub is in the sacred groves dominated by *H. integrifolia* trees. Values of H',

ENS and I/D for shrubs have been observed highest in the sacred groves dominated by *M. emarginata* trees (Table 4.5). Shrub diversity is zero with ENS value of 1, in the sacred groves dominated by *A. sericea, D. melanoxylon* and *H. integrifolia* trees, indicating dominance of a single shrub species in these sacred groves.

Table 4.5. Variations in diversity variables of tree saplings and shrub species in sacred groves dominated by different tree species in Rajasthan. Values are mean±SE of multiple replicates.

SNo.	Dominant tree species	Tree sapling (nos. per 125 m^2)			Shrubs (nos. per 125 m^2)		
		H'	ENS	1/D	H'	ENS	1/D
1	*A. catechu*	0.30±0.30	1.41±0.41	1.35±0.35	0.64±0.64	2.3±1.3	2.15±1.15
2	*A. indica*	0.00±0.00	0.00±0.00	0±0	1.03±0.00	2.80±0.00	2.63±0.00
3	*A. leucophloea*	0.11±0.11	1.15±0.15	0.92±0.26	0.46±0.38	2.29±1.18	1.69±0.86
4	*A. nilotica*	0.00±0.00	0.00±0.00	0.25±0.25	0.99±0.15	2.77±0.43	2.37±0.34
5	*A. pendula*	0.21±0.08	1.35±0.15	0.72±0.15	0.43±0.1	1.76±0.22	1.34±0.23
6	*A. senegal*	0.33±0.33	1.47±0.47	0.94±0.94	0.98±0.45	2.94±1.24	2.57±0.96
7	*A. sericea*	0.65±0.00	1.91±0.00	1.83±0.00	0.00±0.00	0.00±0.00	0.00±0.00
8	*A. tortlis*	0.00±0.00	0.00±0.00	0.50±0.50	0.30±0.3	1.41±0.41	1.34±0.34
9	*B. monosperma*	0.38±0.14	1.51±0.21	1.42±0.20	0.64±0.32	2.21±0.75	1.92±0.63
10	*C. decidua*	0.03±0.03	1.04±0.04	0.48±0.17	0.69±0.17	2.29±0.38	1.94±0.42
11	*D. melanoxylon*	0.00±0.00	0.00±0.00	0.00±0.00	0.00±0.00	0.00±0.00	0.00±0.00
12	*E. camaldulensis*	0.54±0.02	1.72±0.03	1.56±0.04	0.62±0.62	2.21±1.21	1.18±1.18
13	*E. officinalis*	0.00±0.00	0.00±0.00	0.00±0.00	1.05±0.00	2.85±0.00	2.48±0.00
14	*F. bengalensis*	0.00±0.00	0.00±0.00	0.00±0.00	0.58±0.32	1.96±0.58	1.59±0.86
15	*H. integrifolia*	0.00±0.00	0.00±0.00	1.00±0.00	0.00±0.00	0.00±0.00	0.00±0.00
16	*M. emarginata*	0.00±0.00	0.00±0.00	0.25±0.25	1.13±0.39	3.7±1.03	2.91±0.76
17	*M. parviflora*	0.00±0.00	0.00±0.00	0.00±0.00	0.6±0.31	1.98±0.53	1.65±0.35
18	*P. cineraria*	0.00±0.00	0.00±0.00	0.29±0.18	0.36±0.26	1.85±0.67	1.58±0.63
19	*P. juliflora*	0.06±0.04	1.08±0.06	0.87±0.12	0.48±0.14	1.90±0.30	1.59±0.29
20	*P. roxburghii*	1.55±0.00	4.70±0.00	4.01±0.00	1.03±0.00	2.81±0.00	2.4±0.00
21	*S. oleoides*	0.13±0.07	1.17±0.09	0.75±0.19	0.34±0.12	1.55±0.23	1.32±0.26
22	*T. grandis*	0.30±0.21	1.53±0.39	0.94±0.49	0.53±0.21	1.90±0.41	1.60±0.46
23	*T. undulata*	0.35±0.35	1.50±0.50	1.00±1.00	0.56±0.09	1.75±0.15	1.63±0.18
24	*T.arjuna*	0.23±0.23	1.29±0.29	0.69±0.69	0.41±0.41	1.64±0.64	1.48±0.48
25	*W. tinctoria*	1.30±0.00	3.67±0.00	3.38±0.00	0.98±0.00	2.66±0.00	2.49±0.00
26	*Z. rotundifolia*	0.00±0.00	0.00±0.00	0.25±0.25	0.28±0.28	1.52±0.52	1.13±0.52

5. AGRO-CLIMATIC ZONES AND DIVERSITY VARIABLES

All diversity variables for trees of girth class >30 cm and 10-30 cm differ significantly except for the population of latter category. Values of tree population, species richness, H' and ENS for both girth classes have been observed lowest in the sacred groves in IA (Arid western plain) agro-climatic zone of Rajasthan. The highest population is observed in the sacred groves in IIIA (Semi arid Eastern Plain) for trees with girth class >30 cm and in the sacred groves in IVA (Sub humid Southern Plain) agro-climatic zone for tree population of girth class 10-30 cm. In general, diversity indices for both tree girth classes increases with increase in rainfall from arid to dry sub-humid region. This indicates positive impact of rainfall on tree diversity. Species richness for both the girth classes tree have been recorded highest in the sacred groves in IVB (Humid Southern plain) agro-climatic zone. This also shows that groves in relatively high rainfall zone accommodate greater number of tree species as compared to the groves situated in relatively dryer parts of the state. The ENS is also highest in the sacred groves available in IIIB (Flood Prone Eastern Plain) agro-climatic zone. The value of 1/D has also been highest in the sacred groves available in V (Humid Southern Eastern Plain) zone for trees of >30 cm girth class and in the sacred groves available in IIIB for the trees of 10-30 cm girth class (Fig 4.3).

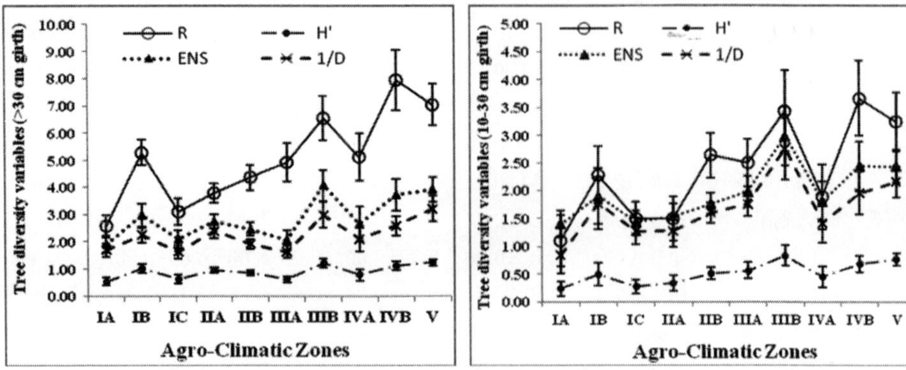

Figure 4.3. Variations in tree diversity in sacred grove situated in different agro-climatic zones of Rajasthan. Error bars are ± 1SE.

Except tree sapling population, all other diversity variables vary significantly among the sacred groves of different agro-climatic zones. Population of tree saplings in the sacred groves has been recorded highest in V agro-climatic zone, whereas the lowest values of 1.25 individuals per 125 m^2 area has been observed in IIIB. The highest and lowest species richness (number of species) for the tree saplings are in the sacred groves of IVA and IIA (Irrigated North Western plain) agro-climatic zones. Shanon-Weiner diversity index (H') and ENS values are highest in the sacred groves in IVB, whereas lowest diverse sacred groves are in IA zone. Simpson's reciprocal diversity index (1/D) is highest in IIB and lowest in IIIB agro-climatic zone (Fig 4.4, left panel).

Variations in shrub diversity variables between agro-climatic zones are not significant (P>0.05) except for shrub population, which is highest in IC (Hyper-arid partly Irrigated Zone) agro-climatic zone and lowest in IVA agro-climatic zone. Species richness of shrub species is lowest in IB (Irrigated North Western plain) agroclimatic zone, whereas lowest number of shrub species is recorded in IIIB (Flood Prone Eastern Plain) zone. Agroclimatic zone IVB shows highest shrub diversity in terms of both H' and ENS value, whereas IIB zone indicates lowest shrub diversity. However, Simpson reciprocal diversity index shows highest shrub diversity in IA zone and lowest shrub diversity in IB zone (Fig 4.4, right panel). Most of these diversity variables follow almost similar pattern though varies in values.

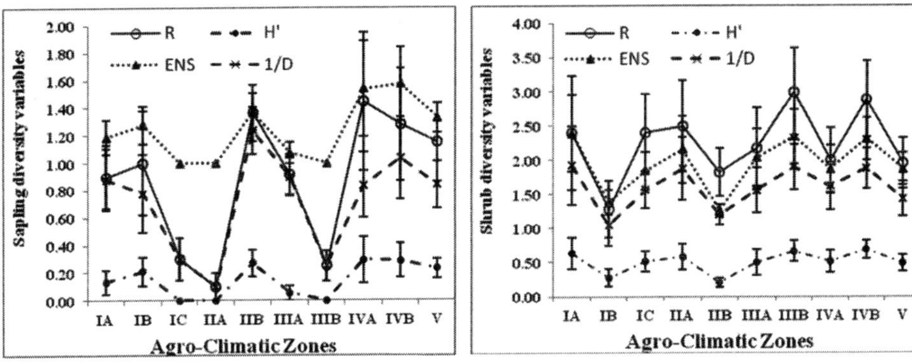

Figure 4.4. Variations in tree sapling (left panel) and shrubs (right panel) diversity in sacred groves situated in different agro-climatic zones of Rajasthan. Error bars are ± SE

6. BASAL AREA OF VEGETATION

Though protected and taken care of by local communities, anthropogenic pressure in the form of grazing, branch removal and even pollarding or tree removal can be seen in many of the groves. Literature indicates a sharp decline in tree density and basal area from the undisturbed (2103 trees ha^{-1} and 26.9 m^2 ha^{-1}) to the moderately disturbed (1268 trees ha^{-1} and 18.6 m^2 ha^{-1}) and finally to the highly disturbed (852 trees ha^{-1} and 7.1 m^2 ha^{-1}) groves, where density–girth curves depict a successive reduction in number of trees in higher girth classes from the undisturbed to the moderately and highly disturbed groves (Mishra et al., 2004). Basal area varies significantly for the studied 123 sacred grove in Rajasthan also, where basal area for trees of >30 cm girth has been recorded highest (35.82 m^2 ha^{-1}) in Baba Rahupeer Dargah, Sri Ganganagar and lowest (0.13 m^2 ha^{-1}) in Pabu Ji Ka Oran, Jodhpur. The basal area of tree of 10-30 cm girth class ranges from 0.91 m^2 ha^{-1} in Dhundhali Mata Ki Banni in Pratapgarh district to 0.01 m^2 ha^{-1} in Bihari Ji Ka Madir Oran in Jhunjhunu district. Though contribution of this category is almost zero in eight sacred groves situated in Barmer, Bhilwara, Jhunjhunu, Jodhpur, Karauli, Nagaur and Udaipur districts, the basal area of tree saplings varies from 2.63 m^2 ha^{-1} in Dev Narain Ji Ka Oran, Kota to almost zero in many of the sacred groves. For shrubs, it varies from 27.07 m^2 ha^{-1} in Hanuman Ji Ka Oran, Alwr to almost zero in Prathurinath Ji Ka Oran, Banswara.

6.1 Spatial variations in basal area

Among the districts of Rajasthan, average basal area for trees of >30 girth category ranges between 0.36 m^2 ha^{-1} in Dungarpur and 13.47 m^2 ha^{-1} in Sri Ganganagar (followed by Jalore with 12.82 m^2 ha^{-1} and Sawai Madhopur with 9.34 m^2 ha^{-1}). For trees of girth class 10-30 cm, the basal area varies from 0.01 in Bhilwara, Bikaner, Jhunjhunu and Rajsamand to 0.44 in Banswara district (Fig 4.5A). This indicates significantly low contribution of trees of this category in total basal area as compared to the tree of >30 cm girth class category. Basal area for tree saplings has been observed almost nil in the arid districts Churu, Jhunjhunu and Sikar. In rest of the districts, the basal area for the seedlings in sacred groves ranges from 0.02 m^2 ha^{-1} in Alwar, Dausa and Sawai Madhopur to 1.20 m^2 ha^{-1} in Baran district. Variation in basal area of shrubs in the sacred groves of different districts is also significant, where it varies from 0.25 in Jodhpur to 11.08 in Barmer district. This indicates that basal area of all components varies significantly among the sacred groves of different districts in Rajasthan.

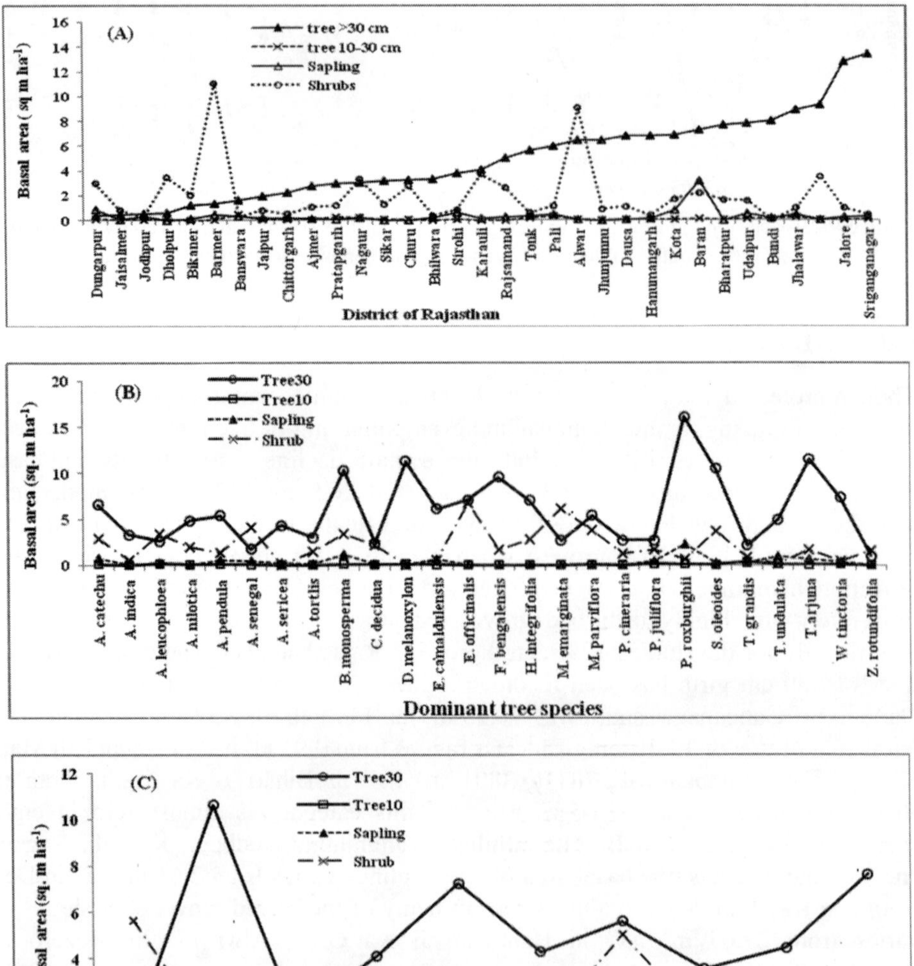

Figure 4.5. Changes in basal area of different plant habits in sacred groves of different district of Rajasthan (A) influenced by dominant tree species (B) and agro-climatic regions (C).

6.2 Effect of tree dominance

For the trees of >30 cm girth class and tree saplings basal area varies significantly both due to dominant tree species and agro-climatic zones, whereas basal area of trees of 10-30 cm girth category and that of shrubs does not show significant variation due to both dominant tree species and agro-climatic zones (Fig 4.5). Basal area due to shrub appears lesser than the basal areas of trees of >30 cm girth category. It is lowest for trees of 10-30 cm girth category in most of the sacred groves dominated by various tree species. Interestingly the pattern between basal area of trees of >30 cm girth class and shrub is almost in reverse in most of the cases, whereas basal area of tree saplings follow the pattern of trees (>30 cm girth). Basal area due to saplings is also greater as compared to that of the trees of 10-30 cm girth category. This indicates reduced recruitment of the sapling to trees of 10-30 cm girth class (Fig 4.5B). Among the dominant tree species, highest (16.14 m^2 ha^{-1}) and lowest (0.99 m^2 ha^{-1}) basal area have been recorded in the sacred groves dominated by *P. roxburghii* and *Z. rotundifolia*, respectively. For trees of 10-30 cm girth, the corresponding basal area is 0.39 m^2 ha^{-1} under *T. grandis* and almost zero under *E. officinalis*. Basal area of tree saplings (ranging from 0.01 to 2.30 m^2 ha^{-1}) is observed highest under *P. roxburghii* and almost zero under *A. indica*, *E. officinalis*, *D. melanoxylan*, *F. bengalensis* and *M. parviflora*. Range of basal area of shrubs is 6.12 m^2 ha^{-1} under *M. emarginata* and almost zero under *Anogeissus sericea* and *D. melanoxylan*.

6.3 Effects of agroclimatic zones

The highest and lowest basal areas of trees >30 cm girth class have been 10.62 m^2 ha^{-1} and 0.92 m^2 ha^{-1} in IB and IA zone, respectively, whereas the corresponding values for trees of 10-30 cm girth class are 0.17 m^2 ha^{-1} and 0.02 m^2 ha^{-1} in IVA and IIB. For tree saplings highest basal area is recorded in V zone and lowest in IIIB zone (Fig 4.5C). Shrubs indicate highest basal area in IA and lowest basal area in IB agro-climatic zone. These results reveal low tree density and basal area in the sacred groves of Rajasthan as compared to the reported values of tree density and basal area of 379.5-750.0 individuals ha^{-1} and 42.9-100.4 m^2 ha^{-1} in western Ghat (Bharati and Devi Prasad, 2015); 999 stems ha^{-1} and 31.7 m^2 ha^{-1} in sacred forests of Chittoor, Andhra Pradesh (Rao et al., 2011), 387 stems ha^{-1} and 23.62 m^2 ha^{-1} in the Kaan forests of Shimoga, and 384-566 stems ha^{-1} and 46.3-99.7 m^2 ha^{-1} in Valparai, Annamalai hills, Tamil Nadu. The most important factors for this variability appear to be climatic and anthropogenic.

7. ANIMAL DIVERSITY

Sacred groves are not only store house of floral diversity but they also harbour many animal species. Among the species harboured by a set of small groves shows variability among the species, which are endemic to endangered in nature. The studies on sacred groves relating to faunal diversity includes bird diversity (Kangah-Kesse et al., 2007), primates (Baker et al., 2009), butterflies (Gaude and Janar-thanam, 2015), arthropods (Parihar, 2015) and surface dwelling faunas (Imam et al., 2016). Thus smaller groves should be given as much importance as larger groves while designing conservation strategies (Padmanabhan, 2005; Chandrakar et al., 2014).

Reports are available on ground dwelling insects from 53 sacred groves of Purulia district (Imam et al., 2016). The 33 species of butterfly belonging to 31 genera accounts for about 13% of the species have been recorded from Goa (Gaude and Janarthanam, 2015). Similarly 13 faunal species have been reported from relatively smaller sacred groves in Balasore, Odisa (Mohanta et al., 2012). Besides, several species of honey-bees are also nest in large trees like *Ficus religiosa, H. integrifolia, A. indica* etc., in sacred groves. The communities have their faith and traditions linked to these rich biodiversity spots in the landscape. A wide range of animals are also observed in the sacred groves of Rajasthan. Some of the sacred groves have been observed very much important in terms of wild life, particularly for birds (Annexure V). Some of the wild animals associated with these sacred groves are:

7.1 Birds

Sacred groves provide habitat, water and nest-sites for wildlife and birds. Khajjur (*Phoenix* spp) and other groves are some important roosting places for bird including weaverbird (*Ploceus philippinus*) (Parihar, 2015). Several cavity-nesting birds excavate their nest in these trees. Important birds observed or reported from the sacred groves in Rajasthan are: Indian sand goose (*Pterocles exustus*), grey quil (*Coturnix coturnix*), spotted owlet (*Athene brama*), spotted dove (*Streptopelia decocto*), common babbler (*Turdoides caudatus*), red-vented bulbul (*Picnonotus cafer*), grey shrike (*Lanius exubitor*), grey partridge (*Francolinus pondicerianus*), white vulture (*Gyps bengalensis*), koel (*Eudynamys scolopaceus*), sun-bird (*Leptocoma zeylonica*), sparrow (*Passer domesticus*), myna (*Acridotheres tristis*), parrot (*Psittacula krameri*), crow (*Corvus splendens*), eagle (*Aquila hastate*), pigeon (*Columba livia*), pea fowl (*Pavo cristatus*) and Alexandrine parakeet (*Psittacula eupatria eupatria*) (Plate 4.2).

7.2 Mammals

Sacred groves are important habitats not only for the wild mammals but also for the domestic animals, whose survival depends on this important community resource. Parihar (2015) has recorded about 9 species of mammals in sacred groves (Orans) in western, Rajasthan. Various mammals recorded in the sacred groves of Rajasthan include chinkara (*Gazella gazella*), blue bull (*Boselaphus tragocamelus*), desert fox (Vulpes vulpes pusilla), Indian fox (*Vulpes bengalehsis*), mongoose (*Herpestes edwardsi*), jackal (*Canis aureus auveus*), desert cat (*Felis silvetri oranata*), hedghog (*Hemiechinus auritus callaris*), porcupine (*Hystrix indica indica*), five-striped (*Funambulus pennati*), Indian desert gerbil (*Meriones hurrianae*), Indian gerbil (*Tetera indica*), desert hare (*Lepus nigri collis*), fruit bats (*Aproteles bulmerae* and *Pteropus giganteus*), cow (*Bos taurus*), hyena (*Hyaena hyaena*), rat (*Rattus norvegicus*), deer (*Axis axis*), wild cat (*Felis chaus*), squirrel (*Funambulus palmarum*), wild pig (*Sus scrofa*), wolf (*Canis lupus*), Samber deer (*Rusa unicolor*), wild dog (*Cuon alpinus*) and bear (*Melursus ursinus*).

7.3 Reptiles

Besides birds and mammals, a number of reptiles are also visible in the sacred groves (Bhatt et al., 2012; Parihar, 2015). Common reptiles present in the sacred groves of Rajasthan include house gecko (*Hemidactylus flarviviridis*), agma (*Agma agilis*), viper (*Echis carinatus*), earth snake (*Eryx johnii*), varanus-pata goh (*Varanus bengalensis*), sand fish (*Ophiomorus tridactylus*), ajgar or python (Python *molurus molurus*), king kobara (*Ophiophagus hannah*), chameleon (*Chamaeleo zeylanicus*), varanus-chandan goh (*Varanus griseus*) and Indian spiny-tailed (*Uromastix hardwickii*).

Many of these animal species are associated with some of the important plant species like *Ficus benghalensis*, which has been observed keystone species in many sacred groves because of supporting many birds and insects also. *Ficus benghalensis* provides a niche for the large number of birds and plants (King et al., 1997). In addition to these animals, several honey combs are also present in trees like *Ficus benghalensis*, *Ficus religiosa*, *Holopetelia integrifolia, Azadirachta indica* etc.

8. SPECIES OF IUCN LIST

Many species available in the sacred grove are of rare and threatened categories because of different types of threats (Iyer, 2010; Joshi and Shringi, 2014). Some of the species found in the sacred grove are of have been categorized in different categories of IUCN list (Table 4.6).

Table 4.6 Red lists of some of the species found in the sacred groves of Rajasthan.

SNo.	Species	Local name	Family	Habit	Red List category
1	*Acacia catechu* (L.f) Willd.	Khair	Mimosaceae	Tree	Invulnerable
2	*Adina cordifolia* (Willd) ex Roxb.	Haldu	Rubiaceae	Tree	Vulnerable
3	*Ailanthes execlsa* Roxb.	Ardu	Simaroubaceae	Tree	Vulnerable
4	*Alangium salvifolium* L.f.Wang.	Ankol	Alangiaceae	Tree	Rare
5	*Alysicarpus vaginalis* (L.)D.C.	----	Fabaceae	Herb	Invulnerable
6	*Boswellia serrata* Roxb.	Salar	Burseraceae	Tree	Rare
7	*Celastrus paniculata* Willd.	Malkagni	Clastraceae	Climber	Rare
8	*Cordia dichotoma* Forst.	Lasora	Ehreteaceae	Tree	Vulnerable
9	*Crateva nervosa* DC.	Varana	Capparaceae	Tree	Rare
10	*Dalbergia latifolia* Roxb.	Safed Shisham	Fabaceae	Tree	Invulnerable
11	*Feronia limonia* L.	Kaith	Rutaceae	Tree	Invulnerable
12	*Gloriosa superba* L.	Kalihari	Liliaceae	Climber	Endangered
13	*Mallotus philippensis* Lam.	Sindhuria	Euphorbiaceae	Tree	Rare
14	*Manilkara hexandra* Roxb.Dub	Khirani	Sapotaceae	Tree	Invulnerable
15	*Melhania futtetyporensis* Munro	-	Sterculeaceae	Shrub	Rare
16	*Mimosa hamata* Willd.	Bander ki roti	Mimosaceae	Shrub	Invulnerable
17	*Morinda tomentosa* Heyne.	Aal	Rubiaceae	Tree	Vulnerable
18	*Nyctanthes arbortristis* L.	Harsingar	Nyctagenaceae	Tree	Vulnerable

SNo.	Species	Local name	Family	Habit	Red List category
19	*Sarcostemma viminale* L.	Sambher bel	Asclepiadaceae	Climber	Endangered
20	*Soyamida febrifuga*	Rohin	Meliaceae	Tree	Rare
21	*Tecomella undulata*	Rohida	Binoniaceae	Tree	Rare
22	*Terminalia alata* Heyne. Ex Roth.	Sadad	Combretaceae	Tree	Rare
23	*Terminalia bellerica* Gaertn.	Baheda	Combretaceae	Tree	Invulnerable
24	*Wrightia tinctoria* R.Br.	Dhudhi	Apocynaceae	Tree	Invulnerable

9. STATISTICAL RELATIONS

Population, species richness (S), Shanon-Winer diversity index (H'), effective number of species (ENS) and Reciprocal Simpson diversity index (I/D) of both trees of >30 cm girth and 10-30 cm girth have been observed positively related to both long term average rainfall (RF1) as well as rainfall averaged for 2003-2012 (RF2). For tree saplings and shrubs only the diversity ($r=0.252$, $P<0.05$, $n=123$) and population ($r=-0.218$, $P<0.05$) show significant correlations with RF2, indicating the positive effect of rainfall of 2003-2012 period on tree saplings and negative effects on shrub population (Fig. 4.6). The relationship of rainfall of 2003-2012 appears more strong in case of trees of >30 cm girth and relatively weaker in case of trees of 10-30 cm girth. Rainfall of 2003-2012 also shows better relationship with tree saplings S, H' and ENS and basal area of trees of 10-30 cm girth as compared to the long term average rainfall as has been observed in other study (Klopfenstein et al., 2015). Thus rainfall of latter period play a significant role in adding the number of trees in 10-30 cm girth category and further to >30 cm girth category.

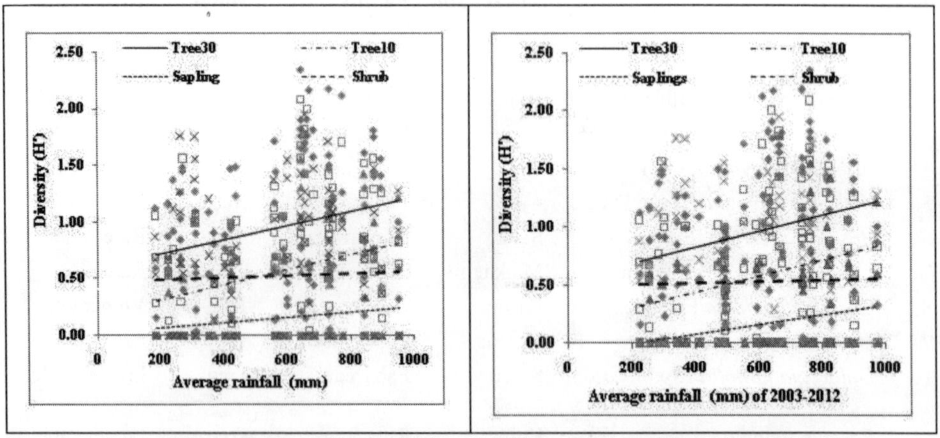

Figure 4.6. Trend lines for diversity variables of different plant habits and rainfall of long term average (left) and 2003-2012 average (right). Tree30: trees of girth class >30 cm at 1.37m height; tree10: trees of 10-30 cm girth; sapling: plants of 3-10 cm girth.

Species richness, H' and ENS of trees of >30 cm girth have also shown positive correlations to the same parameters of the trees of 10-30 cm girth category. ENS and S values of both tree categories exhibit beneficial effects on the number of species as well as diversity of trees saplings (Table 4.7). But population of trees of both girth categories does not show any effects on the tree sapling population and diversity. There is a clear link between biodiversity of native primary forests and water availability. Preservation and retention of native forests as sacred groves dominated by relic species appears beneficial in the form of better status of soil moisture and ground water. Thus it needs to be protected as integral to the cultures of the region (Koradia, 2000).

Table 4.7. Correlation coefficients indicating relationships between different plant, topography and soil variables

	Variable	RF1	RF2	Altitude	Tree H' (>30 cm)	Tree H' (10- 30cm)	Shrub H'	Sapling H'
Tree >30 cm girth	Population	0.234	0.303	NS	NS	0.181	NS	NS
	Richness	0.390	0.413	0.315	0.806	0.474	0.238	NS
	H'	0.244	0.244	NS	-	0.495	0.195	NS
	ENS	0.250	0.243	0.212	0.950	0.472	0.185	NS
	1/D	0.216	0.167	NS	NS	0.324	NS	NS
	Basal area	NS	NS	NS	0.842	0.379	0.220	NS
Trees 10-30 cm girth	Population	0.229	0.242	NS	-	0.205	NS	-
	Richness	0.286	0.281	0.273	0.485	0.912	0.281	0.218
	H'	0.265	0.249	0.228	0.474	-	0.265	0.219
	ENS	0.39	0.225	0.291	0.500	0.951	0.261	0.219
	1/D	0.234	0.202	0.227	0.174	0.445	NS	NS
	Basal area	0.239	0.244	0.274	0.478	0.923	0.209	0.184
Sapling	Population	NS	NS	NS	NS	-	NS	0.565
	Richness	NS	0.220	0.219	NS	0.218	NS	0.897
	H'	NS	0.253	NS	NS	0.219	NS	-
	ENS	NS	0.256	NS	NS	0.219	NS	0.971
	1/D	NS	NS	NS	NS	NS	NS	NS
	Basal area	NS	NS	NS	NS	0.184	NS	0.743
Shrub	Population	-0.218	-0.177	NS	NS	NS	0.561	NS
	Richness	NS	NS	NS	0.238	0.281	0.919	NS
	H'	NS	NS	NS	0.195	0.265	-	NS
	ENS	NS	NS	NS	0.185	0.261	0.964	NS
	1/D	NS	NS	NS	NS	NS	0.338	NS
	Basal area	-0.188	NS	NS	0.220	0.209	0.938	NS
Altitude		-0.335	0.354	-	NS	0.228	NS	NS

Only significant values are provided. NS-not-significant (P>0.05)

10. CONCLUSIONS

Sacred groves of Rajasthan vary widely in their floral and faunal diversity. These sacred groves function as important refugia for many plant and animal species. Under anthropogenic pressure or when the area of the grove is small, a strong selection and hierarchy are taking place among the species available in the groves. There is a great variation in species from grove to grove. In the 123 studied sacred groves, there are 26 types of tree species observed dominating in the area depending upon adaptability to the particular locality and climatic conditions. The local variations reflect variability due to edaphic and environmental conditions, topographic differences, climatic variations, degrees of human interference and elements of climax vegetation. However, dominant species in a particular sacred grove influence many key processes of ecological importance.

A total number of 131 woody species belonging to 48 families have been identified from the studies of 123 sacred groves of Rajasthan. Dominant species with decreasing number of sacred groves are in following order: *A. pendula*>*P. juliflora*> *Salvadora oleoides*> *Capparis deciduas*>*P. cineraria*>*Tectona grandis*> *Acacia leucophloea*>*Acacia nilotica*>*Butea monosperma*>*Zyziphus rotundifolia.* It has been observed that under condition of stress, single species dominance takes place in the community. Rapid changes in the plant communities are also happening due to changes in land-use pattern. In addition, *Gloriosa superba* and *Asparagus racemosus* have also been found in some of the groves, which are medicinally important tuber-bearing climbers and appears to be threatened. Uprooting these threatened plants for medicinal uses will make them disappear from the sacred groves. Sometimes the biomass in groves is lesser than that of adjoining vegetation, particularly in western of Aravalli, it is attributable to the decline in the number of climax species. However, it is higher in eastern Aravalli. The loss of climax species can be expected as a result of the dominance of certain secondary species, affecting the regeneration of the climax species.

Sacred groves dominated by *P. roxburghii*, *W. tinctoria*, *E. officinalis*, *B. monosperma*, *M. parviflora*, *A. indica*, *A. catechu* and *Tectona grandis* indicate higher vitality in terms of diversity and regeneration status in the sacred groves and can be promoted under restoration of the groves. In arid region *P. cineraria* and *Salvadora* spp. are effective in maintaining diversity of the groves. Likewise sacred groves of IB, IIB, IIIB and IVB agro-climatic zones show better floral diversity and thus ecological status. However, it is clear that with the decrease in soil water availability (increase in dryness) and disturbance there is corresponding decrease in the endemic elements of vegetation leading to invasion by the other species. Further the sacred groves available in west of Aravalli are relatively less diverse in nature and lower in soil nutrients as compared to that of the east Aravalli regions and the groves available in the east of Aravalli. At least 19 birds, 25 mammals and 10 reptiles including the Alexandrine parakeet have been observed or said visited these sacred groves. For example Kalu Das ji Ki Dungari, Dausa is full of Alexandrine parakeets and could be a bird watching place, where almost each tree of *Anogeissus pendula* has 1-3 numbers of holes as residence of these parakeets. While developing strategies for restoration of degraded groves, consideration should be given to the species selection supporting wild animals also.

Significantly low values of diversity and the basal area for the trees of 10-30 cm girth even lower than the tree saplings indicate lower rate of natural regeneration and recruitment of plants from saplings to trees of 10-30 cm girth in most of the sacred groves. This needs to be taken into consideration to improve conversion of seedlings of indigenous species to saplings and the tree status. Further, in most of the sacred groves the climax species particularly *Anogeissus pendula* is degenerating because of very poor regeneration and seedling recruitment.

Pandanus odoratissimus

Terminalia arjuna

Randia domentorum

Crateva nurvala

Aegle marmelos

Ephedra foliata on Opuntia

Trema orientalis-wild fruit for birds

Plate 4.1. Some trees and shrubs species occurring in sacred groves.

Plate 4.2. Kalu Das Ki Dungari sacred grove, Dausa, Alexandrine parakeet and holes in *A. pendula* trees (left panels) and Hanuman Ji ka Oran, Alwar, a group of birds (Red ring parakeet, Ring Dove, Pigeon and House Sparrow) and an old tree of *Cordia myxa* (right panels).

5

SOIL STATUS AND SOIL CARBON STOCK OF SACRED GROVES

A well conserved, old stand of forests including the sacred groves provide multiple benefits by conserving biodiversity and regulating climate change effects. Additional significance of sacred groves is probable carbon storage. Soil physico-chemical properties, nutritional status and soil carbon stock depends upon parent material of the soil, climatic conditions, topography and types and composition of vegetation or forests. Different components of forests like old-grown trees, natural regeneration and under-storey species, mycorrhizae or free-living fungi and bacteria also influence soil properties and nutritional characteristics. This chapter describes soil physico-chemical properties, i.e. soil pH, electrical conductivity, soil bulk density, soil organic carbon, soil carbon stock (density) in 0-30 cm soil layer and their altitudinal and spatial variations. Most of the soil variables increase with increase in altitude except soil pH and electrical conductivity, which indicates a decreasing trend and no relation with altitude, respectively. Soil carbon stock in top 0-30 cm soil layer varies between 52.24 tons ha^{-1} in Devnarayan Ji Ka Oran in Chittorgarh and 1.51 tons ha^{-1} in Pabu Ji Ka Oran in Jodhpur district. Alkaline soil conditions, low nutrient status and low soil organic carbon and carbon stock are indicative of degradation of these sacred groves under varying human and livestock pressures. The studies related to sacred groves are taking momentum because of probable benefits under REDD+. Thus based on carbon accounting incentives of different forms may encourage the communities to maintain natural regeneration payments for climate services and biodiversity conservation of these sacred groves.

1. BACKGROUND

The soil quality parameters vary across the landscapes based on the variation in parent material, climatic conditions, topography and different types of vegetation or forest under different species of trees (Yousefi et al., 2008; Ayres et al., 2009). Fire regime, frequency of insect outbreaks, changes in precipitation patterns and land-use change alter the relative abundance of tree species, which in turn may alter the soils (Silva et al., 2013). Varying composition and productivity of forests among sites also influence soil properties. Forest ecosystems with low soil nutrient availability are characterized by competitive effects of the growth-limiting resource between several players, i.e. various components of vegetation, like old-growth trees, natural regeneration and under storey species, mycorrhizal fungi, free-living fungi and bacteria. Plant and animal residues derived soil organic matter is the food source of soil organisms, playing an important role in breaking down nutrients into a form which can easily be taken up by the plants and help regulate carbon and nitrogen cycles. Soil organic carbon (SOC) is important for all three aspects of soil fertility, i.e. chemical, physical and biological. Traditionally, conserved forests or sacred groves serve as a natural habitat for many endemic, rare, primitive and economically valuable organisms representing micro-level biodiversity. For example, native soils of the pristine sacred groves of Meghalaya, India reveal dominance of Bacillus, Paenibacillus, Lysinibacillus and Viridibacillus that show plant growth promoting traits including phosphate solubilization, 1-aminocyclopropane-1-carboxylate deaminase and catalase activity, production of auxin, siderophores, hydrogen cyanide and ammonia. Further, bacteria native to the pristine niche of sacred groves indicate better plant growth promoting activities than the isolates from disturbed forest (Lyngwi and Joshi, 2015; Lyngwi et al., 2015).

The sacred forests have additional benefits in terms of probable carbon stores. Perennial grasses, shrubs and trees constantly maintain and develop their roots and woody biomass that help increasing the associated carbon in addition to providing vegetative cover to the soils. In dry areas where soil moisture restricts plant growth, the characters enabling the survival and growth of the plants under the low precipitation and high temperature regimes through opportunistic strategies (Ogle and Reynolds, 2004) and govern carbon distribution in the soil profile, influence carbon storage by way of primary productivity and soil carbon stabilization (Schenk and Jackson, 2002). A study in western Rajasthan (Singh, 2015b) finds spatial variations in soil organic carbon influenced by rainfall, edaphic factors and vegetation status, showing lowest SOC in Baitu block of Barmer and highest value in Aburoad block of Sirohi district (Fig 5.1A). It has been also influenced by land use changes, where SOC is highest in forest lands and lowest along the road side (Singh, 2015b) Carbon is also available in the form of inorganic carbon, e.g. lime or carbonates in the soils in the drier areas. However, land use and management typically have a larger impact on soil organic carbon. Being an important carbon pool on the Earth's surface, increasing the global storage of carbon in soils is one important method to reduce atmospheric carbon dioxide.

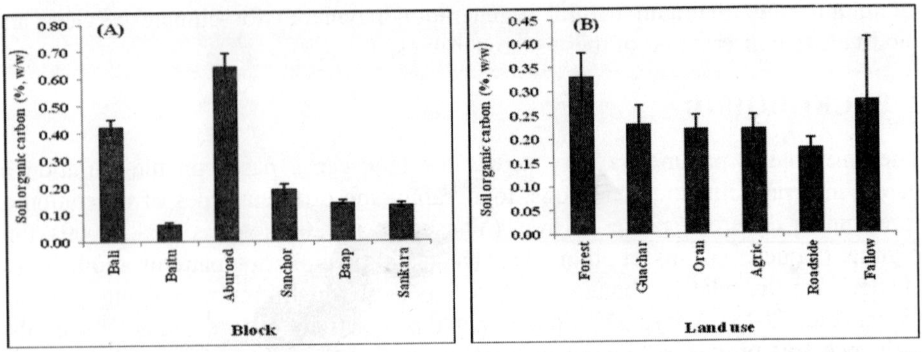

Figure 5.1. Changes in per cent soil organic carbon under spatial (A) and land uses (B) changes in western Rajasthan. Error bars are ±1SE of multiple replications. Source: Singh (2015).

1.1 Protecting natural habitat

About 4 billion hectares of forests and 5 billion hectares of natural grasslands, globally, are a massive reservoir of carbon both in vegetation above ground and in root systems below ground. World's forests store about 300 billion tons of carbon in the living parts and are absorbing as much as 30% (2 peta grams carbon year^{-1}) of annual global anthropogenic CO_2 emissions. Such accumulated carbon is roughly 30 times the annual amount of emissions created by burning of the fossil fuels. As forests grow, carbon accumulates in woody tissues and soil organic matter. Most of the literatures indicate that the net rate of carbon uptake is greatest when forests are young, and slows down with time, because old forests sequester carbon for a long time but provide essentially no net uptake. A well-stocked forest typically sequesters carbon at a maximum rate between age 10 and 20–30. At an age of 30 years, about 200 to 520 tons CO_2-equivalent are sequestered per hectare in forests with productivity ranging from low to high (Australian Green House Office, 2001). After this age, rate of sequestration slows gradually until maturity is attained at about 80 to 100+ years and flattens out from then because the growth is balanced by decay of the tree. Though debatable a nature report indicates that mass growth rate increases continuously with tree size for most of the species. Thus large and old trees do not act simply as senescent carbon reservoirs but actively fix large amounts of carbon compared to the smaller trees (Stephenson et al., 2014). Further, the accelerated production of dead wood or coarse woody debris and the attendant maintenance of the growing space for live carbon accumulation also expand the carbon sink in old forests (McGarvey et al., 2015).

Conserving old forests serves many political objectives including protection of biodiversity and climate mitigation by conserving forest carbon stock and avoiding GHGs emission to the atmosphere. The techniques that increase both the amount of wood produced and the carbon stock retained in the forest need prioritization. Replacing dying or low-productivity stands, protecting young sprouts from damage after harvest, planting tree mixes that are more resilient, and optimizing fertilizer use and tree growth by adding nitrogen-fixing species in afforestation programmes will all contribute to climate-change mitigation. Effective forest management like fertilization and use of N-fixing vegetation leads to enhancement in soil carbon and nitrogen. Thus, conservation of old stands including the sacred groves provide multiple benefits by conserving biodiversity and regulating climate change effects. Different forms of incentives may also encourage the local communities to maintain natural regeneration payments for climate services, biodiversity conservation and forest fire control (Johnson and Coburn, 2010).

2. SOIL CHARACTERISTICS

Many studies demonstrate the effects of environmental conditions on soil characteristics in forest ecosystems (Ashagrie et al., 2007; Christenson et al., 2009; Li et al., 2014), shrubland (An et al., 2010), cultivated stands (Gartzia-Bengoetxea et al., 2009), crop production systems (Haynes, 1999; Huang et al., 2010; Beedy et al., 2010), land use change (Bettina et al., 2005; Leifeld and Kögel, 2005) and overgrazed pasture lands (Fallahzadeand Hajabbasi, 2011). Sacred groves and less disturbed non-sacred groves generally show high porosity and lesser values of soil bulk density, which are inversely related in most of the cases except in areas of extreme clayey soil content (Chandy et al., 2014). The soils of the sacred groves with sandy loam texture are found to be ranging in different soil properties like pH from 4.5 to 5.6, total nitrogen from 0.43 to 0.96%, available phosphorus from 0.01 to 0.07% and exchangeable potassium from 0.01 to 0.05%, whereas soil moisture content ranges from 30 to 72% in these grove in north east region in India. Relatively high organic matter content reduces soil bulk density and increases porosity in the soils and thus soil fertility. Likewise, high values of electrical conductivity indicate the presence of calcium, magnesium, chloride and sulphate ions in the soils that depend upon the parent material from which the soils are derived (Chandy et al., 2014).

2.1 Soil physico-chemical properties

Average altitude of the sacred groves studied is 338 m from mean sea level (msl). It ranges from 105 m msl in Khuvas Mataji Ka Oran in Barmer to 880 m msl in Ubeshwarji Oran in Udaipur. Gravel content varies significantly among the sites. It has been observed highest (i.e., 85.2%) in Nahar Singhji Mata ka Oran in Chittorgarh district. Eleven sacred groves have been found to have no gravel content and appear to be better in soil conditions. In the rests, it is lowest in Karni Mata Ji Ka Oran, Bikaner. The average value of gravel content is 29.5% across the study sites. It shows an increasing trend with increase in altitude. Soil bulk density ranges from 1.25 g cm^{-3} in Devnarayanji ka Oran, Chittorgarh district to 1.71 g cm^{-3} in Pabujika Oran, Jodhpur district. Average value of soil bulk density is 1.49 g cm^{-3} for the state. Almost 50% sacred groves indicate above average soil bulk density is probably influenced by compaction or erosion. A large spatial variability in soil bulk density appears to be due to varying disturbance regime and soil compaction in the grove as soil bulk density increases with increase in grove disturbance. The forest interiors with lowest disturbances shows low soil bulk density, whereas forest clearing particularly in exteriors indicates greater soil bulk density indicating greater soil compaction (Cardelús et al., 2013). Greater soil compaction due to cattle grazing often correlates with lower regeneration because roots have more trouble in penetrating the compacted soils (Trouse and Humbert, 1961; Willatt and Pullar, 1984; Skinner et al., 2009). The high light demanding species are more successful along the forest edge, whereas more shade tolerant species shows greater success in the interior of the forests.

Soil pH of the sacred groves has been observed slightly alkaline in reaction with an average value of 7.48 across the sites. Among these, soils of about 37.4% sites show slightly acidic behaviour. Soil pH has been recorded highest (8.74) in Gogaji Ka Oran in Bikaner district and lowest in Muchkund Oran in Dholpur district. This is probably due to variations in rainfall, concentrations of basic cations in parent material and soil organic carbon influencing soil pH and electrical conductivity. The highest electrical conductivity (EC)- a measure of soil salt content, has been observed 1.94 dSm^{-1} in Musain Mataji Ka Oran in Baran district, whereas the lowest value of EC is found in Gogaji Ka Oran (0.02) in Churu district. There are no variations in soil EC with altitude (Fig 5.2A).

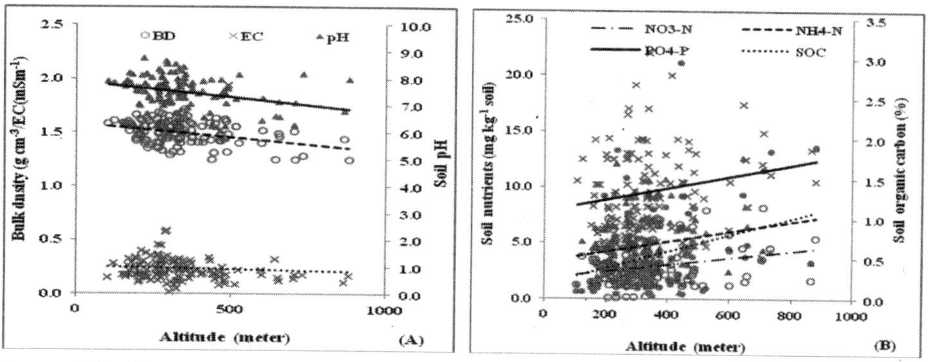

Figure 5.2. Change pattern in soil variables of sacred groves of Rajasthan with varying altitude.

Both soil bulk density and soil pH shows decreasing trends with increase in altitude. Average value of EC across the sites is 0.23 dSm^{-1}, which indicates normal soil condition. Available salts in parent materials from which the soils are derived play important role in governing soil physico-chemical properties. However, they are also governed by relationships between factor and the vegetation types. For example, relations of *Tamarix* spp. and *Seidlitzia rosmarinus* have been observed with EC and soluble minerals, whereas *Artemisia sieberi* has been found related to the soil texture and that of *Astragalus* spp. and *Stipabarbata*found related to soil lime concentration (Yousefi et al., 2008). The decrease in soil pH has been observed associated with decrease in soil bulk density and increase in SOC and soil available NH_4-N, NO_3-N and PO_4-P with increase in altitude. This appears to be due to alkaline nature of the soils. However, there are reports which show a decrease in soil pH, associated with lower availability of plant nutrient, particularly Ca^{2+}, Mg^{2+}, NH_4^+ and precipitation and loss of HPO_4^{2-}, which is also linked with higher soil bulk density (Cardelús et al., 2013). Studies indicate that increase in soil bulk density and soil pH across the gradient are also associated with decrease in carbon and nitrogen, which indicates an increasing trends in soil degradation. Though degradation in sacred groves is throughout but it is relatively greater in exterior region and in lower altitude as compared to those in interior and at higher altitude areas because of grazing in the surrounding areas of the groves (Cardelús et al., 2013).

2.2 Soil nutrients

Soil nutrients availability is very much related to their utilization by diverse plant species. Increased phosphorus availability promotes the coexistence of more tree species within a community and reduces genetic diversity of some of the species because of reduced average niche breadth (Xu et al., 2016). Forest ecosystems with low availability of soil nutrients are characterized by competitive effects of the growth-limiting resource among several factors like different components of vegetation, i.e. old-growth trees, natural regeneration and under storey species and microbial strains. Soil available nitrate nitrogen (NO_3-N), ammonium nitrogen (NH_4-N), and phosphate phosphorus (PO_4-P) varies widely among the sacred groves studied in Rajasthan, India. All these nutrients increase with altitude of the sacred groves along with soil organic carbon (Fig 5.2B). Available NO_3-N has been observed highest (8.20 mg kg^{-1} soil) in the soils of Kundeshwar Mahadev Ka Oran in Udaipur district, whereas the lowest concentration (0.09 mg kg^{-1}) has been observed in the soil of Viratra Mata Ji ka Oran, Barmer. Concentration of NH_4-N is observed highest (13.55 mg kg^{-1}) and lowest (0.54 mg kg^{-1}) in the soils of Ubheshwarji Oran in Udaipur district and Devnarayanjika Oran in Chittorgarh district, respectively. The concentration of PO_4-P has been recorded maximum in the soil of Sidheswar Mahadeoji Ka Mandir (22.00 mg kg^{-1}) in Sirohi district and its lowest concentration has been observed in the soil of Devnarayanji ka Oran (2.32 mg kg^{-1} soil) in Kota district. This indicates importance of sacred groves in maintaining soil fertility (Singh and Garg, 2014). However, availability of these soil nutrients is very much positively related to soil organic carbon (Fig 5.3). Increased climatic adversities like low rainfall, high temperature and sandy soils in western Rajasthan affect soil nutrients availability and soil organic carbon negatively. The average values of NO_3-N, NH_4-N and PO_4-P across all sacred groves studied have been recorded 2.94 mg kg^{-1}, 4.76 mg kg^{-1} and 9.65 mg kg^{-1}, respectively. The high variability in soil nutrients of different groves are not only the effects of climatic differences but also the human interferences in the form of removal and high intensity grazing resulting in trampling and compaction of the soils. Increased grazing intensity generally shows depletion in soil organic carbon (SOC). Changes in soil has been linked to increased grazing intensity by lowering of the carbon: nitrogen ratio, an increase in soil bulk density, a decrease in soil moisture content, and transfer of phosphorus from less intensively grazed areas toward animal night pens (Yuan and Hou, 2015). Study of Sahu et al. (2012) also indicates the influence of species on soil nutrients and properties, where species diversity is observed to be positively correlated with organic carbon and phosphorus, and is negatively related to soil nitrogen, EC and pH.

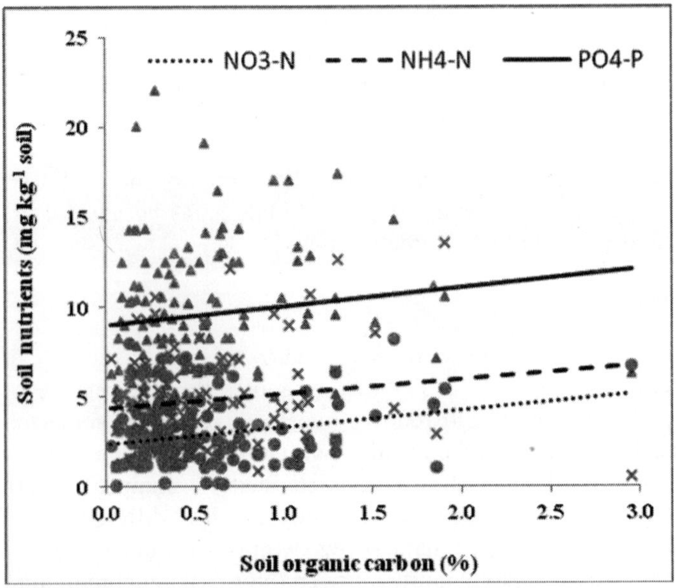

Figure 5.3. Soil available NO_3-N, NH_4-N and PO_4-P in the soil of sacred groves of Rajasthan increase with increase in soil organic carbon.

Altitudinal variations in different soil chemical and physical properties have also been observed in other studies, where greater values of organic matter and cation exchange capacity (CEC) are at higher altitude and total K is at the lower altitude. Soil pH, total P, K, CEC and slope also affect diversity, where decrease in soil pH and total P increase the Shannon index, and increase in CEC and slope enhance species richness of the herbs.

2.3 Soil organic carbon and density

The actual amount of SOC present in soil depends on a number of factors like rainfall, air temperature, vegetations, soil types, topography etc. (Singh, 2005; Singh, 2011; Tripathi et al., 2013). Increasing the SOC pool is a major challenge, especially in developing countries, where crop productions are low, climate is harsh, and water and nutrients are limited. For example, Follett et al. (2005) have observed a sequestration rate of 1.0 to 1.9 Mg C ha^{-1} yr^{-1} in an irrigated Vertisol in Central Mexico, where aboveground crop residue carbon produced and the amount of SOC sequestered shows significant correlation. Likewise soil C pool rate of 2- 3 Mg C ha^{-1} yr^{-1} has also been recorded in sodic soils in northern India under agroforestry practices (Garg, 1998). A SOC sequestration rate of 3.37-6.58 tons ha^{-1} y^{-1} (average value of 4.93 tons C ha^{-1} y^{-1}) has also been observed in 0-40 cm soil layer under afforestation and rainwater harvesting during 65 months period of restoration of degraded hills, where a wide variation in rate of carbon sequestration is due to topography as well as rainwater harvesting devices (Singh, 2013).

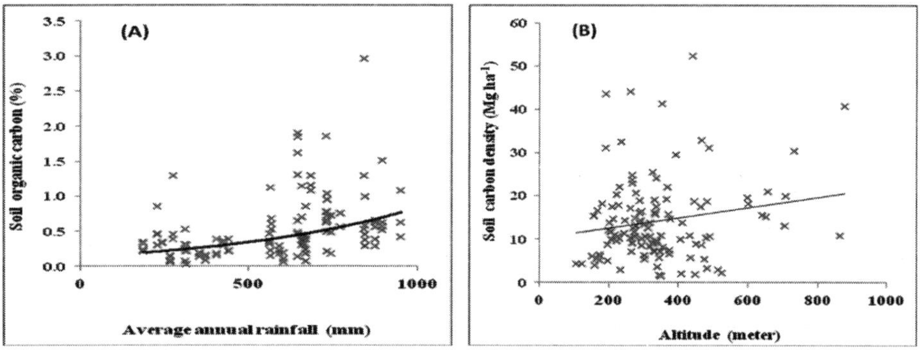

Figure 5.4. Per cent soil organic carbon increases exponentially with increase in annual rainfall (A), and soil carbon density increased with altitude of the sacred groves of Rajasthan.

Per cent soil organic carbon (SOC) varies significantly among the groves with an average value of about 0.53% across the sites in Rajasthan. Devnarayanji Ka Oran, Chittorgarh shows highest (2.95%) value of SOC as well as soil carbon density (52.24 tons ha^{-1}) in top 0-30 cm soil layer. The lowest values of SOC (0.03%) and soil carbon density (1.51 tons ha^{-1}) are in Pabuji Ka Oran, Jodhpur. The lowest SOC in Pabu Ji Ka Oran appears to be due to climatic adversities coupled with grazing pressure from the surrounding villages. Similar situations are there for most of the groves in western Rajasthan. Almost 60% sacred groves show below state average per cent soil organic carbon. This indicates that most of the groves are under varying degree of degradation. Soil organic carbon and soil carbon density both are strongly influenced by physiographic and edaphic factors as they show increasing trend with increase in altitude (Fig 5.1B) and vary with soil texture (Giri et al., 2007; Pala et al., 2013). However, a strong spatial variation in soil organic carbon concentration and its density is highly related to rainfall and disturbances through animal grazing, which appear dominant factors affecting these soil variables. Extent of degradation of forest and grazing lands has also been shown by a study carried out in Jodhpur district of Rajasthan, where average organic carbon stock (both live biomass in above- and below-ground and SOC in top 1 m soil layer) ranges from 16.16 Mg ha^{-1} in pastureland to 26.83 Mg ha^{-1} in sacred groves with spatial variation from 14.43 Mg ha^{-1} in Bilara to 25.67 Mg ha^{-1} in Balesar forest range (Singh, 2015a). The soil organic carbon concentration in the sacred groves across Rajasthan indicates exponential increase with increase in annual rainfall (Fig 5.4). Reduced disturbances at higher altitude and corresponding greater vegetation status influences SOC positively as compared to the area at lower altitude and affected by grazing and other disturbances.

Table 5.1. Changes in soil organic carbon stock (tons ha^{-1}) under spatial and land uses changes in western Rajasthan. Source: Singh (2015a).

Land use	**Soil carbon stock (tons ha^{-1}) in different Block (Panchayatsammiti)**					
	Bali	**Baitu**	**Aburoad**	**Sanchor**	**Baap**	**Sankara**
Forest	20.68	1.94	30.20	6.22	5.44	5.45
Oran	15.44	2.35	26.02	6.88	5.55	6.56
Pastureland	16.67	3.08	27.39	7.80	5.68	5.26
Agriculture	19.89	3.84	30.01	9.69	6.30	5.21
Roadside	12.90	1.69	22.86	8.92	8.24	5.75
Fallowland	19.94	2.71	39.50	5.62	5.63	2.84

A other study conducted (Singh, 2015b) in different land uses in different districts of western Rajasthan indicates significant variation in soil carbon stock between the blocks (Panchayat samiti) in six different districts reflecting significant spatial variations, whereas it did not differ between the land uses (Table 5.1). However, soil carbon stocks vary significantly due to land use changes and have been observed to be highest under forest in general, less under shifting cultivation and the smallest under continuous cultivation. However, carbon/stocks are observed correlated to total annual rainfalls and latitude at regional level and soil type, hill-slope, distance to the water sources and the slope angle at the local level (Chaplot et al., 2010; Singh et al., 2013). Soil carbon stock shows a decreasing trend with increase in number of livestock per household in Rajasthan indicating that increased grazing pressure negatively affects soil carbon stock (Singh, 2015b). Beneficial effects of rainfall are through increased soil water availability, which favours vegetation growth and biomass production and subsequently soil organic carbon enrichment improving soil structure. Sustaining carbon uptake however, relies on growth and development of vegetation, which requires careful balance with biomass removal through grazing or harvesting affecting carbon sequestration (Derner and Schuman 2007; Thomey et al., 2014). The average values of SOC and soil carbon density in 0-30 cm soil layer are 0.54% and 17.09 tons ha^{-1}, respectively. Such high spatial variation in soil organic carbon and carbon density appears to be much affected by vegetation as well as climatic variability (Erfanzadeh et al., 2014), as many species have high potential of carbon sequestration and could be utilized accordingly in such type of programmes.

3. SPATIAL VARIATIONS IN SOIL CHARACTERISTICS

Physiographic and physico-chemical properties of the soils among different districts vary significantly indicating their spatial heterogeneity. Average altitude of sacred groves in different districts varies from 154.0 m (msl) in Jalore to 690.3 m in Udaipur district (Table 5.2). Soil pH ranges between 6.55 in Banswara and 8.63 in Bikaner depending upon the elemental constituents of the parent materials and salinization under increased aridity. Electrical conductivity is observed lowest (0.05 mSm^{-1}) in Churu represented by soils of sandy texture, whereas its highest value of 0.90 mSm^{-1} in Baran district, which has been observed dominated by clayey soils. Soil bulk density ranges from 1.64 g cm^{-3} in Jodhpur to 1.33 g cm^{-3} in Chittorgarh, whereas average values of SOC in different districts of Rajasthan varies from 0.12% in Ajmer to 1.59% in Chittorgarh. This indicates an inverse relationship between soil bulk density and soil organic carbon, particularly in Chittorgarh district in the state. Soil bulk density has been observed showing negative relationships with Si, C, $CaCO_3$, organic matter content, total macro and total micro nutrient content, but shows positive relationship with sand content indicating that it not only the texture and organic matter content but nutrient concentration also that affects the bulk density of soils (Chaudhari et al., 2013).

Table 5.2. Soil characteristics of sacred groves of in different districts of Rajasthan. Values are mean ± SE of multiple replicates.

SNo	District	Altitude	pH	EC (dSm^{-1})	Bulk density (g cm^{-3})	SOC (%)
1	Ajmer	478.7±19.0	7.34±0.09	0.18±0.05	1.52±0.04	0.12±0.03
2	Alwar	328.2±55.0	7.82±0.10	0.24±0.03	1.38±0.02	0.59±0.15
3	Banswara	267.3±1.5	6.55±0.33	0.16±0.02	1.39±0.02	0.71±0.20
4	Baran	371.3±60.3	7.69±0.09	0.90±0.53	1.49±0.02	0.41±0.08
5	Barmer	164.8±20.8	7.72±0.12	0.17±0.01	1.59±0.01	0.10±0.02
6	Bharatpur	671.0±8.0	7.36±0.32	0.12±0.01	1.49±0.01	0.33±0.01
7	Bhilwara	409.7±29.5	7.50±0.12	0.24±0.03	1.41±0.01	1.18±0.06
8	Bikaner	266.7±24.3	8.63±0.06	0.26±0.04	1.50±0.01	0.34±0.01
9	Bundi	304.0±5.5	7.70±0.26	0.30±0.04	1.35±0.05	0.69±0.07
10	Chittorgarh	455.0±13.1	6.98±0.39	0.17±0.03	1.33±0.06	1.59±0.72
11	Churu	315.0±6.7	6.73±0.18	0.05±0.01	1.56±0.01	0.17±0.01
12	Dausa	346.3±14.0	7.49±0.13	0.26±0.01	1.44±0.01	0.38±0.41
13	Dholpur	661.7±49.7	6.96±0.91	0.26±0.04	1.46±0.04	0.46±0.16
14	Dungarpur	221.3±10.8	7.16±0.25	0.20±0.03	1.42±0.03	1.09±0.39
15	Hanumangarh	189.3±4.8	7.83±0.16	0.28±0.02	1.50±0.04	0.71±0.29
16	Jaipur	377.3±15.7	7.75±0.20	0.20±0.01	1.54±0.02	0.50±0.31

SNo	District	Altitude	pH	EC (dSm^{-1})	Bulk density ($g\ cm^{-3}$)	SOC (%)
17	Jaisalmer	255.3±17.0	8.02±0.05	0.35±0.02	1.63±0.02	0.28±0.03
18	Jalore	154.0±15.7	7.87±0.02	0.21±0.03	1.57±0.02	0.13±0.02
19	Jhalawar	340.6±13.9	7.67±0.26	0.14±0.02	1.54±0.05	0.54±0.12
20	Jhunjhunu	379.3±36.0	6.85±0.37	0.16±0.03	1.63±0.01	0.25±0.07
21	Jodhpur	261.8±25.7	7.63±0.37	0.29±0.03	1.64±0.02	0.21±0.05
22	Karauli	574.5±341.5	7.07±0.43	0.25±0.04	1.43±0.06	0.56±0.15
23	Kota	288.8±10.3	7.83±0.09	0.37±0.03	1.46±0.04	0.72±0.09
24	Nagaur	326.5±5.3	8.09±0.29	0.27±0.02	1.55±0.05	0.29±0.09
25	Pali	293.2±17.0	7.14±0.03	0.16±0.02	1.55±0.01	0.34±0.11
26	Pratapgarh	434.5±69.4	6.66±0.33	0.17±0.03	1.40±0.02	0.81±0.23
27	Rajsamand	637.0±17.6	6.98±0.21	0.15±0.02	1.49±0.02	0.60±0.04
28	Sawaimadhopur	267.7±35.2	7.57±0.05	0.25±0.01	1.51±0.02	0.62±0.03
29	Sikar	464.3±16.7	6.65±0.25	0.18±0.08	1.60±0.01	0.28±0.06
30	Sirohi	323.0±54.5	8.00±0.15	0.17±0.01	1.49±0.04	0.23±0.03
31	Sriganganagar	180.3±11.1	7.50±0.18	0.20±0.02	1.56±0.04	0.52±0.12
32	Tonk	326.0±13.1	7.34±0.06	0.24±0.01	1.63±0.04	0.16±0.05
33	Udaipur	690.3±51.5	7.37±0.27	0.19±0.02	1.36±0.04	1.05±0.25

Soil nutrients like NH_4-N, NO_3-N and PO_4-P and soil carbon storage in 0-30 cm soil layer also indicate significant spatial variations (Table 5.3). Availability of NO_3-N ranges between 0.18 mg kg^{-1} and 6.64 mg kg^{-1} in Sawaimadhopur exhibiting its low values in the state. Likewise availability of NH_4-N varies from 1.35 mg kg^{-1} in the sacred groves of Barmer to 6.95 mg kg^{-1} in sacred groves of Bundi districts. This also reflects low availability of nitrogen in the sacred groves of different districts of Rajasthan, when considered the available nitrogen rating values of low (<272 kg ha^{-1}), medium (272 to 544 kg ha^{-1}) and high (>544 kg ha^{-1}) reported by Arora (2002). Soil available PO_4-P ranges from 5.55 mg kg^{-1} in Jodhpur to 18.15 in Sirohi district. Categorization of soil available phosphorus indicates that <12.40 kg ha^{-1} is low, 12.40 to 22.40 kg ha^{-1} is medium, and >22.40 kg ha^{-1} phosphorus availability is high (Arora, 2002). Thus soil available phosphorus in the studied sacred groves of these districts is medium to high. Soil carbon storage (density) also indicates significant spatial variations ranging from 3.36 tons ha^{-1} in Ajmer to 28.13 tons ha^{-1} in Bhilwara district.

Table 5.3. Soil nutrients and carbon density in sacred groves of different districts of Rajasthan. Values are mean±SE of multiple replicates.

SNo	District	NO_3-N (mg kg^{-1})	NH_4-N (mg kg^{-1})	PO_4-P (mg kg^{-1})	SOC (tons ha^{-1})
1	Ajmer	1.18±0.04	3.78±0.67	6.77±1.18	3.36±1.03
2	Alwar	2.17±0.16	5.19±1.79	9.12±0.89	17.31±4.42
3	Banswara	1.68±0.44	3.49±0.56	11.76±0.77	9.71±1.59
4	Baran	1.47±0.33	8.63±0.76	10.40±1.30	12.23±1.92
5	Barmer	1.06±0.24	1.35±0.09	7.51±1.01	4.48±0.54

SNo	District	NO_3-N (mg kg^{-1})	NH_4-N (mg kg^{-1})	PO_4-P (mg kg^{-1})	SOC (tons ha^{-1})
6	Bharatpur	6.64±0.50	3.85±0.82	10.23±2.23	12.70±1.05
7	Bhilwara	1.96±0.12	6.54±2.32	11.92±1.19	28.13±3.22
8	Bikaner	1.23±0.63	3.77±0.63	8.41±0.66	13.02±1.01
9	Bundi	2.99±0.27	6.95±1.27	15.29±1.99	11.52±1.48
10	Chitaurgarh	6.52±0.11	4.04±1.78	8.35±1.22	24.77±14.43
11	Churu	2.62±0.38	5.29±1.01	13.19±1.05	7.88±0.82
12	Dausa	3.53±0.35	2.84±0.23	7.98±1.97	7.99±0.61
13	Dholpur	4.48±0.88	2.80±0.92	7.90±2.31	14.68±4.24
14	Dungarpur	1.73±0.35	3.53±0.49	9.85±1.66	20.76±6.09
15	Hanumangarh	3.35±0.72	3.15±0.68	7.29±1.53	23.46±10.43
16	Jaipur	3.15±1.10	4.74±1.16	9.93±0.42	19.28±11.09
17	Jaisalmer	1.61±0.21	2.15±0.21	5.72±0.69	13.16±1.31
18	Jalore	3.73±0.38	4.87±0.91	12.31±1.16	5.30±0.53
19	Jhalawar	4.65±0.94	4.42±0.64	7.74±0.91	16.51±3.20
20	Jhunjhunu	2.54±0.76	4.23±1.68	8.93±1.88	11.97±3.49
21	Jodhpur	2.82±0.43	6.06±0.35	5.55±0.38	7.70±1.91
22	Karauli	4.63±1.23	4.16±2.01	9.47±1.92	19.20±5.22
23	Kota	1.60±0.19	6.51±0.67	10.20±1.72	19.22±1.96
24	Nagaur	3.15±0.69	3.30±0.57	8.12±0.90	10.93±2.63
25	Pali	2.32±0.26	2.54±0.19	6.03±0.43	13.33±1.11
26	Pratapgarh	4.07±0.64	4.16±1.51	7.71±0.85	23.69±6.89
27	Rajsamand	1.66±0.25	6.45±1.09	10.41±1.20	18.12±1.66
28	Sawaimadhopur	0.18±0.03	6.08±0.43	14.97±0.74	9.35±0.30
29	Sikar	2.74±0.47	5.83±0.27	10.98±2.02	13.22±2.75
30	Sirohi	3.65±0.30	8.79±1.19	18.15±2.91	7.06±2.50
31	Sriganganagar	2.99±0.40	4.04±0.57	8.21±0.84	18.72±4.29
32	Tonk	2.94±0.19	6.84±1.26	7.51±0.52	4.75±1.61
33	Udaipur	5.06±0.76	6.32±1.51	12.26±1.33	17.72±4.42

4. EFFECTS OF DOMINANT TREES ON SOILS

Types of tree species influence soils by providing organic matter of varying chemical composition which in turn contributes to different levels of organic matter depending upon the variations in species and its composition. Preferences of species change soil conditions. For example, in a study in Sudan indicates *Faidherbia albida* tree species dominates loamy sand and sandy loam, whereas *Balanites aegyptiaca* dominates at the sandy clay loam soil. Sandy clay loam at sedentary plain and sandy loam of lower hill slopes have been observed dominated by *Albizia amara* and *Acacia senegal*, but *Acacia seyal* and *Acacia senegal* dominate clay plain area. *Boswellia papyrifera* and *Albizia amara* dominate the sandy loam of stony hill slope (Desougi et al., 2016). Simpson diversity index shows sandy loam of lower terraces (0.15) as the most diverse site, whereas the upper terrace areas harbour the lowest diversity (Desougi et al., 2016). Likewise, red alder (*Alnus rubra*) promotes productivity and long term sustainability through its higher litter nutrient concentration and accelerates nutrient cycling, while vine maple leaves (*Acer circinatum*) increase N, P, K, Ca, Mg, and Zinc (Zn) and decompose much quicker than conifer litter. Further, soil microbial or arthropod communities do adapt to types of litter produced within stands; litter in stands of origin usually decompose faster than if the same litter is introduced to different species (Tripathi et al., 2013).

In the study conducted in Rajasthan, all soil variables have been observed varying significantly between the sacred groves dominated by different tree species, except for electrical conductivity (EC) and NO_3-N and PO_4-P concentrations. Gravel content varies significantly between 1.99% in the sacred groves dominated by *Anogeissus sericea* and 69.32% in sacred grove dominated by *P. roxburghii.* This indicates the preferences of the species towards different soil conditions as *A. sericea* prefer rocky and gravelly area. Soil pH is observed highest (8.51) in the sacred grove dominated by *Z. nummularia* (*rotundifolia*), indicating its adaptability to such conditions of the soils. One can see that most of the sacred groves or Orans in Western Rajasthan are dominated mostly by *Z. rotundifolia* probably because of its adaptability to the environmental conditions and social acceptance among the local people as well. The lowest soil pH is recorded in the sacred groves dominated by *E. officinalis* trees (Table 5.4). There are two probability of low soil pH under *E. officinalis*. The first one is its preference to such soil conditions, and the second one is its soil ameliorative characteristics because of acidic litter reducing soil pH. Studies conducted involving different tree species under afforestation indicates highest soil organic carbon increase and lowest soil pH under *E. officinalis* (Khoshoo, 2008; Singh et al., 2008). The results of Sann et al. (2016) also highlight remarkable associations of species such as *Acacia catechu, Dalbergia paniculata, Terminali aoliveri* and *Millettia multiflora* with soil texture, where *Acacia inopinata* is associated with a high soil pH of about 9–10, whereas *Terminalia tomentosa* is observed associated with soil hardness.

Table 5.4. Soil characteristics under the influence of dominant tree species in sacred groves of Rajasthan. Values are mean ± SE of multiple replicates.

SNo.	Dominant tree species	Physico-chemical		
		Gravel	pH	EC (dSm^{-1})
1	*A. catechu*	41.18±4.16	7.82±0.09	0.39±0.06
2	*A. indica*	17.75±0	8.14±0.00	0.22±0.00
3	*A. leucophloea*	38.03±9.05	7.59±0.19	0.22±0.04
4	*A. nilotica*	28.24±10.08	7.55±0.11	0.3±0.06
5	*A. pendula*	45.19±3.89	7.26±0.12	0.31±0.07
6	*A. senegal*	41.11±15.88	8:23±0.06	0.17±0.02
7	*A. sericea*	1.99±0.00	7.08±0.00	0.17±0.00
8	*A. tortilis*	2.50±2.5.00	6.84±0.66	0.16±0.02
9	*B. monosperma*	25.25±6.63	7.57±0.15	0.27±0.11
10	*C. decidua*	6.27±2.58	7.71±0.20	0.26±0.02
11	*D. melanoxylon*	38.32±0.00	8.43±0.00	0.11±0.00

SNo.	Dominant tree species	Physico-chemical		
		Gravel	pH	EC (dSm^{-1})
12	*E. camaldulensis*	39.95±2.93	8.16±0.18	0.26±0.09
13	*E. officinalis*	12.54±0.00	6.38±0.00	0.26±0.00
14	*F. bengalensis*	43.02±13.08	7.19±0.56	0.21±0.05
15	*H. integrifolia*	40.75±0.00	7.68±0.00	0.23±0.00
16	*M. emarginata*	9.66±8.98	7.08±0.13	0.11±0.04
17	*M. parviflora*	45.74±14.69	7.82±0.20	0.20±0.04
18	*P. cineraria*	17.93±9.54	7.5±0.30	0.23±0.04
19	*P. juliflora*	34.46±5.53	7.50±0.09	0.21±0.01
20	*P. roxburghii*	69.32±0.00	7.96±0.00	0.16±0.00
21	*S. oleoides*	13.90±4.86	7.51±0.15	0.18±0.02
22	*T. grandis*	47.91±8.01	6.70±0.23	0.18±0.01
23	*T. undulata*	28.3±0.99	7.86±0.06	0.2±0.02
24	*T. arjuna*	19.78±8.35	6.88±0.83	0.33±0.25
25	*W. tinctoria*	45.97±0.00	6.78±0.00	0.11±0
26	*Z. rotundifolia*	11.73±5.81	8.51±0.14	0.3±0.04

Electrical conductivity ranges between 0.11 dSm^{-1} in *P. cineraria* dominated sacred groves and 0.39 dSm^{-1} in *A. catechu* dominated sacred groves. Relatively greater electrical conductivity in *Acacia catechu* dominated area appears to be due to soil characteristics of the area with relatively better silt content where there are chances of accumulation of salts along with water from the other areas and its tolerance to salt concentrations (Ramoliya et al., 2004). Soil pH and acidity are very much related with proton-producing components of the soil N cycle like nitrification, which is positively correlated with species-level variability in N concentrations. Accumulation of mineral substrates through dead materials changes soil features and consequently influences the plant growth and the competition between plant species. Concentrations of soil nutrients namely NO_3-N, NH_4-N and PO_4-P vary widely due to dominance of trees (Table 5.5).

Table 5.5. Soil nutrients under the influence of dominant tree species in sacred groves of Rajasthan. Values are mean ± SE of multiple replicates.

S.No.	Dominant tree species	Soil nutrients (mg kg^{-1} soil)		
		NO_3-N	NH_4-N	PO_4-P
1	*A. catechu*	2.35±0.01	5.49±1.71	8.99±4.01
2	*A. indica*	5.30±0.00	2.86±0.00	9.08±0.00
3	*A. leucophloea*	3.34±1.00	2.84±0.43	8.83±1.21
4	*A. nilotica*	2.15±0.42	4.51±1.51	8.71±2.78
5	*A. pendula*	3.21±0.42	6.26±0.58	11.00±0.81
6	*A. senegal*	4.30±0.27	7.38±3.15	16.57±5.44
7	*A. sericea*	3.12±0.00	3.23±0.00	5.66±0.00
8	*A. tortilis*	2.88±1.16	5.73±1.13	10.38±0.03

S.No.	Dominant tree species	Soil nutrients (mg kg^{-1} soil)		
		NO_3-N	NH_4-N	PO_4-P
9	*B. monosperma*	1.75±0.37	8.34±1.12	10.88±2.05
10	*C. decidua*	2.61±0.29	3.93±0.53	7.93±0.69
11	*D. melanoxylon*	3.65±0.00	3.20±0.00	6.50±0.00
12	*E. camaldulensis*	3.49±1.96	10.34±3.22	6.42±4.10
13	*E. officinalis*	3.45±0.00	0.82±0.00	6.40±0.00
14	*F. bengalensis*	4.97±1.62	3.31±0.69	11.05±2.54
15	*H. integrifolia*	4.23±0.00	2.82±0.00	11.80±0.00
16	*M. emarginata*	1.81±0.72	4.21±1.27	9.72±2.61
17	*M. parviflora*	3.59±2.02	4.74±1.01	11.61±1.88
18	*P. cineraria*	2.93±0.52	3.94±0.86	7.54±1.29
19	*P. juliflora*	2.65±0.48	5.18±0.57	9.04±0.83
20	*P. roxburghii*	4.46±0.00	3.05±0.00	14.06±0.00
21	*S. oleoides*	2.63±0.39	3.75±0.48	9.26±0.80
22	*T. grandis*	2.90±0.71	4.18±0.93	9.48±1.16
23	*T. undulata*	2.82±1.57	3.9±2.40	10.02±4.22
24	*T. arjuna*	3.15±0.80	3.24±1.23	7.88±1.36
25	*W. tinctoria*	1.73±0.00	3.3±0.00	13.34±0.00
26	*Z. rotundifolia*	1.24±0.44	3.47±0.54	7.94±0.66

The highest and lowest concentrations of NO_3-N have been observed in sacred groves dominated by *A. indica* (5.30 mg kg^{-1} soil) and *Z. rotundifolia* (1.24 mg kg^{-1} soil), respectively. It probably depend upon the litter quality and quantity in which latter one is generally more in favour of *A. indica* as compared to *Z. rotundifolia*. Concentration of NH_4-N ranges from 0.82 mg kg^{-1} soil in the sacred groves dominated by *E. officinalis* to 10.34 mg kg^{-1} soil in sacred groves dominated by *E. camaldulensis*. Though quantity of litter available under *E. camaldulensis* is relatively high but its rate of decomposition is reported to be very low (Bala et al., 2010). In such a condition highest amount of NH_4-N under this appears to be due to contributions by the other species. Soil available PO_4-P varies from 5.66 mg kg^{-1} soil in *A. sericea* dominated sacred groves to 16.57 mg kg^{-1} soil in *A. senegal* dominated sacred groves (Table 5.5). Establishment and occurrence of *A. sericea* generally on calcareous soil or the soil with very shallow depth may be the reason of low PO_4-P availability under this species, but increased concentration of PO_4-P under *A. senegal* appears to be due to parent material and enhanced microbial growth and its utilization by *A. senegal* trees (Githae et al., 2011). Such changes in soil nutrients lead to species with low growth rates and low biomass loss rates being replaced by species with high potential growth rates and high biomass losses. The plant properties responsible for reduced biomass loss rates appear to result in the litter produced being poorly decomposable, whereas the litter from plants with high potential growth rates decomposes more easily (Berendse, 1998).

Soil bulk density ranges from 1.32 g cm^{-3} in the sacred groves dominated by *E. officinalis* to 1.63 g cm^{-3} in the sacred groves dominated by *Diospyros melanoxylon*. The sacred groves dominated by *A. sericea, A. tortilis, C. decidua, D. melanoxylon,*

M. emarginata, *P. cineraria*, *S. oleoides*, *T. undulata* and *Z. rotundifolia* show above average soil bulk density, whereas other sacred groves show below average (horizontal line on the bars) soil bulk density with relatively better soil conditions (Fig 5.5, top). Further, the sacred groves dominated by above mentioned tree species are mostly situated either in arid region or the dryer parts of the semi-arid region and are affected by compaction. Soil organic carbon varies from 0.16% in the sacred groves dominated by *T. undulata* to 1.32% in sacred groves dominated by *E. camaldulensis*, which indicates the effects of dominant tree species in soil carbon concentration added through leaf litter and other dry materials.

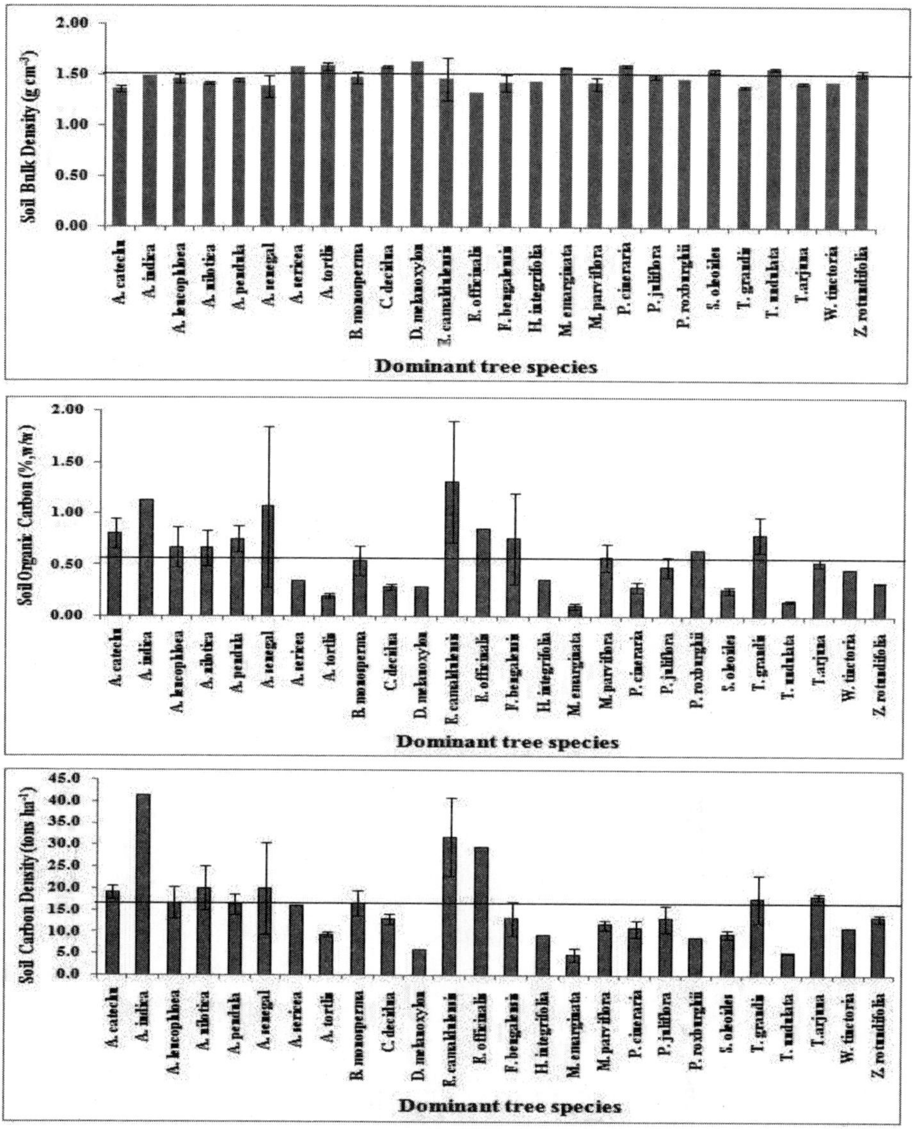

Figure 5.5. Changes in soil bulk density (top), per cent soil organic carbon (middle) and soil carbon density in 0-30 cm soil layer (lower) in sacred grovess dominated by different tree species. Error bars are ±1SE.

While comparing the data of average for all sites, the sacred groves dominated by *Acacia catechu*, *Azadirachta indica*, *Acacia leucophloea*, *A. nilotica*, *Anogeissus pendula*, *Acacia senegal*, *E. camaldulensis*, *E. officinalis*, *F. benghalensis*, *P. roxbughii* and *T. grandis* show above average SOC and appear to be beneficial in enhancing soil carbon because of addition of relatively greater amount of litter. This indicates that these species are relatively better in sequestering soil organic carbon at a higher rate; though it varies between the species (Jandl et al., 2007; Du et al., 2015). The study of Xuluc-Tolosa et al. (2003) also indicates that the earlier successional phase of a forest shows higher decomposition rates than later stages of succession resulting in high rate of nutrient turnover. Sacred groves dominated by other species under study show below average per cent SOC, which are mostly available in relatively dry areas or are affected by degradation (Fig 5.5, middle). This indicates that sacred groves situated in arid regions possess relatively lower SOC as compared to those situated in semi-arid region or east of Aravalli. Soil carbon density measured in 0-30 cm soil layer (after correction of gravel content) indicates highest ($P<0.05$) carbon density in the sacred groves dominated by *A. indica* (41.36 tons ha^{-1}). The lowest carbon density of 4.83 tons ha^{-1} has been observed in the sacred groves dominated by *M. emarginata* trees. Thus spatial variation in soil carbon density (ratio of highest to lowest) is 8.56 fold indicating strong influence of tree species under dominance and their contribution in the form of added litter and fine root turn over. However, variations in soil conditions from sandy to clay loam and rainfall also influence soil carbon storage (Giri et al., 2007).

When comparing with the average carbon density across all sites, sacred groves dominated by *Acacia catechu*, *Azadirachta indica*, *A. nilotica*, *Acacia senegal*, *E. camaldulensis*, *E. officinalis*, *T. grandis* and *Terminalia arjuna* have been observed above state average soil carbon density. The other sacred groves indicate below average soil carbon density value in 0-30 cm soil layer (Fig 5.5, bottom). This suggests strong effects of dominant species together with vegetation traits such as plant species composition and soil conditions influencing soil carbon storage (Erfanzadeh et al., 2014; Gao et al., 2014; Sariyildiz et al., 2015). Two main governing processes influencing soil carbon storage are the quantity and quality of the litter input and the decomposition of soil organic matter in the form of carbon output. While increasing the productivity of the forest and thereby increasing the carbon input to the soil enhances soil carbon storage, minimizing the disturbances in forests and the associated soil reduces the risk of carbon losses. Likewise, establishment of mixed species forests can also increase the stability of the forest and avoid high rates of soil organic matter decomposition (Jandl et al., 2007). Carbon storage increases with stand age, i.e. greater in older than in younger stands, vary with forest zone, stand age, and forest type. This is indicated by soil carbon contribution to total forest carbon storage ranging from 62.5% in young to 30.1% in over-mature stands in boreal and from 70.1% in young to 26.0% in over-mature in temperate forests (Wei et al., 2013).

5. SOIL CHANGES IN AGRO-CLIMATIC REGION

Effects of agro-climatic zones on soil properties of the sacred groves under study has been observed significant ($P<0.05$) on all recorded soil variables except NH_4-N, which is not significant ($P>0.05$). Electrical conductivity ranges between 0.18 dSm^{-1}

in IIB (Transitional Plain of Luni Basin) and 0.38 dSm^{-1} in IVB (Humid Southern plain), whereas soil bulk density varies from 1.38 g cm^{-3} in IVB to 1.61 g cm^{-3} in IA (Arid Western plain) agro-climatic zone. Sandy soils in Arid Western Plain, which is mostly dominated by sand dunes, shows highest soil bulk density indicating positive influence of sand content on increased soil bulk density (Berendse et al., 1998). Percent soil organic carbon has been observed highest (1.13%) in IVA (Sub humid Southern Plain), whereas the lowest amount of soil organic carbon has been recorded in IA agro-climatic zone (Fig 5.6). This indicates that increased aridity negatively affects soil organic carbon though it also relates to rainfall, temperature, elevation, clay and silt contents, land use, rainfall and soil conditions (Liu et al., 2011; Venkanna et al., 2014).

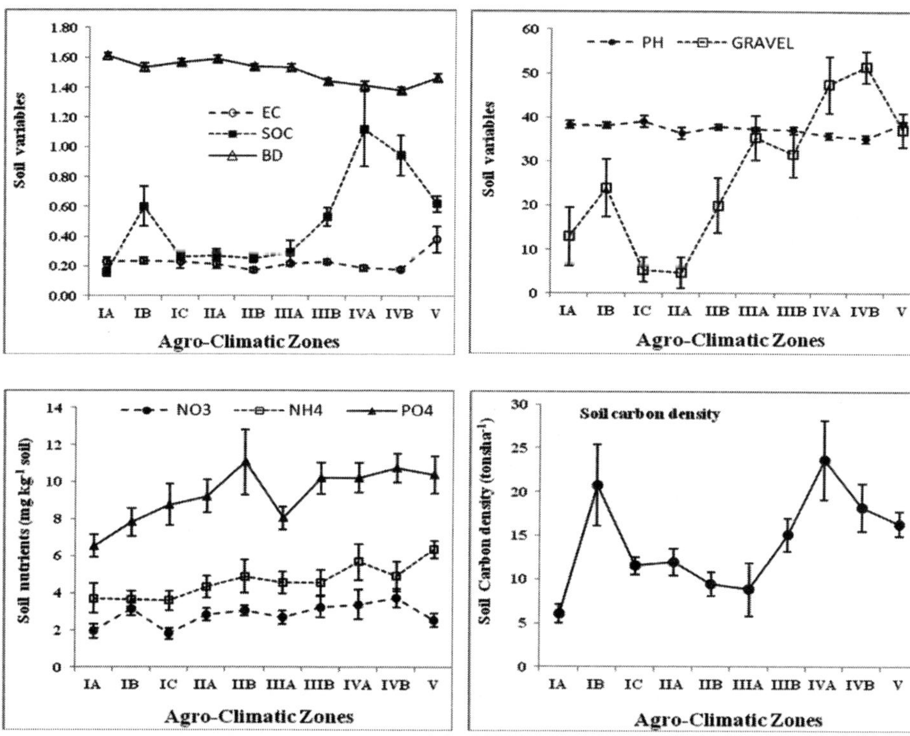

Figure 5.6. Variations in soil characteristics in sacred groves in different agro-climatic zones of Rajasthan. Error bars are ±1SE.

Lowest and highest values of soil pH in the soils of sacred groves of different agro-climatic zones are 7.04 in IVB and 7.82 in IC (Hyper-arid partly irrigated zone), respectively. Gravel content shows wide variations between the agro-climatic zones being highest ($P<0.01$) in IVB (51.48%) and lowest in IIA (Internal drainage dry zone) agro-climatic zone (Fig 5.6, top right). Lowest concentrations of SOC, PO_4-P and soil carbon density are in the soils of sacred groves in IA agro-climatic zone; whereas NO_3-N and NH_4-N concentrations are lower in the soils of sacred groves in IC zone. These observations are obvious because of arid condition and soil of sandy texture in these zones, influencing soil characteristics and soil nutrient availability under low vegetation cover, low litter input and enhanced rate of SOC decomposition caused by warmer soil under open canopy (Chen et al., 2016). SOC and related soil carbon density in top 0-30 cm soil layer has been observed highest in the soils of sacred groves situated in IVA agro-climatic zone due to relatively better vegetation conditions in this agro climatic zone. The increased irrigation facility in IB zone and its influences on increased soil water availability enhances vegetation status and thus soil carbon density in this zone. Availability of NO_3-N, NH_4-N and PO_4-P are highest in IVB, V and IIB agro-climatic zone, respectively. Such spatial heterogeneity in soil nutrients appear to be due to differences in vegetation types because the main source of soil nutrients in natural ecosystems is litter input.

However, it is also influenced by parent materials and its mineral constituents from which the soils have been derived. The spatial heterogeneity of soil nutrients have been reported random in the dense, uniform vegetation of relatively humid regions, and in the patchy vegetation of relatively arid regions. Concentrated distributions of soil nutrients beneath vegetation patches are often observed and referred to as "islands of fertility" in dry areas (Xu et al., 2012). Protecting the natural forests or groves enhances vegetation cover, height and above- and belowground biomass, soil organic carbon and total nitrogen concentration and a decrease in soil pH, electrical conductivity and soil bulk density. According to a study in China, fencing promotes vegetation recovery in 6 years, soil pH in 8 years, soil organic carbon in 16 years, total nitrogen in 30 years and total phosphorus concentrations in 19 years (Li et al., 2014).

6. STATISTICAL RELATIONS

Soil characteristics are very much influenced by rainfall and the vegetation growing on it. Rainfall of 2003-2012 shows better relationship with soil bulk density, per cent SOC, NO_3-N, NH_4-N and PO_4-P as compared to the long term average rainfall. Long term average rainfall show negative relationships with soil pH ($r=-0.252$, $P<0.01$) and soil bulk density ($r=-0.541$, $P<0.01$) and positive relationships with gravel content of soil ($r=0.597$, $P<0.01$), per cent SOC ($r=0.406$, $P<0.01$), soil available NH_4-N ($r=0.278$, $P<0.01$) and PO_4-P ($r=0.274$, $P<0.01$). A diverse community of tree species is observed to be beneficial (positive correlations) for shrub species, SOC and soil carbon density; whereas both tree and shrub diversity show negative correlations with soil bulk density and appear beneficial in improving soil conditions (Table 5.6). Unfortunately, there is no significant correlation of soil nutrients with diversity variables, except PO_4-P, which has been observed positively correlated with effective number of species (ENS) and reciprocal Simpson's index (1/D). Both per cent soil organic carbon and soil carbon density show negative correlation with population but link positively with diversity variables of the trees of girth class > 30 cm. Restoring woody vegetation on cleared land, croplands by reconverting them to rangelands, vegetation to bare soils, and restoring soil stability could all increase carbon sequestration as well as its storage both in soils and vegetation (Brooker et al., 2013). A significant positive correlation between SOC and H' of trees of >30 cm girth indicates beneficial impacts of older trees in soil organic carbon accumulation, thus their conservation and protection enhance soil carbon sequestration. Tankou et al. (2013) have observed that decrease in soil pH and total soil phosphorus enhances Shannon index, whereas increase in CEC and slope enhance species richness of the herbs. In cases of shrubs, species richness increases with decreases in total P and soil pH and increase in K.

Table 5.6. Correlation coefficients indicating relationships between different plant diversity, topography and soil variables.

	Variable	pH	SOC (%)	BD	NO_3-N	NH_4-N	PO_4-P	Carbon density
Tree >30 cm girth	Population	NS	-0.277	0.204	NS	-0.234	NS	-0.246
	Richness	NS	0.352	-0.436	0.183	NS	NS	0.311
	H'	NS	0.242	-0.321	NS	NS	NS	0.270
	ENS	NS	0.258	-0.357	NS	NS	NS	0.278
	1/D	NS	0.219	-0.324	NS	NS	NS	0.225
	Basal area	NS	NS	NS	NS	NS	NS	NS
Trees 10-30 cm girth	Population	-0.250	NS	NS	NS	NS	NS	NS
	Richness	NS	NS	-0.301	NS	NS	NS	NS
	H'	NS	NS	-0.255	NS	NS	NS	NS
	ENS	NS	NS	-0.273	NS	NS	0.190	NS
	1/D	NS	NS	-0.248	NS	NS	0.192	NS
	Basal area	-0.186	NS	NS	NS	NS	NS	NS
Sapling	Population	NS	NS	NS	NS	NS	NS	NS
	Richness	NS	NS	NS	NS	NS	NS	NS
	H'	NS	NS	NS	NS	NS	NS	NS
	ENS	NS	NS	NS	NS	NS	NS	NS
	1/D	NS	NS	NS	NS	NS	NS	NS
	Basal area	NS	NS	NS	NS	NS	NS	NS
Shrub	Population	NS	NS	NS	NS	NS	NS	NS
	Richness	NS	NS	-0.255	NS	NS	NS	NS
	H'	NS	NS	-0.262	NS	NS	NS	NS
	ENS	NS	NS	-0.217	NS	NS	NS	NS
	1/D	NS	NS	-0.222	NS	NS	NS	NS
	Basal area	NS	NS	NS	NS	NS	NS	NS
Altitude		-0.337	0.245	-0.333	0.376	NS	NS	NS
SOC		NS	-	-0.731	0.248	NS	NS	0.841

Significant relationships have also not been observed between basal area of trees of >30 cm girth and the soil variables, but basal area of the trees of 10-30 cm girth shows negative relationship with soil pH ($r=-0.186$, $P<0.05$). This indicates that increase in population of trees of 10-30 cm category shows more favourable effects on soil properties under high rate of litter turn over (Du et al., 2015). Positive relationship of per cent soil organic carbon with rainfall ($r=0.406$, $P<0.01$), altitude ($r=0.245$, $P<0.01$), tree of >30 cm girth, species richness, H', ENS, 1/D, gravel content ($r=0.440$, $P<0.01$), NO_3-N ($r=0.246$, $P<0.01$) and soil carbon density ($r=0.841$, $P<0.01$) in 0-30 cm soil layer indicates the beneficial effects of soil carbon on both soil improvement and vegetation development. It is also indicated by an increasing trend in carbon accumulation in the soil with increase in rainfall, species diversity and number of species (Fig 5.7).

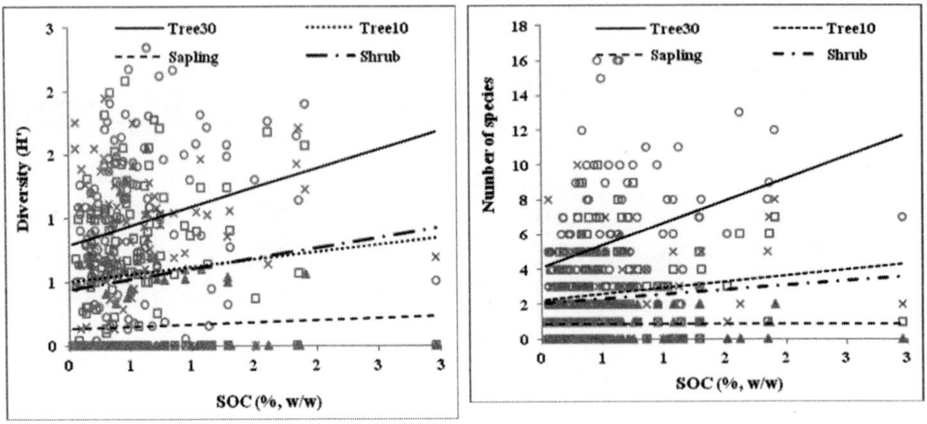

Figure 5.7. Relationships between SOC and Effective number of species of different plant habit

7. CONCLUSIONS

Local variations in dominant species appear to be governed by edaphic and environmental conditions, topographic differences, climatic variations, varying degrees of human interference and elements of climax vegetation. There are strong spatial variations in altitude and soil characteristics indicated by their values in the sacred groves distributed throughout Rajasthan, between different districts and agro climatic regions. Dominant species play significant role in improvement of groves and sequestration of carbon. This is indicated by high amount of soil organic carbon accumulation in the sacred groves dominated by *A. indica*, *A. senegal*, *A. nilotica*, *E. officinalis*, *T. grandis* and *T. arjuna*, which can be promoted under future afforestation programme in managing sacred groves. Soil organic carbon, soil carbon storage (density) and soil nutrients increase with altitude, where chances of grazing and trampling are relatively less. However, soil carbon storage increase with rainfall, tree of >30 cm girth, species richness and tree diversity, which suggests the importance of diverse community of trees of older age in enhancing soil carbon sequestration. Increase in aridity is harmful, both for SOC and soil nutrients storage, which is indicated by their low values in agro climatic zone I under sandy soil and increase in soil temperature. These both increase rate of soil organic matter decomposition due to low vegetation and open canopy. Changes in quantity of SOC, NH_4-N, NO_3-N and PO_4-P and soil carbon density with increased rainfall or irrigation facilities indicate favourable conditions of the sacred groves available in east of the Aravalli in Rajasthan. This is supported by greater soil carbon density in agro-climatic zone IB and IV as compared to the other zones. Increase in vegetation cover, height and above- and below ground biomass, soil organic carbon and total nitrogen concentration and a decrease in soil pH, electrical conductivity and soil bulk density are beneficial effects of protection and conservation of sacred groves.

For sustaining the goods and services, grove eco-systems must recover after disturbances and should not degrade further. In such process the surviving and recruited plants grow more quickly in response to the reduced competition, and carbon productivity and live basal area recover to the original levels within a few years (Carlson et al., 2012; Hansen, 2014). Effective management may stabilize sacred groves from degradation and deforestation that could be offset directly through carbon financing by avoiding its degradation and deforestation (McNeely et al., 1990; Devagiri et al., 2012). This appears substantially cheaper in ensuring groves cover as compared to restoring the sacred grove ecosystem after complete damage. Further, revenue for groves or forest conservation offer potential to achieve financing for ecosystem based adaptation on the scale required to ensure biodiversity security under climate change scenario.

Promoting woody perennials including grasses, increasing the number of groves under reforestation and afforestation of indigenous species, restoring degraded groves by conserving soil and water, and effective management of grazing and livestock will be more beneficial in restoring ecological services including carbon sequestration and soil improvement, of the sacred groves. However, the most effective mean in maintaining the services of the groves involve participation of local people, which can only be done by ensuring their day to day needs.

6

LIVELIHOODS FROM AND THREATS TO SACRED GROVES

As an integral part of life since ancient times, the local communities respect and protect sacred groves due to sacredness associated with taboos. These are considered remnants of the past dominant forest vegetation which once existed in the locality. Sacred groves provide varying ecological, social and economic benefits; like enhancing water availability, place of grazing, fuel wood, fodder and plants of medicinal values. They also work as communally shared commons and are used to perform important social ceremonies in indigenous societies. Several cultural festivals are performed in these sacred groves like social gatherings, marriage, after death rituals etc. However, sacred groves are facing various threats, which include construction of temples, shrines, residential and commercial buildings, etc. in and around the groves, fragmentation of the groves, encroachment of the grove land for agriculture purposes, exploitation of resources by locals, removal of vegetation, grazing of cattle, tourism activities, replacement of indigenous tree species by the exotic species, etc. Such types of anthropogenic interventions lead degradation of the sacred groves and if continued, these cultural heritages may become lost in due course of time. Out of 123 sacred groves studied in Rajasthan, 6 are in the control of state forest department (Rajasthan) and the rests are under the control of village Panchayat, in which one has been recently converted sacred grove from pastureland. This chapter highlights the benefits accrued from and threats to the sacred groves, for development of effective management strategies in conservation and management of these community resources.

1. BACKGROUND

Human activities have been continuing ever since men started cultivation and exploitation of natural resources for their livelihood. Moreover, sacredness and social restrictions make the sacred forests patches important reservoirs of biological diversity and preserve unique floral and faunal species (Gupta and Sharma, 2014). Having left undisturbed and unexploited for the years, these sacred groves even now remain a treasure of biodiversity and gene pool, conserving many endemic and endangered plants and animals of economic and scientific importance which does not exist even in the adjoining forest areas at some of the places (Chandrashekara, 2011; Gaude and Janarthanam, 2015). A majority of people are attached with the function of sacred groves to spiritual purposes; however many regard the grove for its environmental benefits (Hembram, 1983). The local inhabitants believe in the sanctity of the groves and deities and this make the people aware of the spiritual function of the groves (Bhasin, 1999). People residing in the adjoining area of the grove recognize that trees have an economic value by providing fruit, flower, fodder and a medicinal value as well as providing shade and wood for building houses and making farm implements (Sukumaran and Raj, 2010; Kumari, 2011). In addition they contribute to the protection of the waters sources like ponds, rivers and streams, collection of wild honey, control of soil erosion, and protection of buildings against storms. For example, the environmental importance attached to the grove by Aye community in Nigeria has been attributed to the roles of the grove in providing more dependable source of water for the community (Oyelowo et al., 2014). Thus sacred groves are not only the home of the local flora and fauna; but they are also a real gene pool (Eaton et al., 2000). Ancient secrets of herbs and traditional medicine, primitive practices of sorcery and magic are culturally associated with the groves, which provide multiple social, cultural (Nath, 1960; Godbole et al., 1998; Jeeva et al., 2005; Anthawal et al., 2006), economic (Kalam, 1996; Kandari et al., 2014) and environmental benefits (Hangarge et al., 2012; Tewari and Rao, 2015).

2. BENEFITS OF SACRED GROVES

Various ecosystem services like fuel wood, fodder, small timber, water source, livelihood opportunities, soil conservation, lowering of temperature and NTFP collection are derived from sacred groves (Gokhale and Pala, 2011). Their important role in maintaining hydrological cycle of the area favours local livelihood (Appell, 2013). Scared groves are the area, where one can find a water body in western Rajasthan. These sacred groves are utilized for multiple purposes depending upon the need of the local people. People use the sacred groves as grazing land for their cattle and collect minor produces for their day to day livelihood (Table 6.1).

Table 6.1. Benefits accrued from sacred groves by the people of adjoining villages in Rajasthan.

SNo.	Type of requirements	Number of sacred groves	Percent grove
1.	Water availability	113	91.9
2.	Fuel wood collection	38	30.1
3.	Grazing for livestock	94	76.4
4.	Dry wood collection	39	31.7
5.	Fruit and fodder collection	17	13.3
6.	Drinking water for human beings	1	0.8
7.	Fodder and other material collection	1	0.8
8.	Place of rituals and social gathering	123	100
9.	Place of meeting and decision making	all	100
10.	None	13	10.6

2.1 Conservation and water availability

Forests are known for their role in water conservation and soil moisture retention because of thick humus and litter layer, maintaining micro-climate and reducing evapotranspiration which in turn, maintain hydrology of the area (Appell, 2013). Sacred groves that have greater ecological and cultural roles in the rural landscapes since past years, are being increasingly targeted for development of hydroelectric projects or are being reclaimed for cultivation and other alternative land uses by the local communities (Ray et al., 2014). Sacred groves help conserve soil and water, improve soil stability and act as soil binder.

Many plants like vetiver grass (*Vetiveria zizanioides*), *Desmostachya* spp., and *Eucalyptus* spp. are maintained to bind the soil thereby preventing soil erosion. Most of the sacred groves are found to be associated with some water bodies including pond, Talab, Nadi, spring, stream, deep well etc. These water bodies not only provide drinking water for wild and domestic animals but also help in supply of water to the human habitation. In some cases these water bodies act as aquifer and help charging ground water, which is utilized for drinking water or for irrigation. Some of the sacred groves have more than one water body, if bigger in size. The study of Chaudhary et al. (2015) in Meghalaya, India shows that sacred groves increase resource in the form of NTFPs to the local people, provide water for the rice fields, working as a source of two minor streams flowing into the nearby villages and sources of springs, a source of bees for local apiculture and contribute nutrients to the fields in its vicinity.

The study conducted in Rajasthan also indicates that sacred groves work as a catchment area for many water bodies as most of the groves (91.9%) have at least one water body though some sacred groves had many water bodies in the form of spring, tanks, ponds and wells (Table 6.1). They also work as water supply source for the springs passing through it and promote agricultural activities in the nearby areas. However, due to negligence most of the water bodies are under deteriorated conditions. One such example is the 'Khadin System of Cultivation' in western Rajasthan, where water coming from uplands, mostly forests, sacred groves or other similar areas, is collected in farmlands down slope areas and conserved for cultivating agricultural crop (Plate 6.1A).

2.2 A place of grazing

Pastoralism represents the next development of human society after hunting due to domestication of animals like horse, cow, sheep and goat. As compared to the agriculture which developed later, pastoralism did not involve much labour. This is the land where grazing animals appear relaxed and peaceful. Further, the centuries old deforestation practices have driven by the need for fuel and agricultural and grazing land that has resulted in radical alteration in many natural landscapes (McCann, 1997; Nieuwhof, 2006). Sacred grove has a strong association with animals, the rearing of livestock and the use of the grove as a feed corridor in the dry areas and periods of drought. The groves also serve as a barrier which prevents livestock from causing crop damage in fields. These attributes of the sacred groves are of considerable economic value to most people living nearby the groves and contribute greatly to the maintenance of the sacred groves (Sarfo-Mensah et al., 2010). By tradition of a kind of taboo, the trees on the sacred areas are protected from being cut down. However, people usually lope or cut the branches of the trees for their livestock (Anthawal et al., 2006; Sharma and Devi, 2014).

Animal husbandry is an important support to rural people in dry areas and second occupation in Rajasthan. People use sacred groves/Orans as grazing ground together with the pasturelands. At sometimes demarcation between pastureland and Oran is very confusing except a small temple put aside in the area in case of Oran. Almost 76.4% of the sacred groves are utilized for the animals and ruminants grazing intensively. Sometimes the grazing needs of the local peoples are accommodated by the management committee of the sacred grove. For example Viratra Mata Oran covers about 1859 ha area and is ruled by the Oran Prabandhak Samiti involving 7-8 villages, which impose a penalties of Rs 500 to 1000 under violations of formulated rules. Earlier the Samiti was dominated by Rajputs, but now other casts have also joined, and their needs have been accommodated. However, people sometimes debranch the tree for the grazing animals. One can see the trimming of *Prosopis cineraria* canopy almost at same height from the soil surface in the area and many sacred groves in western Rajasthan. This is done by the Camel-the ship of the desert (the population of which is dwindling now because of varying threats) feeding on this important multipurpose tree. Likewise other trees are also pruned and grazed in different localities. However, overcrowding of the animals is almost making the soil surface devoid of surface vegetation (Plate 6.1B).

2.3 Fuel wood collection

As an institution sacred grove is perhaps the best example of indigenous traditional resource use practices promoting conservation of biodiversity, where removal of any living thing from sacred grove is a taboo. Despite of such social restriction, people collect fuel wood from these areas. The tribal community of Kerala usually collects canes and firewood indiscriminately from the grove areas in their vicinity (Warrier et al., 2015). Overexploitation of the resources by them has lead to its retrogression to a great extent. During the survey 38 sacred groves are still being used for collection of fuel wood. The collection is mostly of fallen dry twigs and branches. However, at some places people have pruned and pollarded some plants for their collection after drying. For example, people use to even cut some trees despite of protection made by the temple personals in Dev Bani of Shivji Mahraj, Garh Bassi, Alwar-said the head priest of the temple.

2.4 Dry wood collection

Earlier timber collection from sacred grove was only allowed for repairing of the temple, located in the sacred grove. Tree cutting for personal and commercial purpose was restricted in the grove area. This is true in many sacred groves available across the world (Gadgil and Vartak, 1976; Roy Burman, 1995; Godbole et al., 1998; Behera and Pradhan, 2015). Malhotra et al. (1998) have also reported the existence of numerous groves in Kerala from which plants and animals are not harvested. In some of the groves, one can observe lying dead wood in plenty and people use to collect these dry woods only at the time of death of a person for cremation. Dead wood is also collected while performing some rituals and social gathering (Plate 6.1C). Dead woods are part of the reasons people use to enter in the sacred grove and its collection is allowed for cooking food at present. Demand for fuel wood used for cooking food (langar) for the pilgrims and devotees are also fulfilled by the sacred groves. But unfortunate part is that people use the sacred groves for commercial purpose. The groves from where the fuel wood is extracted, local communities derive certain direct economic benefits from the sacred groves (Oyelowo et al., 2014). The study conducted in Rajasthan also indicates dead wood collection from almost 30.1% sacred groves, and in some of the regions it is more intense and the people use to cut the tree branches and pollard tree and leave these cur materials for drying so that these dried wood could be collected latter on as fuelwood (Table 6.1).

2.5 Collection of minor produce

In addition to fruits, fodder, and fuel wood, sacred groves are important source of a variety of medicinal plants used in curing different ailments and bites (Dev and De, 2012; Behera and Pradhan, 2015). Diverse vegetation growing in the sacred groves provides a variety of non-wood produce. Fruits and berries produced by the trees and shrubs are used by the wild as well as domestic animals. The respondents have pointed out some of their benefits from the groves. Most of the communities around sacred groves generally agree that the groves serve as a source of food, income and medicine. A total of 120 medicinal plants, widely used for the treatment of various ailments, have been reported from the sacred groves of Manipur (Khumbongmayum et al., 2005a & b). The roles and benefits of the sacred groves in the conservation of the regional medicinal plant have been emphasized in several studies from different part of world and conservation, utilization and assessment of the implications of exploitation of the medicinal species have become an important task (Pandey, 1998a & b; Jeeva et al. 2005 & 2006; Rampilla et al., 2015). People use to collect these produce for their personal uses and sometimes considered famine foods. Fruits of *Zizyphus* spp., *Salvadora* spp., *Capparis decidua*, *Diospyros melanoxylan* etc. are collected by the local villagers and utilized immediately or stored after drying for long term uses (Plate 6.2A).

Some of the tree species have multiple uses and have considerable value for the traditional medicinal use among the local community. For example Khakhra (*Butea monosperma*) is used as brushes for white washing, whereas its leaves are used for preparation of plates and 'Dona' (bowl). They are cut into strips and used to make a juice, which is mixed with butter milk, gum is eaten by women, root and bark are used in heeling blood in stool and dried and powdered flowers heel blood in urine. Lac insect is also found on its leaf. Branches of Salar (*Boswellia serrata*) or stem of Khajoor (*Phoenix sylvestris*) are used to make a 'Mandap', which plays an important role in a marriage ceremony or ‘Yagya’, respectively. Gum of *B. serrata* is used to make dhoop (incense sticks), whereas that of Gugul (*Commiphora wighttii*) is used in many medicines. Pipal (*Ficus religiosa*) helps in treatment of Guinea worms, whereas different parts of Neem (*Azadirachta indica*) are used in various ailments. Many other herbs growing naturally in the sacred groves are collected and utilized by the villager in the form of medicine as a part of traditional knowledge or with the help of local 'Baidya'. Some of them are *Ephedra foliata*, *Barlaria* spp., *Moringa concanensis* etc. Prospects of honey collection have also been observed in some groves.

2.6 A place of social gathering and rituals

Sacred groves provide sacred physical space, which is communally shared as commons, and used to observe important social ceremonies in indigenous societies. Several cultural festivals are performed in these sacred groves/Orans (Plate 6.2B). These sacred groves also provide a meeting place on various occasions including social gatherings, marriage, after-death rituals, etc. In some of the sacred groves, 'Mela' are also organized, during which people of different villages, regions or the states gather together. For example, lakhs of people gather together not only from Rajasthan, but also from Gujarat and Haryana at Ramdeora in Jaisalmer. Such gatherings are associated with the agricultural calendar. About sixty percent of the celebrations are concentrated after the traditional harvest time or the period when cultural practices related to agriculture are almost negligible and the people are more available (Malhotra et al., 2001). Study conducted in Chhatishgarh indicates that people follow many rituals, for example tribes of Jashpur region celebrate together Sarhul festival, by singing and dancing in a group on the maturity of the crops (Avinash and Yadav, 2014). At the time of drought or less rain in the village during whole year, villagers pray to the deity Bheema dev. Every year, during January-February two-fairs are conducted by the villagers and are famous by the name as Punni mela and Madai mela. Bastar Dushhera begins with the ritual known as the 'Kachangadi' in which a virgin girl swings in the bed made up of thorns. They also use to make many food items or dishes in these festivals (Avinash and Yadav, 20014). Other seasonal events are linked to the religious festivals or the birthday of the priest or prophet. The cultural event triad like pilgrimage, religious celebration, and popular festivals are rooted in religious rituals with their economic and social aspects in the region.

3. THREATS TO THE SACRED GROVES

Extraction of more and more resources from the forest area due to less income and inadequate marketing support together with lack of communication results in overexploitation of available resources. Studies reveal that these conservation sites face different levels of threats from urbanization, over-exploitation of natural resources, fragmentation, expansion of agriculture, religious tourism, lack of awareness, sanskritization, invasiveness etc. (Ormasby, 2013; Priya and Sharma, 2014; Nwankwo and Agboeze, 2016; Kemeuze et al., 2016). Increased biotic pressure on the existing sacred groves in the form of cuttings, over grazing, hunting and forest fires, which are neglected by the local people, leads to gradual degradation. There are dilution of traditional values, such as expansion of agriculture and settlement areas, unplanned development, illegal tree felling, poaching, and weaken and decrease cultural, biological and ecological values of these sacred groves (Madeweya et al., 2004). Further, human activities that were previously taboo, such as dead wood collection, biomass gathering, lopping of tender branches and green leaves for goats, creation of footpaths, cattle grazing, mining of sand and clay, brick-making and increased collection of wild fruits, vegetables, medicinal plants, fruit-eating bats and fireflies are affecting the ecology of the sacred groves.

The number of sacred groves in various locations and regions depend upon the magnitude of the varying threats which are area specific. Over the past few decades many people have lost their right of forest management to the government. There are several reasons for this decrease in terms of area and number. Primary cause is the lack of documentation as a result of which it becomes difficult to recognize the importance and conserve and maintain these sacred groves. Considering the varying levels of threats, many researchers have worked upon the conservation aspects of sacred groves through socio-cultural practices (Gadgil and Vartak, 1975; Khiewtam and Ramakrishnan, 1989; Tiwari et al., 1998; Sinha and Maikhuri, 1998; Bhagwat et al., 2005; Deb, 2007; Jaryan et al., 2010; Anthwal et al., 2010). Some common threats identified during the survey of the sacred groves of Rajasthan are being described below:

3.1 Disappearance of the traditional belief

Traditional belief is fundamental to the concept of sacred groves throughout the world. These systems and their rituals are now considered mere superstition (Ugwu, 2014). Lack of awareness in terms of long term future benefits has also resulted in the destruction of these sacred groves. Various religious beliefs, influx of large number of pilgrims and tourists and conversion to other religions axing of plants for monetary benefit are contributing consistently to the degradation of these well flourished areas of vegetation to a mere crunch of trees in many parts of the state. The rituals are now known to very few people, mostly belonging to the older generation. In some of the regions with larger sacred groves the traditional rituals are still performed in accordance with the customary beliefs, but in smaller groves the traditional rituals are no longer performed. Though silently but the traditional values appear to be gradually disappearing with the recent advent of modernization, urbanization and people's changing aspirations. As a result, the violation of cultural norms and taboos no longer carries heavy consequences, and the sacred groves are degrading.

> Erosion in people's beliefs in nature worship, religion and social values are major threats to the sacred forests, which are conserved patches of forests that lay undisturbed. Fear of Gods and social taboos prevented people from harvesting resources from these sacred forests though there are no physical borders. Thus change in social and religious values are now posing a danger to these sacred forests.
>
> **Bhattacharya (2014)**

3.2 Urbanization and developmental interventions

Though some of the groves are well protected, still most of the groves are subjected to moderate to high anthropogenic pressure in the form of construction activities, urbanization and modernization. The sacred groves in sub-urban with sufficient area are facing the threat of urbanization (Priya and Sharma, 2014). The areas situated near the urban or village vicinity are more prone to urbanization or encroachment. More over the other threat to most of the sacred groves is construction activities within the temple premises itself due to which large number of trees are being removed.

Sacred groves in many parts of India and the state of Rajasthan have been destroyed due to construction of roads, railways tracks, dams etc. Thus the recent threat to sacred groves/Oran is witnessed from the process of modernization, industrialization, greed of land mafias and construction of building. These all result in loss of cultural and ecological importance among the younger generation of local people. Out of 123 sacred groves, eight sacred groves provide place of cremation, whereas 16% sacred groves have been utilized for a mixed activities like construction of temple, Dharmashala and road as well as railway tract (Plate 6.3). Fortunately, 15% of the sacred groves have been found in safe condition and with full sanctity (Table 6.2).

Table 6.2. Types of utilization of sacred groves of Rajasthan.

SNo.	Type of land uses	Sacred groves (nos.)	Sacred groves (%)
1	Construction of temple or mosque	57	46.3
2	Railway track	1	0.8
3	Construction of Road or highway	19	15.5
4	Place for cremation	8	6.5
5	Construction of both Temple and Road	12	9.8
6	Construction of temple, Road and railway track	1	0.8
7	Construction Dharamshala, shops and Road	2	1.6
8	Construction Dharamshala, Temple and Road	3	2.4
9	Plantation activity	2	1.6
10	No change	18	14.6
Total		123	100

3.3 Grove fragmentation

Fragmentation through canal, railway or roads or even a walking trail, implementation of electric cables and poles leads to many small patches in the groves, which are more prone to encroachments (Vipat and Bharucha, 2014; Rajesh, 2016). A fragmented habitation and destruction made in it caused by development of infrastructure like roads are also threat to biodiversity, because roads act as barriers by impeding animal movements and restricting space use (Chen and Koprowski, 2016). Out of 123 sacred groves studied in Rajasthan, one is observed bisected by railway tract, 19 sacred groves are observed bisected by both railway and highway, whereas there are many motorable pathways in many sacred groves (Table 6.2). Multiple crossing and a network of roads and the railway lines fragment the groves into smaller isolated patches of some trees of plants, which ultimately lose their entity in absence of effective pollination and seed setting and dispersal as well. Ultimately it affects regeneration of the climax species observed in the sacred groves. An assessment carried out by Daye and Healy (2015) on fragmentation at the landscape level between 1995 and 2010 indicates trends of simplification and reduction in heterogeneity in sacred groves. In this, large increase (by 81.4%) in mean patch size is attributable to the loss of most small patches to other land-uses indicated by the decrease in patch number by 44.3% and reduction in patch size coefficient of variation by 9.1%. The landscape-level changes are also associated with changes in individual patch attributes with a large reduction in mean edge density, which have been reduced by 22.8% along with slight reduction in mean shape index. Small size and increasing edge density indicate high vulnerability, especially if an erosion of traditional cultural values reduces their protection (Daye and Healey, 2015). The negative effect of the small size and isolation of many sacred forests present a threat to the viability of population of many of the species. This can only be countered by establishment and protection of a network of other forest patches within the landscape providing the habitat to support metapopulation of tree species available in these sacred groves (Hanski, 1998).

3.4 Sanskritisation

In the past, only small temples used to exist under trees or in the grove, but now big temples are taking their place along with some residential and commercial complexes. In some sacred groves, trees are cleared for making big temples, developing parks, resting places for the devotees and pathways. The green area of sacred groves is shrinking due to conversion of the grove area by people for the developmental purposes as well as to accommodate the big gatherings during festivals and the 'Melas'. Similar conditions have also been observed in Rajasthan, where many groves are suffering due to '**Sanskritisation**' and transformation of the primitive forms of nature worship into formal temple or mosque worship. This has also changed the shape of the sacred groves. Out of the 123 sacred groves surveyed, 57 sacred groves are under construction of temples and mosque and are now transforming the nature of primitive worship into a modern type of pilgrimage (Table 6.2).

3.5 Increased effects of waste disposal

In the process of modernization and **Sanskritisation,** there are increasing fluxes of pilgrimage to the temples established in the sacred groves. Green patches are cleared of from the grove areas to accommodate the influx of the people during seasonal festivals. Although, tourism represents both possible benefit to groves through recognition and valuing of the site that can lead to funding and conservation support, but it also poses a threat to sacred sites due to the negative consequences of possible over-use and the consequences of solid waste and ecological impacts. Such anthropogenic pressure on the area in the form of water requirement, demand for fuel wood used for cooking food for the pilgrims and devotees leads generation of huge amount of solid waste (Priya and Sharma, 2014). Although increased tourist and pilgrimage generates alternative livelihood at local level, but corresponding increase in the load of solid waste results in the problem of waste disposal (Poddar, 2014). It is now becoming common problem in tourist places even in forest area. In case of sacred groves or places, Viratra Mata grove in Barmer, Baba Ramdev Temple grove in Pokharan, Jaisalmer, Sawanria Ji Seth Dham, Dausa, Ranakpur in Pali are few examples among many, where problem of solid waste disposal deteriorate water and environmental quality. For example, a large fair is held at Ramdevra, Jaisalmer from Bhado Sudi 2 to Bhado Sudi 11 (Aug - Sept), which is attended by lakhs of devotees who come in large groups from far and wide.

3.6 Over-grazing and fuel wood collection

Increasing human and livestock population and corresponding increase in collection of firewood, food and fodder by the local people result in deterioration of the grove condition. Nomads are seen freely collecting firewood and grazing their cattle. Initial investigations of sacred groves in the Konkan region of northern Western Ghats also reveal that these sites face threats of development and agriculture encroachment, increased resource use, cattle trampling and renovations of temples that lead to damage of the sacred groves environment (Blicharska et al., 2013). Overgrazing also prevent regeneration of many tree species and appears to be a common problem existing in many sacred groves and may lead these holy forests to disappear in due course of time (Deil et al., 2005). Enhanced intensity of grazing, lopping and removal of biomass also result in dwindling of the groves. There are usually 5 types of disturbances in the sacred groves. These are grazing, cutting, mining, movements of vehicles and dumping of the garbage (Table 6.3).

Table 6.3. Types of disturbances and exploitation of resources in the studied sacred groves in Rajasthan.

SNo.	Type of exploitation	Sacred groves (Nos)	Sacred groves (%)
1.	Grazing	104	84.6
2.	Grazing and cutting	4	3.3
3.	Grazing and Mining activity	1	0.8
4.	Vehicle movement	6	4.9
5.	Dumping Garbage and Grazing	1	0.8
6.	No disturbance	7	5.7
Total		123	100

Seven sacred groves observed are found intact without any disturbance, i.e. 6% of the total sacred groves, whereas 85% sacred groves are under pressure of grazing of varying intensity. In some of the sacred groves there is severe grazing affecting ground vegetation and thus regeneration. About 3% of sacred groves are affected by both grazing and cutting of fodder or fuel wood (Plate 6.1C). Studies indicate that grazing followed by shrub encroachment influence patch-level effects of shrubs on plant density, biomass and similarity in species composition between shrub under-stories and the open areas, that ultimately affects soil stability, nutrient cycling, and infiltration (Soliveres and Eldridge, 2014). Uncontrolled grazing in dry areas reduces soil carbon stock (Singh, 2015b), but no grazing also deteriorate plant-soil system (Reeder et al., 2004). Plants in natural plant communities are generally damaged and forage production is reduced with increase in animal stocking rates under grazing and trampling by the domestic animals (Ellison, 1960; Belsky, 1986; Painter and Belsky, 1993; Holechek et al., 1999). This also leads to simplification of the plant communities, establishment of woody vegetation in grasslands, and regression to earlier succession stages or conversion to invasive dominated communities in the grove (Reisner et al., 2013).

Removal of dry woods and litter along with the seeds from the grove floor results in retrogression by hindering the natural regeneration process (Warrier et al., 2015). Sometime people pollard trees or debranch the trees for their animals as fodder, and after drying of these materials they transport it to use as fencing material or fuel-wood. Trees of smaller girth class are easy to cut and transport as head loads as the most of the groves are very close to the road and human settlements. In some of the studies, majority of the cut trees are in the girth classes of 30-60 and 60-90 cm GBH (Mala et al., 2013). This indicates the preference of lower girth class wood by the villagers to meet the household fuel wood demand and also the demand of poles for construction purposes. Sundriyal et al. (1994) have observed similar condition in Eastern Himalaya forests, where a range of 30-90 cm GBH is preferred by the villagers for commercial consumption of fuel wood and construction of the huts during tourist season. Sometime people collect timber and firewood for sell-out in market and many industries are ready to pay for it. In many sacred groves timber and firewood collection is the main occupation for the local people (Choudhary, 2016). Thinning of the woody layer and change in the microclimate of the grove may impair regeneration processes of the tree species on the one hand, and help in colonization and establishment of shade-intolerant shrubs and annuals on the other hand (Mishra et al., 2004).

3.7 Human greed

Sacred forests often exist as isolated patches of natural forest even after conversion of the surrounding matrix to different forms of land-use. Encroachments of sacred groves by local communities or different government departments, and migration and immigration of people have contributed to the extinction of some of the sacred groves. One such problem is the mining activities deteriorating the environmental condition of the sacred groves. Though pure mining activity has been observed at one site, but excavation of soils, sand and murram (calcium carbonate aggregates) have been observed in many of the sacred groves and appears to be the common problem. Because of sanctity visit of sacred groves by the local people is also common, vehicular movement causing disturbances have been observed in 6 sacred groves.

Encroachment of sacred groves for varying uses is one of the major threats for the continued existence of sacred groves. About 64.2% sacred groves have been found intact with minimal encroachments in Rajasthan under shifting of agriculture field boundary in the grove and results in shrinkage in the grove area. Major encroachments are for agriculture purpose, which account for almost 27% of the total number of sacred groves surveyed (Table 6.4). Next is for both construction of houses and agriculture activities. About 3% sacred groves have been encroached upon for construction of houses only.

Table 6.4. Type of encroachments observed in the studied sacred groves of Rajasthan.

SNo.	Type of encroachments	Sacred groves (nos.)	Sacred groves (%)
1.	Agriculture	33	26.8
2.	Construction of House/ Dhani	3	2.4
3.	Both Agriculture and house or Dhani	8	6.5
4.	No encroachment	79	64.2

Like railways and roads, encroachments for agriculture and other activities also lead to fragmentation of the sacred groves into smaller patches and simultaneously loss to entity. Encroachment has led to the shrinkage of some of the largest groves in India. Greed for certain plants like sandal (*Santalum album*) in Southern India and Teak (*Tectona grandis*) in Rajasthan has resulted in decrease in their population. A study in Jodhpur district (Chaudhary et al., 2002) also indicates varying degree of encroachment of Oran and pasture, where half of the surveyed oran/pasture lands have marginal (0-20%) encroachments and 25% have more than 20% encroachments. More than 90% and up to 100% encroachments have also been reported in some sacred groves and pastureland in the above-mentioned study (Chaudhary et al., 2002).

3.8 Species in afforestation programme

Inappropriate selection of species under strengthening plantation of sacred groves has also affected the ecology and overall diversity of the sacred groves. Indigenous species are being neglected in population strengthening of the grove. Local species have also been observed replaced with exotic ornamental and avenue trees like *Araucaria cookie* (Priya and Sharma, 2014). One such example is a sacred groves dominated by *Eucalyptus camaldulensis*. Further, plantation of *Acacia tortilis* in some of the sacred groves particularly in western Rajasthan does not appears appropriate. For example, plantation of *Acacia tortilis* and *Acacia nubica* has been done in Kolu Pabu Ji Ka Oran, Jodhpur. Rather, indigenous species like *Prosopis cineraria*, *Cordia* spp., *Zizyphus* spp., *Bauhinia racemosa* and *Salvadora* spp. should be promoted in sacred groves of western Rajasthan.

3.9. Invasion by exotic weeds/species

The indigenous species found in the sacred groves are now threatened by the introduction of many exotic weeds. Some of the common exotic weeds are *Chromolaena odorata,* floss flower (*Eupatorium odoratum*), shrub verbenas (*Verbena officinalis*), Lantana (*Lantana camara*), Vilaiti Tulsi (*Hyptis suaveolens*), Parthenium (*Parthenium hysterophorus*), Climbing hempweed (*Mikania micrantha*) and Vilayati Babul (*Prosopis juliflora*) etc. These play a major role in degradation of sacred groves (Singh, 2014; Warrier et al., 2015). Invasion of *L. camara* and *P. juliflora* in the sacred groves of Rajasthan has become a serious problem for the ecology of these sacred groves. We have observed dominance of *P. juliflora* in almost 16 sacred groves. However, *P. juliflora* of girth class >30 cm have been observed in 58 (47%) sacred groves. While considering the second dominant tree *P. juliflora* has been observed in almost 47% of the sacred groves of Rajasthan (Table 6.5). Saplings of *P. juliflora* have also been observed in 38 sacred groves. Among the shrubs, the invasive *L. camara* has been observed in 11 sacred groves mostly in humid areas of Rajasthan under the study (Table 6.5).

As compared to the total population of trees in >30 cm and 10-30 cm girth class population of *P. juliflora* varies under dominant tree species (Fig 6.1). It ranges from 0.25 tree per 1.25 ha under *M. emarginata* dominated sacred grove (with total 66.5 trees per 1.25 ha) to 49.31 tree per 1.25 ha under *P. juliflora* dominated sacred grove (with total 73.3 trees per 1.25 ha). Average population of *P. juliflora* of >30 cm girth class is 6.03 tree as compared to 75.1 tree per 1.25 ha for all tree across state.

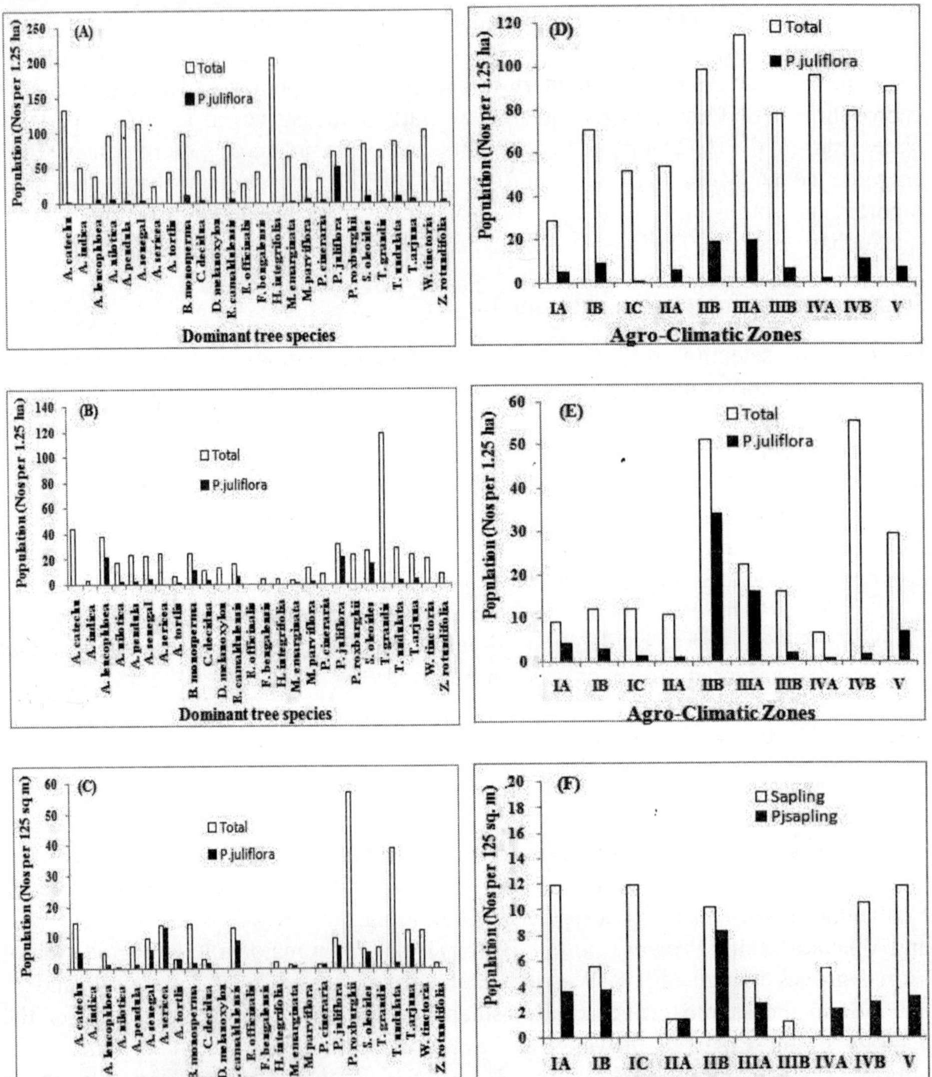

Fig 6.1. Population of *P. juliflora* of >30 cm girth (top), 10-30 cm girth (middle) and saplings (bottom) with respect to total population of all species influenced by dominant tree species and agro-climatic zones in Rajasthan.

For trees of 10-30 cm girth class, population of *P. juliflora* varies from 0.17 tree in *T. grandis* dominated grove (as compared to total 118.5 tree) to 22.0 in *A. leucophloea* dominated grove (as compared to 38.0 number of total trees). In this category, average population of all tree species and *P. juliflora* are 21.4 trees and 5.7 trees, respectively (Fig 6.1, left middle panel). In case of saplings, population of *P. juliflora* saplings ranges between 1 and 48 individuals per 125 sq m. In this the highest population of *P. juliflora* sapling has been recorded in Neelkanth Madeo Oran, Dungarpur. Among the dominant tree species saplings of *P. juliflora* is significantly high with *A. sericea* followed by *T. arjuna* dominated sacred groves (Fig 6.1 left panels).

Table 6.5. Extent of distribution of *P. juliflora* (trees of >30 cm girth, 10-30 cm girth and saplings) and *Lantana camara* in sacred groves of Rajasthan.

SNo.	Invasive species	Total grove	Sacred groves number
1	*P. juliflora* tree >30 cm girth	58	1-3, 6, 8, 11-15, 18-19, 22, 24-26, 28-30, 38, 40-41, 43, 45-46, 51, 56-59, 61, 68, 69, 70, 75, 77-78, 80, 82-83, 85, 89, 90-92, 98, 100-102, 106-108, 110-113, 115, 120
2	*P. juliflora* tree 10-30 cm girth	57	1-3, 6, 8, 11, 13-15, 20, 22, 25-26, 28-30, 40-41, 43, 45-46, 49, 51, 54, 56-59, 61, 68, 69, 70, 75, 76, 77-78, 81, 82-83, 85, 89, 90-92, 97-98, 101-102, 106-108, 110, 112-115, 120
3	*P. juliflora* saplings	38	1-2, 14-15, 22, 24, 29, 43, 45-46, 49, 51, 57-58, 68-71, 75-77, 81-82, 85, 88, 90-92, 97-98, 106-108, 110, 112-115
4	*Lantana camara*	11	1, 9, 11, 33, 44-45, 94-96, 116, 123

Among the agro-climatic region, population of *P. juliflora* having >30 cm girth varies from 0.6 to 19.8 trees per 1.25 ha as compared to the total tree population of 51.8 tree and 113.9 tree in IC and IIIA agro-climatic zone, respectively. Population of *P. juliflora* trees of 10-30 cm girth ranges from 0.56 tree to 34.0 tree as compared to the total tree population 6.67 trees and 51.32 trees per 1.25 ha in IVA and IIB, agroclimatic zone, respectively (Fig 6.1 right panels). The saplings of *P. juliflora* have been observed highest in IIB and almost nil in the sacred groves of IC and IIIB agro-climatic zones.

Among the shrubs *Lanata camara* confined only to the sacred groves of Ajmer, Alwar, Pratapgarh, Banswara, and Udaipur districts. Population of *L. camara* varies from 1 to 68 individuals per 125 sq meters. Among the dominant tree species its population is highest with *E. camaldulensis* followed by *W. tinctoria* (Fig 6.2, left panel), whereas almost no *L. camara* has been found to be associated with the tree species of the western Rajasthan. Among the agro-climatic zones, highest population of *L. camara* has been observed in the sacred groves of IVB agroclimatic zone, whereas the sacred groves of IA to IIB zones have not shown *L. camara* shrub.

3.10 Policy implication

Conflicts among the sacred grove managers have also resulted in loss of biodiversity in certain sacred groves, when policy decisions have been made to benefit certain minority sections of the village society, against the traditions of the sacred grove. Moreover, area and legal status of several sacred groves has not been clearly defined (Chandrakanth and Romm, 1991). This results in lack of effective legislation against offenders. Further, lack of co-ordination between the revenue officials and other parallel departments including forest department and their callous attitude towards the groves is also affecting proper management. Some vested interests take political mileage also. In Kranataka, Devarakaadus sacred groves have been given the status as protected forest after 1888 and are declared as belonging to the department; thus no single family could claim absolute right over these forests (Latitha, 2015). Despite of this no action is being taken on the land grabbers and more worse is the regularization of the encroachments by the Governments in many of the cases. Many protected forests have been misused, abused and worst vandalized resulting in loss of entire ecosystem in many areas. Thus policies should be developed for managing these threats and in addition it should gear improvement and enhance capacity of local communities who are capitalizing their cultures and religions in conserving important biodiversity of the sacred groves particularly in sacred groves situated at distant place from the main forests.

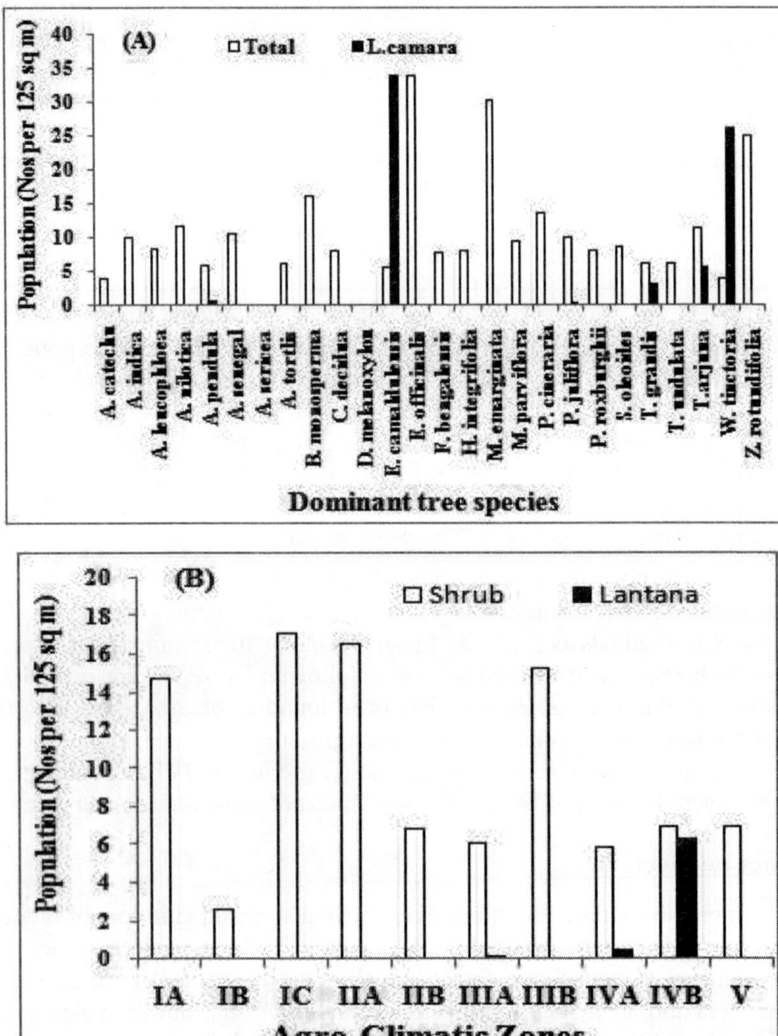

Figure 6.2. Changes in the population of *L. camara* in the sacred groves dominated by different tree species (top) and in different agro-climatic zones (bottom) of Rajasthan.

4. CONCLUSIONS

Sacred groves provide varying ecological, social and economic benefits. These sacred groves are place of water availability, grazing, fuel wood and fodder collection and work as communally shared commons and use to observe important social ceremonies in indigenous societies. Several cultural festivals are performed in these sacred groves, which also provide a meeting place on various occasions including social gatherings, marriage, after death rituals etc. Various threats to the sacred groves are construction of temples, shrines, residential and commercial buildings, etc. in and around the groves, encroachment of the grove land for agriculture purposes, exploitation of resources by locals, removal of vegetation, grazing of cattle, tourism activities, replacement of indigenous tree species by the exotic species, etc.

Modern age of liberalization, privatization and globalization has made community to deviate from the customs which were used to guide principle in protecting and maintaining the sacred groves. Sanctity of life has no place in market-linked livelihood systems, whereas modern education has no respect for local traditions. As a result this ancient and widespread institution is now weakening in terms of both cultural and biological integrity. Increasing pressure from human population and livestock and encroachments by the individuals and government organizations are the major threats. Developmental activities, agricultural encroachment, increased resource use, cattle trampling and renovations of temples has lead to the damage of these valuable habitats (Blicharska et al., 2013). Over-exploitation and unscientific collection of plants for varying uses, eroding community values and lack of faith in these institutions in the younger generation are other threats (Yadav et al., 2010; Rao and Sunitha, 2011).

Human induced disturbances lead to thinning of the woody layer and change in the forest microclimate and impair regeneration processes of the indigenous tree species and help colonization and establishment of shade-intolerant shrubs and annuals (Mishra et al., 2004). The degradation of sacred groves is also in the form of loss of species and loss of rich cultural heritage of area. Invasion of sacred groves by *P. juliflora* and *L. camara* are now becoming major threats as population of these invasive species is increasing every year and requires special attention to adopt control measures. Moreover, lack of policy and any legal status to the sacred groves also promote offenders for illegal activities in this ancient treasure. Though threats to the sacred groves vary from one region to another or even from one grove to the other, but most common threats are:

(i) Loss of traditional beliefs;

(ii) Rapid urbanization and developmental interventions (dams, large-scale mining works, road construction, implementation of cables and poles, tourism, and industrialization);

(iii) Commercial forestry;

(iv) Transformation of the primitive forms of nature worship into formal temple worship;

(v) Invasion by exotic weeds like *Lantana camara* and *Prosopis juliflora*;

(vi) Increasing livestock and fuel wood collection; and

(vii) Fragmentation and perforation of sacred groves.

Such type of anthropogenic activities should be prevented by proper conservation and protection measures of such areas. People involved in the conservation of sacred groves should be encouraged and incentives should be provided to them (Sharma and Devi, 2014), so that this rich traditional conservation institution could be saved for the betterment of future generations.

Plate 6.1A. Sacred grove near Kaladungar Jaisalmer (left) working as area of water resource for cultivating agriculture crop and supporting Salvadora-Tamarix association down slope area (right).

Plate 6.1B Grazing animals in Kolu Pabuji Oran, Jodhpur (left) and drinking water (right) at the adjoining site in the Oran.

Plate 6.1C. Wood collection and storage (left) and Neem tree lopping for fodder to livestock (right).

Plate 6.2A. Collection of fruits of *C. decidua* (left) and harvesting of *S. munja* for thatching (right).

Plate 6.2B. Sacred groves as place of performing rituals (left) and place of discussion (right).

Plate 6.2C. Encroachment for house construction (left) and fragmentation of *Aegle marmelose* grove (right) are some of the important threats to the sacred groves.

Over grazing at Karani Mata Oran

Construction of road

Encroachment for agriculture

Mining activity

Construction of Dharmashala

Cutting for fuel wood

Plate 6.3. Varying types of disturbances in sacred groves of Rajasthan.

7

PEOPLE PERCEPTION AND MANAGEMENT STRATEGIES

With the passage of time, there are considerable changes in the extent, status, vegetation structure, religious beliefs and taboos and even people's perception towards the sacred groves. Management through local institutional structure based on respected priests and the integrity of the local village institution has now being altered under societal changes and modernization. Apathy among the local people, family and village heads, forest department officials, government and panchayats are also leading to gradual loss in sanctity of the sacred groves. People perceptions are also changing due to modern development and the economic benefits which the land opportunists are offering them. However, perception and awareness increase with the educational level. Sacred groves are perfect system of conservation, where an entire ecosystem is protected and conserved for the sustainable development. A holistic understanding of the current status, structure and function of sacred grove is essential for formulating strategies for sacred grove conservation and ensuring ecological role of this valuable ancient institution. Renewal of community traditions, upholding socio-cultural beliefs, reintroduction of indigenous plant and animal species, removal of human induced obstruction to land and water flow as well, removal of invasive species, recharging of ground water and ensuring alternative livelihoods of local communities are some of the strategies for effective conservation and management of sacred groves.

1. BACKGROUND

Sacred groves are usually not virgin forests, if the area is simply a forest and low in species diversity. Reasons for disappearance of sacred groves are many fold. Factors contributing to pressure on sacred groves include cultural change and natural resource demands. The structural variability and the floristic richness of sacred groves are happen under constant disturbance by people in the form of digging, burning, cutting etc. and their livestock like grazing, defecation, etc. (Deil et al., 2005; Manikandan et al., 2011). Management through local institutional structure based on respected priests and the integrity of the local village panchayat has now being altered through a widening economic divide in the traditional community and the new societal changes (Vipat and Bharucha, 2014). In a study related to the human impact on sacred groves and biodiversity in west Bengal, 28% respondents believe that change in people's attitude and socio-cultural practices are the main reasons for degradation of biodiversity and sacred groves, whereas rest respondents are in opinion that habitat degradation, forest fragmentation, exploitation of natural resources are other direct human impacts (Prasad et al., 2015). Many sacred groves have been grabbed or intruded by the migrants from the outside, land grabbers and some have been converted into small village units by communities.

The indifference in local people, family and village heads, forest department officials, government and panchayats add to the woes of the groves and result in gradual loss of pristine purity and glory of the sacred groves (Mahato, 2015; Latitha, 2015). While frequency of rituals conducted in association with the sacred groves is declining, there is economic pressure to extract resources from sacred groves or to reduce the sacred grove size, particularly for coffee production or other production systems (Ormsby, 2013). Modification in the social constitution plays an important role in declining the grove system. The nature centric worship places in the sacred groves of many areas have already been replaced by temples or by solid structures in the name of so called modernization. As the deity has been shifted to the temple, the adjacent areas have become irrelevant for the people resulting in instantly occupisation for other uses (Kar et al., 2015). Modern development is altering the people perception due to the economic benefits that land speculators are offering them, though perception and awareness increases with the educational level of the respondents, but highest familiarity with biodiversity and environmental problems is observed among 18-25 year-olds age group and it decreases with older age group (Prasad et al., 2015). This emphasizes community participation for successful preparation of conservation strategy, as local communities play an important role for protection and conservation of these sacred groves.

Studies on sacred groves relating to plant conservation provide mixed results, where some shows their effectiveness while the others do not (Asokan et al., 2015). Most of the field studies usually suffer from overrating rare species, and show insufficient consideration of socio-economic data (Manikandan et al., 2011; Fournier and Sanou, 2013). Studies on sacred groves particularly in India, present two opposing images on the social institutions (Gadgil and Vartak, 1976; Boojh and Ramakrishanan, 1998; Latitha, 2015). One group represents sacred groves as the archetype of community-based traditional conservation systems and highlights the conservation ethic of communities that protect sacred groves. In this regard several cases have been reported where conservation ethic has resulted in conservation of valuable ancient resources (Ramakrishnan, 1998; Sarfo-Mesah et al., 2010; Manikandan et al., 2011). Karanth (1998) put an opinion about the alternative concept of 'sustainable landscapes' in combination with the ideas of the emerging discipline of ecological economics that provides useful tools for protecting the sacred groves in which the wildlife could survive in coming centuries. The recent empirical studies appear more critique to the conservation ethics attributed to these social institutions, where many case studies argue that sacred groves may not necessarily be a better conservation model, because communities maintaining sacred groves may not necessarily have a conservation ethic or objective (Roy Burman 1992, 1997; Vipat and Bharucha, 2015). The ethnic conservation may be simply a by-product of different factors for which the sacred groves are maintained and the sacred groves are inequitable and unsustainable institutions. Likewise, the beliefs that promote plant cover conservation do so only to a certain extent and regardless of the species content, which raises the concern that alien species may upset the established communities (Fournier and Sanou, 2013). Because of economic and social issues arise out in recent decades; the traditional methods of managing sacred groves appear less effective. This emphasizes external intervention taking the local people into confidence and pulling out important sacred groves under the 'Protected area Network' to ensure their proper conservation and management.

2. PEOPLE'S PERCEPTION

People response to a programme is backbone of success in proper implementation, maintenance and management of an area including sacred groves (Takahashi et al., 2012). An ethnographic study reveals that while the immediate decisions regarding sacred groves are taken locally, the key influences determining their changing and diverse fate emanate from spheres of property regimes, state policy and an environmental discourse that originate far away the local community. State policies, donor agency and NGO's and recent institutions of self governance and resource management play a vital role in dynamics of the sacred site. While the micro-political analysis of sacred groves uncovers defects in the world of social actors whose actions directly influence the fate of sacred groves, an understanding of current trends influencing them requires a macro-political socio-economic perspective (Gandhi and Vardhan, 1997; Vasan and Kumar, 2006). In a study conducted in Meghalaya, when the residents were questioned about the sacred forest expansion, about 65% of the respondents were in view for expansion of the sacred grove, and only 30% of the interviewees were pessimistic about sacred grove expansion because of limited land availability (Ormsby, 2013).

Personal interactions and discussion conducted in Rajasthan indicate that entire respondents get some sort of direct or indirect benefit from the sacred groves and the product from the areas. Most of the respondents are in favour of conserving these areas so that benefits arise out of it cannot be hampered and its entity can be maintained. The requirements of most of the respondents combined according to the sacred groves and have been enlisted in Table 7.1, which are 15 in number. It is clear from the table that respondents of almost 100 (81.3% respondents) sacred grove sites are in favour of erecting boundary wall for effective protection from encroachments and conservation of the diversity and treasures available in the grove area, very much similar to the other study in Kerala (Prasad and Raveendran, 2012). Findings of Alohou et al. (2016) also indicates a decrease in sacred grove area by more than half during three decades and people have been observed well aware about the extent of threat to the sacred forests and different services provided by the groves including the cultural ones as the most favourable, because the religious groves and secret forests are found providing the most diversified services. In a study in Nigeria also, 79% respondents show favourable attitudes towards the concept of conservation of the fragments of sacred forest through cultural practices, whereas 14% respondents agree to conserve the grove because of their regulating, provisioning, social/cultural and supporting services. In this 71% respondents generate income from various activities carried out within the sacred forests, hence contribute to their livelihood economy. However, changes in religious beliefs and the need for development have shown a negative effect on the conservation attitude of the majority of the respondents resulting in the destruction of sacred forests.

Table 7.1. People requirements in protection and conservation of sacred groves of Rajasthan.

SNo.	Types of management	Sacred groves (Nos. /%)
1.	Demarcation and erection of boundary	100 (81.3%)
2.	Plantation	94 (76.4%)
3.	Development of water facility in sacred rove area	48 (39.0%)
4.	Documentation	20 (16.3%)
5.	Construction of Dharamshala for rituals	14 (11.4%)
6.	Requires government intervention in curbing encroachments	11 (8.9%)
7.	Proper protection and caring	8 (6.5%)
8.	Declaration as a tourist place	2 (1.6%)
9.	Road construction	2 (1.6%)
10	No interference from the government departments	2(1.6%)
11.	Construction of Temple	1 (0.8%)
12.	Eradication of invasive, i.e. *Prosopis juliflora*	1(0.8%)
13.	Development of garden/park	1 (0.8%)
14.	Electrification of the main temple area	1 (0.8%)
15.	Ban on mining Activity	1 (0.8%)

About 76.4% respondents have expressed their view for strengthening the sacred groves through plantation so that its diversity and benefits could be enriched. However, the respondents are mostly in favour of indigenous species of multiple benefits fulfilling the requirement of human and domestic as well as wild life. Though most of the sacred groves are associated with some water body, but erosion from the adjoining upper or uphill area has led siltation and negatively affected the water storage capacity and its life span. It is also reflected in the response of the villagers during the survey. About 39.0% respondents put their demand for developing water bodies in the sacred groves so that drinking water supply for domestic and wild animals could be ensured. Further it will also help charging ground water to sustain future needs. Interestingly respondents of almost 20 sacred grove sites have observed sensitive enough about the sanctity of the sacred groves and aware too about ecological, cultural and economic benefits, that they need full documentation of these sacred groves so that effective management strategies could be framed. Similar views have also been highlighted in recent studies (Choudhury, 2015; Gaikwad et al., 2004; Madeweya et al., 2004; Prasad and Raveendran, 2012; Vipat and Bahrucha, 2014). Tribal communities like *Bhil, Meena, Banjara, Kathodi, Rabaris, Sansi* and *Kanjar* are well aware and perceive and manage even natural disasters and extreme weather events in western Rajasthan. This includes their strategies for early detection of coming events and for coping with these events, as well as their perceptions of their short and long term impacts on biodiversity. However, lack of benchmark makes indigenous knowledge difficult to harmonize and integrate into conventional forecasting system because indigenous knowledge is mainly based on relative local experiences.

About 11.4% respondents insist for construction of Dharmashala so that the people coming to the sacred groves to attend different functions as well as during social gatherings could be benefited from it and its sanctity could be maintained.

Moreover, it will also facilitate tourism. Another 8.9% respondents are in favour of government intervention for effective control measures regarding encroachments including enforcement of laws. People are also in view of effective protection and conservation but they require help from the government for proper caring of the sacred groves. Some respondents do not want involvements of government organization in protecting and conserving these sacred groves. However, crumbling rules and taboos, which use to preserve and protect the groves, must be addressed for effective conservation and management of the groves (Onyekwelu and Olusola, 2014).

Awareness about the invasive species like *P. juliflora* has also been observed as some respondents also put forward their response to eradicate or control this species affecting composition, structure and ecological function of sacred groves and disrupt water flow (Tegegn, 2008; Kent, 2013). This needs to be taken care at different levels. Likewise demands to ban mining activities in the sacred groves area are also there. Further, local peoples are also in support of developing some of the potential sacred groves as spot of tourism and nature watch. However, the level of community involvement in the tourism, and tourism profit sharing may affect the effectiveness of the protection and conservation of the sacred groves, in which transparency in revenue sharing will be required for facilitating positive tourism development at the community level. Likewise other requirements of the local people mentioned in the Table 6.1 may have different impacts in terms of environment.

3. CAUSES OF CONCERNS

Tree worship is not related to any particular religion, but to the entire mankind since time immemorial. Trees thus have occupied a prominent place of pre-eminence and sanctity in every Indian in one way or the other (Ramakrishnan, 2002). The present study reinforces the notion that the sacred groves, though small in size, are important repositories of endemic flora and fauna. However, it is clear that tradition and way of life associated with the sacred groves which were used to be followed strictly earlier have been eroded during the last few decades and the groves are no longer enjoying the same status and privilege as they used to be in the past.

> National Environment Policy 2006 says that sacred groves are nature's laboratories for evolution of wild species and repositories of significant genetic and ecosystem diversity.

Lands of sacred groves are at present no man's land. In absence of effective legislation and action against deter offenders it will be difficult to protect and conserve the ancient system of nature conservation. Community-based land and resource management systems in conservation of grove and the associated biodiversity were recognized by the anthropologists, biologists, conservationists and environmentalists; and restrictions on the access and use of such areas generally reduced or even eliminated human impact and thereby helped protect species in the sacred groves (Christopher et al., 2014). The legal status and total area of several groves have not been clearly defined. Despite of significant role played by sacred groves in conserving biodiversity there is no policy for management, conservation and utilization of these resources. Further there are no special laws for eviction of unauthorized occupants. Absence of a single coordinating agency results in fragmentations as well as duplication of the efforts undertaken by various governmental and non-governmental agencies for conservation and rehabilitation of the sacred groves in some of the cases. Further, almost all sacred groves do not have any kind of physical boundary of demarcation. In absence of such demarcation the extension of the agriculture lands and encroachments for various uses are continued to stay uninterrupted.

Diversion of sacred groves as well as grazing lands is increasing at a very fast rate for other uses. Utilization for construction of various community assets like Schools, Hospitals, Community Halls, Panchayat Bhavans, Public Distribution Systems, and Meeting Halls etc by various government departments and non-governmental organizations are going to hamper all the cultural, social, economic and ecological values of the sacred groves. Construction of houses, *bara*, shops and other commercial establishment by local resident, elected representatives, land-mafias, etc together with slow but steady extension of the boundary of the adjoining agriculture fields are also resulting in reduction in the size of the sacred groves.

The principal cause for fragmentation is population pressure, requirement for biomass and agriculture as well as for transport (Ray et al., 2012). Once the grove interior is exposed, it will lead other species to invade the area and change the floral composition as well as microclimate to a great extent. These changes will facilitate establishment of the invasive species in the grove and leaving the endemic flora on the verge of extinction. Simultaneously shrinkage in the area of grove and change in microclimate would affect the composition of flora and fauna (Ramesh and Ishwar, 2008). Large to medium size animals cannot stay in a small fragment and smaller ones which can stay, they face shortage of resources, shelter etc. affecting the normal life of the biological entity.

There is lack of adequate scientific data about the ecological, economic as well as environmental values of these sacred groves. Though there is projection of about 25000 sacred groves in Rajasthan, but a concerted effort is still lacking to come to the exact number and the area therein in the sacred groves of Rajasthan. In view of climate change also there is very limited data available about the existing carbon stock in both green biomasses as well as in soil of the sacred groves of Rajasthan.

Gradual decline in social structure play significant role in sacred grove degradation. Cultural changes among the young people are so rapid that they no longer believe in the methods their ancestors followed to maintain such fragile ecosystem. The worship places in many of the areas have already been replaced by temples/solid structures in the name of modernization. Once the god/goddess will be shifted to the temple, the surrounding places will become irrelevant to the people, leading to its conversion for other usage and encroachments as well. As a result, the local shelter for plants and animals as well as ecosystem functioning will be completely lost.

Neglecting the maintenance of groves increases the vulnerability of the system under various factors like cattle grazing affecting regeneration, leaf litter and twig collection (affects nutrient cycling and moisture conservation), encroachment for some other purposes, etc. This is going to affect overall functioning of the grove.

Most hidden matter in conservation of sacred grove is that the village people living nearby the sacred groves are relatively poor and their livelihood depends upon the grove. So, any step for the conservation of the sacred groves may not be successful unless viable option is provided to the local people for sustaining their livelihood.

In addition to the anthropogenic problems, frequent drought is inherent characteristic of the climatic conditions of dry areas and Rajasthan in particular. The adverse effects of the deficit rainfall affect natural regeneration and seedling recruitments of indigenous tree species and overall survival and growth of the plantation too if taken in it. Higher mortality of the plants raised under various schemes in rehabilitating sacred groves results in frequent plantation in absence of suitable species and sufficient provision of funds for cultural practices like fertigation, hoeing, irrigation or application of fungicide/insecticides for the raised plantation.

Though a significant amount of literature are generating on sacred groves, systematic research is still required on a variety of sacred places with respect to biological matters such as their size, age, species composition, biodiversity level, and degree of naturalness as well as any historical, social, cultural, religious, economic, political, tenure, and legal matter (Madeweya et al., 2004; Sponsel, 2008). Likewise there is a need for controlled comparison among sacred places and adjacent secular ones of the same size and type of biotic community in order to describe and assess any differences that might arise because of sacred status of an area (Roy Burman, 1996; Sponsel, 2008).

Many noxious invasive weeds are also common to the sacred groves which are harmful for the functioning of this ecosystem (Kemeuze et al., 2016). A study conducted in Jhabua district, Madhya Pradesh show *Ageratum conyzoides, Antigonon leptopus, Argemone mexicana, Cassia tora, Datura innoxia, Echinochloa crus-galli, Lantana camara, Lagascea mollis, Leucaena leucocephala, Parthenium hysterophorus, Opuntia elatior* and *Xanthium strumarium* as the invasive species (Wagh and Jain, 2015). Among these *Ageratum conyzoides, L. camara, P. hysterophorus* and *P. juliflora* are harmful to native flora (Singh et al., 2010; Tripathi and Shukla, 2007; Dogra et al., 2009).

Protecting and conserving sacred groves through adopting natural means can simultaneously help protect cultures, religions, and rights as well as the associated biotic species and ecosystems functioning. Aldo Leopold (1887-1948), a pioneer wildlife biologist and conservationist, stated most succinctly the essence of a viable ecocentric environmental ethic in his essay the “Land Ethic”: "A thing is right when it tends to preserve the integrity, stability, and beauty of the biotic community. It is wrong when it tends otherwise." This ethic very much applies to sacred groves and their role in conserving biological diversity and the functioning of this ecosystem.

4. MANAGEMENT STRATGIES

Sacred groves are perfect system of conservation, where an entire ecosystem is protected and conserved for the sustainable development. In this there are needs of reintroduction of indigenous plant and animal species to the grove, dismantling of the man-made or induced obstruction to the water flow, removal of invasive species and recharging of ground water. This could only be strengthened and promoted by monitoring of the sacred groves through newly established 'Bio-diversity monitoring committees' by formulating local specific rules and regulations for the sustainable management. If their security is ensured through fencing and community awareness programs, these groves can play more important roles as decentralized centers of biodiversity conservation and ecological services. These groves have been once recognized sources of springs and streams and have benefited the village communities. It is now essential to restore sacred grove system and revitalization of community centered management system under joint management regime which can continue to play a greater role in strengthening rural livelihoods by supplying various non-timber forest produce and providing water security, rendering pollination services and may be an ideal local level mitigation measure for global climatic change (Sigu et al., 2001; Busch et al., nil). Restrictions on access and use of such areas generally reduce or even eliminate human environmental impact and thereby help protect species in the sacred groves (Christopher et al., 2014).

Socio-economic, ecological and conservation importance of sacred groves has been recognized and it has been emphasized that immediate conservation of the sacred groves is must (Gadgil and Berkes, 1991). Several approaches and options could be adopted to conserve these scared spaces. Local people have acquired unique knowledge and skills by adapting to the specific constraints and advantages of their fragile and sometimes in-hospitable environment. Though now eroded, by using such knowledge, taboos and cultural ethics and norms, the local people has ensured long-term sustainability and conservation of these sacred sites and grove for varying benefits. In some of the cases, rural poor depend upon biological resources for meeting about 90% of their day-to-day needs. Until and unless viable option is provided to the local grove dependent communities for sustaining their economic condition, any step for the conservation of the sacred groves will not be successful.

Spectrum of researchers and environmentalist often range from the radical to the reformist changes in conservation and management of the sacred groves. One group of researchers believes that humanity can achieve harmony with nature only through radical adjustments. They believe that a complete overhauling of the system including political, economic and industrial is required to achieve a sustainable society. On the other hand the reformist feels that countless campaigns need to be carried out to reform laws and sympathetic law makers need to be elected and the public needs to be won over. Likewise, difference between preservation and conservation has also been defined. While environmental preservation is viewed as the strict setting aside of natural resources to permit damages caused by contact with humans or by certain activities like grazing, fishing, logging, mining etc., the environmental conservation allows for some degree of industrial development but within limits. Enacting regulations and laws both with respect to preservation and conservation of natural resources is on top priority (Burman, 1995; Kandari et al., 2014).

The strategies for biodiversity conservation that link culture and nature are more likely to be effective than those imposed from the top down by government and/or international agencies and those which ignore the traditional beliefs, values, institutions, and practices of local societies (Gadgil et a., 1993; Ramakrishnan, 1998). For achieving radical changes there is need to include better designs and more widely distributed social innovations under planning. One cannot use protest as a mean to achieve sustainability, rather one must use constructivism by making use of already existing tools, models and ideas in an innovative way. In most of the cases, the management of sacred groves is under the Village Panchayats, which unfortunately are highly political institutions with artificially constituted units of communities or villages with divergent agendas and social and economic identities. This leads to conflict development many of the times.

For ecologists, traditional ecological knowledge offers a mean to improve research and also to improve resource management and environmental impact assessment. The two obvious aspects in sacred grove conservation and management include religious beliefs behind their origin, resource use rules and meanings to local communities at one side, and well-preserved state of forests in them on the other side. These two elements could then be extrapolated to present sacred groves as inherently oriented towards biodiversity conservation and communities managing them as conserving communities (Vasan and Kumar, 2006; Kaur et al., 2013). For effective management of the sacred groves there is need to strengthen the existing conservation activities of the villagers and to restore degraded sacred groves following landscape level approach.

4.1 People participation and awareness programme

For the protection and conservation of sacred groves and to maintain their natural identity and sanctity, there is need to uphold traditions and beliefs through people participation. Surrounding village communities need to be educated and guided for sustaining the sanctity of the existing grove. The local people basically are interested in continuation of sacred grove based tradition, but in the way of modernism in life and religion. Here villagers with support from forest department and non government organization can protect the groves at their vicinity under joint action (Kaur et al., 2013). Conservation of sacred groves requires a step-wise approach including comprehensive recognition of the natural and social values, capacity building among local communities and design of locally suited incentives with participatory planning and implementation (Singh, 2010; Blicharska et al., 2013). Grove conservation activities can also be enlisted in the programmes along with the provision of small incentives to the younger ones. These incentives can be made in the form of carbon credit, ecotourism and for monitoring activities depending upon local condition and the requirements. Further, to avoid the scramble of modernization and globalization from causing casting aside of this ancient, decentralized and local community centered conservation practice; conservation of sacred groves can be enhanced by establishing "buffer zones" around the sacred site itself. These zones can help to promote sustainable development and sustain the people livelihood.

Considering the level of threats to the sacred grove the residents need to be sensitized and alternatives to firewood and other needs should be provided to the residents that can decrease the pressure on the sacred grove. Revival of past ceremonies related to the sacred groves as well as through awareness campaigns and education programs highlighting the ecological and spiritual benefits of these sacred forests and reinforcing rules about the use of resources can also help people involvement enthusiastically (Chandrashekara and Sankar 1998; Wild and McLeod 2008). Long term conservation needs an incentive for the effort to be put in by the local people. Some successful models are introduction of medicinal plants and providing information and technical support in the available area of the sacred grove in Konkan Region in Western India; and providing piped water to individual home as additional benefits in Meghalaya (Chaudhary et al., 2015).

4.2 Government policy and planning

There are some inherent environmental principles behind the conservation practices of sacred groves. Apathy of the elders towards decimation of knowledge to the younger generation about the use or importance of the sacred grove, softening over the rituals and taboos etc. are resulting in loss of faith and effectiveness of traditional rules and regulations. These need to be strengthened by upholding socio-cultural aspects. A separate policy may be formulated at state level to ensure conservation, protection and rehabilitation of sacred groves through active cooperation and involvement of the local people while meeting the basic requirements of livelihood like fuel wood, fodder, small timber and ethno-botanical needs of the local residents on sustained basis (Kandari et al., 2014). As women are more enthusiastic regarding conservation and preservation of different plants, a strategy at local and regional level should be adopted with major emphasis on women participation. Existing statutory rules and regulations governing protection and management of sacred groves needs critical review, whereas Acts and rules governing removal of unauthorized occupants from the sacred groves area may be made more stringent and effective to challenge land grabbers, deter poachers and protect shrines. This may bring the grove back to life and their significance.

4.3 Documentation and delineation

Under paucity of knowledge about the extent and distribution of the sacred groves, their edaphic and ecological conditions and social environment in the adjoining areas, any management strategies or management plan is not going to get success. In additions to have a knowledge about the existing species (Sharma and Upadhyay, 2014), there should be detailed surveys to know the number, extent of distribution and present status of the sacred groves by physical approach along with the help of remote sensing technique. Further, the boundary of each sacred grove needs to be physically demarcated by fixing of suitably spaced stone/RCC boundary pillars. Once the boundary is delineated by fixing of boundary pillars, and all available sacred groves have been identified and detailed history card prepared, the periodic changes in their status can be monitored at regular intervals through analysis of the multi-special satellite imageries. Concerned Patwari of the village/area may inspect physical condition of the boundary pillars at least once every year strictly so that any damage or shifting of the boundary pillars by unscrupulous persons can be detected in time and necessary remedial measures could be taken.

4.4 Management of sacred groves and people mobilization

The disappearance and/or degradation of sacred groves not only symbolize the loss of the rich relict flora and fauna but also its rich tapestry of culture associated with the grove. To counter the current threats observed on some sacred groves because of religious and cultural changes as well as from natural resource pressures, there is need of renewal of community traditions (Gowloog, 2009). Further, management of sacred groves or patches/sites through the traditional system is now being challenged by a number of economic and social issues. Thus, following the traditional methods only in conservation and management of sacred groves appears to be less effective in this modern age. For this, complete sacred grove area can be divided into core zone and buffer zone as available in 'Man and Biosphere Reserve' programme. Here, the core zone should be maintained under conservation, whereas buffer zone can be utilized for sustained use. But for the smaller groves, a buffer zone of 10-20 m may be maintained to control encroachments and other illegal occupations. Alternatively, incorporating the sacred sites into conservation networks may enhance the effectiveness of protected areas by covering a wider variety of habitats and by upholding the socio-cultural values of indigenous communities and harnessing their supports (Chandrakanth et al., 2004; Bhagwat and Rutte, 2006).

Mobilization and group formation with the support of the respective village panchayat is necessary for each sacred grove. Besides, grove specific sacred grove management committee should also be formed involving representatives of all stakeholders in the region. Technical support to local communities is essential for the sustainable management of sacred groves. In this regard, some technical organizations may also be involved. Moreover, local youths should be trained for wild plant species identification and assessment, seed collection or propagation, nursery development/ gap filling/planting, weed management, maintenance of sacred grove, selective collection and marketing of non-timber forest produce. For effective protection, village communities should be provided with alternatives for fulfilling fuel and fodder requirement. Information on forest produce from the sacred groves and possibilities of their value addition and marketing linkage may also help people to get linked with the conservation plan by ensuring people livelihood and other benefits of conservation (Ananthakrishnan, 1999). However, there is a need to promote management in an era of market-orientated economy, where communities would need technical help in order to explore the ways in which sacred groves could evolve to meet contemporary needs, i.e. regenerated bio-diversity and biotic composition. For this, adequate documentation of the biological resources of the state could be done implying People's Bio-diversity Registers (PBRs). Further, ecological services rendered by sacred groves needs to be highlighted and people should be made aware and awakened to realize that the conservation of groves is crucial for their sustenance.

Adult education and community development initiatives will augment the existing system of conservation and management towards addressing various safety and security issues of the sacred groves and places; and enhance the relevance in the heritage tourism industry and community development (Nwankwo and Agboeze, 2016). This will also entail sustainable harmonization of traditional and modern measures for addressing various issues of the sacred sites. Each of the conservation measures should complement the weaknesses of other measures. They should be better integrated in the community development initiatives which will promote participation of the local community and make the process more productive and sustainable (Dixit et al., 2007). Further, promoting tourism, cultural heritage of indigenous people, revenue and employment generation may play important roles in the preservation of sacred groves and their biodiversity conservation potential (Onyekwelu and Olusola, 2014).

4.5 Restoration of sacred groves

Sacred groves are no longer free from anthropogenic pressure. In such a condition, restoration activities include planting indigenous species of multiple benefits, protection for seedlings and saplings recruitment, nursery establishment for rare, endemic plants, measures for soil and water conservation etc. However, integrity of the tree-cover adopting landscape matrix needs to be conserved (Bhagawat et al., 2005). If the security is ensured through fencing or boundary wall and the relevant communities are aware for the security of sacred groves, these groves can play more important roles as decentralized centres of biodiversity conservation and ecological functions. Through restoration of sacred groves in accordance with ecological and environmental values and revitalisation of the community centered management system, the groves could continue to play a greater role in strengthening rural livelihood by supplying various non-timber forest produce as well as providing water security, rendering pollination services and as ideal local level mitigation measure for global climatic change (Dande, 2014). To maintain the functional values and attributes of sacred groves, the threatened and indigenous species need to be conserved *in situ* effectively in the grove (Manikandan et al., 2011). For this, external intervention including government organizations taking the local people into confidence is the call of the day.

Because of dominance of large number of secondary or invasive species, the sacred groves cannot be treated as climax species in spite of their rich species diversity and endemism. In such a case, without increase in the area, the vegetation of the groves cannot return to the climax formations by natural dynamics. In such a condition, larger areas could be earmarked for their growth and a network of sacred groves could be established to promote endangered and endemic species together (Sarfo-Mesah et al., 2010). Ecosystem services like pollination, seed dispersal, nutrients cycling and, soil and water conservation are usually operated in larger spatial extent combining different land use types at different times and therefore require an integrated approach at landscape level. For this, important and larger sacred groves could be brought under the 'Protected Area Network' to ensure their proper protection and conservation. The present fragmented conditions of the groves can be used as repositories of endemic plants, soil seed bank and connective corridors for birds and animals in human dominated landscapes. Therefore, it requires combined and holistic approach to conserve the grove tradition at landscape level. In this, landscape approach appears more appropriate due to wider spatial scale as well as in application as it covers thorough knowledge of ecological processes, interactions of grove biota, and issues of landscape ecology, and taking care of grove surroundings.

For rehabilitation of sacred groves through afforestation, perceptions of local people should be taken into care. The local people may be made to understand clearly that the rehabilitation works are to be undertaken for their benefits only and the implementing agency does not have any intention to grab the land. Types of species to be taken into restoration or rehabilitation of the sacred groves should be selected carefully with more emphasis on indigenous flora as well as the liking of the local fauna. For example, *Prosopis cineraria, Tecomella undulata, Cordia* spp., *Salvadora* spp., *Zizyphus* spp, *Azadirachta indica* etc. may be considered for the rehabilitation of the sacred groves in western Rajasthan from utilization point of view and *Commiphora wightii*, *Moringa concanensis*, *Bauhinia racemosa*, *Maytenus emarginata* etc. from conservation point of view. Likewise, *Putanjiva roxburghii*, *Crateva relegiosa*, *Boswellia serrata*, *Acacia catechu*, *Emblica officinalis*, *Mangifera indica*, *Ficus* spp., *Adansonia digiata*, *Mitragyna parviflora*, *Madhuca indica*, *Syzygium cumini* etc. may be considered in the afforestation of the Devabanis of Aravalli and eastern Rajasthan. However, it is better to ensure survival and growth rather than repeated planting.

4.6 Control of over grazing and over exploitation

In order to prevent overgrazing and consequent change in the vegetation composition of the sacred groves, possibilities of adopting the sustainable grazing practices like rotational grazing, deferred rotational grazing or its possibility may be explored. Otherwise, cut and carry system will be more appropriate and can also be adopted. Concerted research programme may also be initiated to assess present as well as potential productivity of such areas. However, in order to conserve representative biodiversity of a particular region, some selected sacred groves' repository of the biodiversity may be protected from any type of interferences in the form of grazing or plant collection.

4.7 Control of invasive species

The expansion of noxious weeds including *P. juliflora* and *Lantana camara* is affecting human health, suppressing indigenous plants, interrupting water flow and decreasing the livestock productivity (Singh et al., 2010; Tripathi and Shukla, 2007; Dogra et al., 2009; Priyanka and Joshi, 2013). The management measures that have been implemented have not been able to yield the desirable results because of the limited spatial scale, cost, and/or improper planning and implementation. Effective strategies for management of these invasive include the engagement of the community and the limiting of the number of vector animals responsible for the spread of these invasive species (Haregeweyn et al., 2013).

4.7.1 Control of Lantana

Lantana camara is a noxious weed expanding and establishing in many regions of the world, including India. Most of the studies reveal that the species have become a menace and has been expanding its range. More than 80% of studies focus on its impact, toxicity and therapeutic uses only. It has many negative impacts including the potential to disrupt the succession cycle and to displace the native biota resulting in decreased biodiversity. Its infestations alter the structural and floral composition of native communities (Sharma and Raghubanshi, 2010). As the density of *Lantana camara* in forest increases, allelopathic interactions increase and hence there is decline in species richness (Day et al., 2003).

Lantana camara can be removed mechanically or physically in several ways, including stick raking, bulldozing, ploughing and grubbing. These techniques are mainly suited to medium-sized infestations and require extensive follow-up, as they invariably lead to re-growth if the rootstock (up to 30 cm depth) is not removed (Love et al., 2009). Another approach for controlling *Lantana* is fire, which is not recommended in forest and thickets because they are highly sensitive to fire. Other control measure of *Lantana camara* is through application of chemicals. Glyphosate is marginally effective as a foliar spray and regrowth is common in such cases. Fluroxypyr (Vista) plus aminopyralid applied twice within 6 months period is found effective, but it is costly. Application of Fluroxypyr as a basal application appears consistently effective. However, use of such chemicals tends to cause harm to the native biota of the ecosystem, thereby affecting food chain, soil health, causing water pollution and giving genesis to ancillary problems. Biological control organisms have also been studied for controlling *Lantana camara* spread. Though less effective, the biological organisms for controlling *Lantana camara* include *Ophiomyia lantanae* (fruit-mining fly), *Calycomyza lantanae* (agromyzid seedfly), *Teleonemia elata* (leaf-sucking bug), *Teleonemia scrupulosa* (leaf-sucking bug) but mostly failed as they have several varieties or forms resulting in complicating the introduction and establishment of exotic insects. Several other host specific insects such as *Diastema tigris* (flower-mining moth), *Salbia haemorrhoidalis* (leaf-floding caterpillar), *Uroplata girardi* (leaf-mining beetle), *Octotoma scabripennis* (leaf-mining beetle) and *Epinotia Lantanae* (flower-mining moth) have been introduced from time to time for the biological suppression of *Lantana camara* but have not been effective in controlling its infestation (Priyanka and Joshi, 2013). In such a circumstance mechanical or physical control of *Lantana camara* can be more effective and involves bulldozing, ploughing and grubbing up 30-40 cm depth. This process can be repeated at frequent interval until the loss of soil seed bank.

4.7.2 Control of P. juliflora

Prosopis juliflora is undoubtedly a significant threat to conservation values of the sacred groves and forests as well and is likely altering ecosystem functions (Belton, 2008). Many efforts have been done to eradicate and control *Prosopis juliflora* from its areas of invasion (Geesing et al., 2004). However, general experiences from America, Asia and Australia have shown that eradication of *Prosopis juliflora*, by the different methods, especially the mechanical and chemical ones is highly expensive and relatively less effective. The magnitude of resilience and distribution of the plant makes *Prosopis juliflora* virtually impossible to eradicate once established. Therefore, it will be more appropriate to work for its eradication during its seed setting, germination, and seedling and at establishment phase. Propagation of *P. juliflora* is mainly by seeds, the dispersal of which is done by the animals feeding on *Prosopis juliflora* pods. In this circumstance control of such animals moving in the sacred groves can be stopped. Alternatively all pods of *P. juliflora* can be collected and grinded to make fodder for the animals. This will provide alternative livelihood to the locals and the spread of the seeds can be controlled.

Mechanical method involves plant removal by machine or mechanically by hand pulling, cutting, hand digging or mechanical uprooting. Stick racking, chain pulling, bulldozer pushing and blade ploughing are some methods to control *P. juliflora*. In this, stick racking (best results are achieved when soil moisture is sufficient to allow machinery to work with minimum strain, but is dry enough so the root system desiccates) and chain pulling (may kill up to 90% of trees in a mesquite infestation) have been found effective in Australia. However, the effectiveness of control may be reduced when either very dense infestations or a high proportion of young trees and seedlings are present. Though larger trees and shrubs are killed by cutting the stem at ground level and spraying or painting the freshly cut stumps with herbicides like Round up, 2,4-D, Glenside Kerosene and diesel oil.

Predators or pathogens are also being used to limit the *Prosopis juliflora* reproduction. Sudanese researchers have found some predator insects that attack the leaves that lead to deterioration of the tree canopy. In Australia, four species of insects viz. *Algarobius bottimeri*, *Algarobius prosopis* (the larvae of these beetles destroy mesquite seeds in mature pods both in the trees and on the ground), the *Prosopidopsylla flava* (a sap-sucking psyllid that causes dieback) and *Evippe* spp. (a leaf-tying moth that causes defoliation) have been introduced as biological control agents against *P. juliflora* (DAFF, Queensland, 2013). Another method that has been used in several countries is burning of the stump after cutting from the ground. In Yemen, for example, the application of kerosene over the stump followed by burning has been used as a way of eradicating the plant. However this only works when the plant is dry (not in stage of flowering) and the root system is not too deep to survive further otherwise re-growth will occur.

4.8 Conversion under compensatory afforestation

While developmental and industrial projects, i.e. erection of dams, mining activities and construction of industries or roads, require diversion of forest land, the Compensatory Afforestation Fund Management and Planning Authority (CAMPA) or Compensatory Afforestation (CA) refers to promotion of afforestation and regeneration activities as a way of compensating for forest land diverted to non-forest uses. In this, the prime task of State CAMPA is regenerating natural forests and building up the institution engaged in this task in the State Forest Department. According to 'Forest (Conservation) Act 1980' and the 'Rules' and 'Guidelines' in these, whenever a forest land is to be diverted for non-forestry purposes, the equivalent non forest land has to be identified for compensatory afforestation. It may be contiguous to or in the proximity of Reserved Forest or Protected Forest, as far as possible. In such a condition the **sacred groves or community forest** may also be considered for diversion purpose under compensatory afforestation and its effective management. This may help in achieving additional 12% area of India under forest cover as projected by Governmwent of India. However, local community may also be involved, not only for the process of transferring this common land to forest but their involvement should also be ensured for effective planning and management of these sacred lands.

5. SYNTHESIS AND FUTURE PERSPECTIVES

Increasing threats through constant disturbance like mining, burning, cutting and removal of trees and livestock grazing, defecation, etc. by the people and religious and cultural changes as well as pressures on existing natural resource are resulting in degradation in sacred grove conditions. Despite of the unintended protection and conservation by the local community or races, the groves are prone to alteration under modernization and vanishing religious sentiments. Lack of sympathy among the local people, family and village heads, forest department officials, government and panchayats add to the miseries of the groves that causes gradual loss of pristine purity and glory of the sacred sites. People's perceptions are also changing towards the groves. There is need of renewal of community traditions and socio-cultural belief through revival of ceremonies related to the sacred groves, as well as through awareness campaigns and education programs that highlight the ecological and spiritual benefits of the sacred forests and reinforcing rules about the use of resources of the sacred groves.

More systematic research work needs to be conducted on the variety of sacred groves or places with respect to biological issues like their size, age, species composition, biodiversity level, exotic species invasiveness, and degree of naturalness as well as any historical, social, cultural, religious, economic, political, tenure, and legal issues. Likewise, controlled comparisons between sacred places and adjacent secular ones of the same size and type of biotic community will also be required for describing and assessing any differences arising out due to sacred status of the place or grove. Protecting sacred groves from such erosion can simultaneously help protect cultures, religions, and rights as well as the associated biotic species and ecosystem functioning. By strengthening traditional institutions and innovating new sacred art movement under legal framework may be able to challenge land grabbers, hold off poachers, protect sacredness and help conserve the biodiversity that will put the sacred grove back to the life and its functioning. However, stringent action through formulation of environmental protection measures, widening of the buffer zone through a network and blocking the developmental corridors which use to channelize anthropogenic pressures need to be considered in preserving, conserving and managing the sacred groves effectively.

8

BIBLIOGRAPHY

Amrithalingam, M. (1998). Sthala Vrikshas of Tamil Nadu. In: The Ecological Traditions of Tamil Nadu, Krishna, N. and Prabhakaran, J. (eds.), C.P.R. Environment Education Centre, Chennai.

An, S., Mentler, A., Mayer, H. and Blum, W.E.H. (2010). Soil aggregation, aggregate stability, organic carbon and nitrogen in different soil aggregate fractions under forest and shrub vegetation on the Loess Plateau, China. *Catena*, 81: 226-233.

Ananthakrishnan, T.N. (1999). Multidimensional links in biodiversity research: an integrated exercise. *Current Science*, 77: 356-358.

Anderson, E.N. (1996). Ecologies of the Heart: Emotion, Belief, and the Environment. New York: Oxford University Press.

Anon. (2004). Annual Report. Arid Forest Research Institute, Jodhpur, Rajasthan, India.

Anthwal, A., Gupta N., Sharma A., Anthwal S. and Ki-Hyum, K. (2010). Conserving biodiversity through traditional beliefs in sacred groves in Uttarakhand Himalaya, India. *Resources, Conservation and Recycling*, 54: 962-971.

Anthwal, A., Sharma, R.C. and Sharma, A. (2006). Sacred groves: traditional way of conserving plant diversity in Garhwal Himalaya, Uttaranchal. *The Journal of American Science*, 2: 35-43.

Appell, G.N. (2013). Sacred groves among the Rungus of northern Borneo: relict areas of biodiversity. http://firebirdfellowships.org/sacred-groves-among-the-rungus-of-northen-borneo-relict-areas-of-diversity.html. Retrieved on 25th March 2016.

Arora CL (2002). Analysis of soil, plant and fertilizers. In: fundamental of soil science, Sekhon, G.S. Chhonkar, P.K., Das, D.K., Goswami, N.N., Narayanasamy, G., Poonia, S.R., Rattan, R.K. and Sehgal, J. (Eds), Indian Society of Soil Science, New Delhi.

Arora, V. (2011). Conserving sacred groves and mitigating climate change. Presented at the National Conference on 'Environment and Biodiversity of India', New Delhi.

Ashagrie, Y., Zech, W., Guggenberger, G., Mamo, T. (2007). Soil aggregation, and total and particulate organic matter following conversion of native forests to continuous cultivation in Ethiopia. *Soil Till. Res.*, 94: 101-108.

Asokan, A., Chouhan, S. and Singh, V. (2015). sacred grove—a nature's gift—as a remedy for human ailments, a biodiversity reservoir for restoring indigenous traits for endangered listed plants—a review. *Open Access Library Journal*, **2**: e1517. http://dx.doi.org/10.4236/oalib.1101517

Australian Green House Office (2001). https://en.wikipedia.org/wiki/Australian_Greenhouse Office.

Avinash, H. and Yadav, K.C. (2014). Festivals, traditions and rituals associated with sacred groves of Chhattisgarh. *Indian J. Tropical Biodiversity*, 22(2): 156-163.

Ayres, E., Steltzer, H., Berg, S., Wallenstein, M.D., Simmons, B.L. and Wall, D.H. (2009). Tree species traits influence soil physical, chemical, and biological properties in high elevation forests. PLoS ONE, 2009; 4(6): e5964, doi: 10.1371/journal.pone.0005964.

Baimas-George, M. (2012). Soil nutrient composition in afromontane forests of northern Ethiopia. Colgate Academic Review: Vol. 8, Article 13. http://commons.colgate.edu/car/vol8/iss1/13.

Baker, L.R., Tanimola, A.A., Olubode, O.S. and Garshelis, D.L. (2009). Distribution and abundance of sacred monkeys in Igboland, southern Nigeria. *American Journal Primatology*, 71(7): 574-586.

Bala, N., Kumar, P., Bohra, N.K., Limba, N.K., Baloch, S.R. and Singh, G. (2010). Production and decomposition of litter in plantation forests of *Eucalyptus camaldulensis* along canal command area in Indian Desert. *Indian Forester*, 136(2): 163-172.

Barbaro, Salvatore, Caracausi, R., Chaix, B., Chisesi, R.M., Cognata, G. and Lorusso, L.C. (2007). Setting up of a protocol of environmental research for preservation and right exploitation of the cultural heritage. *Conservation Science in Cultural heritage*, 7: 267-321.

Basu, R. (2009). Biodiversity and ethnobotany of sacred groves in Bankura District, West Bengal. *Indian Forester*, 135 (6): 765-778.

Batjes, N.H. (1996). Total carbon and nitrogen in the soils of the world. *European Journal of Soil Science*, 47: 151–163.

Baviskar, A. (2014). Mangarbani sacred grove near Delhi. http://www.vikalpsangam.org/article/mangarbani-sacred-grove-near delhi/#.U4R826I_Zko. Accessed on 27th May 2014.

Beedy, T.L., Snapp, S.S., Akinnifesi, F.K. and Sileshi, G.W. (2010). Impact of *Gliricidia sepium* intercropping on soil organic matter fractions in a maize -based cropping system. *Agric. Ecosys. Environ.*, 138: 139-146.

Behera, M.K. and Pradhan, T.R. (2015). Sacred groves of Phulbani forest division of Odisha: socio cultural elements and plant biodiversity. *Indian Forester*, 141(6): 173-176.

Belsky, A.J. (1986). Does herbivory benefit plants? A review of the evidence. *American Naturalist*, 127: 870–892.

Belton, T. (2008). Management strategy for Mexican Thorn (*Prosopis juliflora*) on ascension Island: an assessment of this species and recommendations for management. http://www.issg.org/database/species/reference_files/projul/ Ascension ProsopisPlan.pdf. Accessed on 22nd July 2014.

Bennett, A.B. and Gratton, C. (2013). Floral diversity increases beneficial arthropod richness and decreases variability in arthropod community composition. *Ecological Applications*, 23: 86-95.

Berendse, F. (1998). Effects of dominant plant species on soils during successions in nutrient-poor ecosystems. *Biogeochemistry*, 42: 73-88.

Berendse, F., Lammerts, E.J. and Olff, H. (1998). Soil organic matter accumulation and its implications for nitrogen mineralization and plant species composition during succession in coastal dune slacks. *Plant Ecology*, 137(1): 71-78.

Bettina, J., Yamashita, T., Ludwig, B. and Flessa, H. (2005). Storage of organic carbon in aggregate and density fractions of silt soils under different types of land use. *Geoderma*, 128: 63-79.

Bhagwat, S.A. and Rutte, C. (2006). Sacred groves: potential for biodiversity management. *Frontiers in Ecology and the Environment*, 4: 519-524.

Bhagwat, S.A., Kushalappa, C.G., Williams, P.H. and Brown, N.D. (2005). A landscape approach to biodiversity conservation of sacred groves in the Western Ghats of India. *Conservation Biology*, 19: 1853-1862.

Bhakat, R. and Pandit, P.K. (2004). An inventory of medicinal plants of some sacred groves of Purulia district West Bengal. *Indian Forester*, 130: 37-43.

Bhakat, R. K. and Sen, U.K. (2008). Ethnomedicinal plant conservation through sacred groves. *Tribes and Tribals*, 2: 55-58.

Bhakat, R.K. and Pandit, P.K. (2003). Role of a sacred grove in conservation of medicinal plants. *Indian Forester*, 129: 224-232.

Bhakat, R.K. and Sen, U.K. (2012). Sacred grove as an institution of culture and conservation. *Indian Journal of Biological Sciences*, 18: 38-40.

Bhandari, M.M. (1990). Flora of the Indian Desert. Jodhpur: Printed by MPS Repros.

Bhandary, M.J. and Chandrasekhar, K.R. (2003). Sacred groves of dakshina Kanada and Udupi districts of Karnataka. *Current Science,* 85:655-1656

Bharathi, S. and Devi Prasad, A.G. (2015). Diversity and regeneration status of tree species in the sacred groves of central western Ghats, India. *J. Biodiversity and Environ. Sci.*, 7(2): 10-21.

Bhasin, V. (1999). Religious and cultural perspective of a sacred site – Sitabari in Rajasthan. *Journal of Human Ecology*, 10: 329-340.

Bhatla, N., Mukherjee, T. and Singh, G. (1984). Plants-traditional worshipping. *Indian J. History of Science*, 19: 37-42.

Bhatt, S., Broome, N.P., Kothari, A. and Balasinorwala, T. (2012). Community conserved areas in south Asia. Case studies and analyses from Bangladesh, India, Nepal, Pakistan and Sri Lanka. Kalpavriksh, New Delhi.

Bhattacharya, P. (2014). Sacred Groves at Risk as Faith Erodes. Indian Express. 05th January 2014.

Blicharska, M., Mikusiński, G., Godbole, A. and Sarnaik, J. (2013). Safeguarding biodiversity and ecosystem services of sacred groves – experiences from northern Western Ghats. *International Journal of Biodiversity Science, Ecosystem Services & Management*, 9: 339-346.

Boojh, R. and Ramakrishanan, P.S. (1998). Conserving the Sacred Groves: For Biodiversity Management, Ramakrishnan, P.S., Saxena, K.G. and Chandrasekhar, U. (Eds), UNESCO and Oxford-IBH Publishing, New Delhi. Pp. 253-262.

Booker, K., Huntsinger, L., Bartolome, J.W., Sayre, N.F. and Stewart, W. (2013). What can ecological science tell us about opportunities for carbon sequestration on arid rangelands in the United States? *Global Environmental Change*, 23: 240-251.

Bor, N.L. (1960). The grasses of Burma, Ceylon, India, andPakistan. Pregamon press, London. i-xviii, 1-767 pp.

Boraiah, K.T., Vasudeva, R., Shonil, A., Bhagwat, and Kushalappa, C.G. (2003). Do informally managed sacred groves have higher richness and regeneration of medicinal plants than state-managed reserve forests? *Current Science*, 84(6): 804-808.

Brandis, D. (1897). Indigenous Indian Forestry: Sacred Groves. In: *Indian Forestry Working*. Oriental Institute, England.

Burman, R.J.J. (1995). The dynamics of sacred groves. *Journal of Human Ecology,* 6: 245-254.

Busch, J., Dave, R., Hannah, L., Ashkenazi, E., Cameron, A., Fischman, D., Rasolohery, A. and Schatz, G. (nil). Climate change and the cost of conserving biodiversity in Madagascar.http://www.gripweb.org/gripweb/sites/default/files/disaster_risk_profiles/climate%20change%20and%20the%20cost%20of%20conserving%20biodiversity%20in%20Madagascar.pdf

Byers, B.A., Cunliffe, R.N. and Hudak, A.T. (2001). Linking the Conservation of Culture and Nature: A Case Study of Sacred Forests in Zimbabwe. *Human Ecology*, 29: 187-218.

Campbell, M.O. (2005). Sacred groves for forest conservation in Ghana's coastal savannas: Assessing ecological and social dimensions. *Singapore Journal of Tropical Geography,* 26: 151-169.

Cardelus, C.L., Scull, P., Hair, H., Baimas-George, M., Lowman, M.D. and Eshete, A.W. (2013). A preliminary assessment of Ethiopian sacred grove status at the landscape and ecosystem scales. *Diversity*, 5: 320-334.

Cardinale, B.J., Matulich, K.L., Hooper, D.U., Byrnes, J.E., Duffy, E., Gamfeldt, L., Balvanera, P., O'Connor, M.I. and Gonzalez, A. (2011). The functional rolc of producer diversity in ecosystems. *Am. J. Bot.*, 98: 572-592.

Carlson, C.H., Dobrowski, S.Z. and Safford, H.D. (2012). Variation in tree mortality and regeneration affect forest carbon recovery following fuel treatments and wildfire in the Lake Tahoe Basin, California, USA. *Carbon Balance Manag.*, 7: 7. doi: 10.1186/1750-0680-7-7.

Carret, J.C. and Loyer, D. (2003). Madagascar protected area network sustainable financing: Economic analysis perspective. Presentation to the World Parks Congress, Durban, South Africa.

Castro, P. (1990). Sacred groves and social change in Kirinyaga, Kenya. In: Social change and applied anthroplogy, Chaikan, M. and Fleverts, A. (eds.), Boulden: Westview Press.

CECOEDECON. Undated. Orans: Marubhumi Me Hariyali Ki Chadar (Shil ke Dungri, Chaksu, CECOEDECON).

Chakrabarti, S. (2010). Conservation of orchids by the people of North Eastern India. *NeBio*, 1: 48-52.

Chandrakanth, M.G. and Romm, J. (1991). Sacred forests, secular forest policies and people's actions. *Natural Resources Journal*, 31(4): 741-756.

Chandrakanth, M.G., Bhat, M.G.and Accavva, M.S. (2004). Socio-economic changes and sacred groves in south India: Protecting a community-based resource management institution. *Natural Resources Forum* 28: 102-111.

Chandrakar, K., Verma, D.K., Sharma, D. and Yadav, K.C. (2014). A study on the role of sacred groves in conserving the genetic diversity of the rare, endangered and threatened species of flora & fauna of Chhattisgarh state (India). International Journal of Scientific and Research Publications, 4(1): http://www.ijsrp.org/research-paper-0114/ijsrp-p2552.pdf.

Chandran, M. D. S. and Hughes, J.D. (2000). Sacred groves and conservation: the comparative history of traditional reserves in the Mediterranean area and in South India. *Environment and History*, 6: 169–86.

Chandran, M.D.S., Gadgil, M. and Hughes, J.D. (1998). Sacred groves of the Western Ghats. In: Conserving the Sacred for Biodiversity Management, Ramakrishnan, P.S., Saxena, K.G. and Chandrasekara, U.M. (eds.), Oxford and IBH, New Delhi. Pp. 211-232.

Chandran, S. and Gadgil, M. (1998). Sacred groves and sacred trees of Uttara Kannada. Lifestyle and Ecology. Indira Gandhi National Centre for the Arts and D.K. Printworld, Pvt. Ltd., New Delhi.

Chandrashekara, U. M. (2011). Cultural and conservation values of sacred groves of Kerala, India. International J. *Ecology and Environmental Sci.*, 37: 143-155.

Chandrashekara, U. M. and Sankar, S. (1998). Structure and Funcrion of sacred groves; Case study in Kerala, Pp. 223-235.

Chandy, S., Chandran, M.D.S. and Ramachandra, T.V. (2014). Soil quality –comparative study between sacred and non sacred grove forests. LAKE 2014: Conference on Conservation and Sustainable Management of Wetland Ecosystems in Western Ghats, date: 13- 15th November 2014. http://wgbis.ces.iisc.ernet.in/energy/lake2014/theme8/T7_P4_SAIRA.pdf.

Chao, A., Gotelli, N.J., Hsieh, T.C., Sander, E.L., Ma, K.H., Colwell, R.K. and Ellison, A.M. (2014). Rarefaction and extrapolation with Hill numbers: a framework for sampling and estimation in species diversity studies. *Ecological Monographs*, 84: 45–67.

Chaplot, V. Bouahom, B. and Velentin, C. (2010). Soil organic carbon stocks in Laos: spatial variations and controlling factors. *Global Change Biology*, 16(4): 1380–1393.

Chaudhari, P.R., Ahire, D.V., Ahire, V.D., Chakarvarty, M. and Maity, S. (2013). Soil bulk density as related to soil texture, organic matter content and available total nutrients of Coimbatore soil. *Int. J. Scientific and Research Publications*, 3(2): 1-8.

Chaudhary, H.C., Dimri, B.M. and Choudhary, K.R. (2002). Brief introduction to comprehensive community drought preparedness programme to improve quality of life of women and children in Jodhpur district. In: proceeding of workshop on ' Development of suitable strategy for rehabilitation of Orans and Gauchar in Rajasthan held at AFRI, Jodhpur on 16-17 April, 2002.

Chaudhary, K.L., Nath, R., Dkhar, R., Marak, M.S.R., Bareh, B., Yapa, M., Jyrwa, L.B., Mawkroh, B.M. and Hujon, D. (2015). The Lyngdoh Mawnai sacred grove, West Khasi Hills, Meghalaya: mapping and analysis of disturbance. *Keanean Journal of Science.* Retrieved on 1st April 2016.

Chaudhary, V. (2016). Biodiversity conservation in sacred groves – complete detail. http://natureconservation.in/biodiversity-conservation-in-sacred-groves-complete-detail/. Retrieved on 4th April 2016.

Chen, G., Yang, Y., Yang, Z., Xie, J., Guo, J., Gao, R., Yin, Y. and Robinson, D. (2016). Accelerated soil carbon turnover under tree plantations limits soil carbon storage. *Scientific Reports* **6**, Article number: 19693 (2016), doi:10.1038/srep19693.

Chen, J., Brosofskee, K.D., Noormets, A., Crow, T.R., Bresee, M.K., Le-Moine, J.M., Euskirchen, E.S., Mather, S.V. and Zheng, D. (2004). A working framework for quantifying carbon sequestration in disturbed land mosaics. *Environmental Management* 34: S210–S221.

Chhatterjee, S., Sastry, A.R.K., Roy, B.N. and Lahon, R. (2000). Sacred groves of Sikkim and Arunachal Pradesh. Abstract. National Workshop on Community Strategies on the Management of Natural Resources, Bhopal.

Choudhary, K. (2016). Firewood and Timber Collection and its impact on natural resources. http://natureconservation.in/firewood-and-timber-collection-and-its-impact-on-natural-resources/. Retrieved on 1st April 2016.

Choudhary, K., Singh, M. and Pillai, U. (2008). Ethnobotanical survey of Rajasthan - an update. *American-Eurasian Journal of Botany*, 1: 38-45.

Choudhury, P.R. (2015). Shelve the 'Save Sacred Grove Plan' to save the sacred groves: an appeal to Government of Odisha to revisit sacred grove plan. http://www.orissadiary.com/ShowOriyaColumn.asp?id=59999. Retrieved 5th April 2016.

Christenson, L.M., Lovett, G.M., Weathers, K. C. and Arthur, M.A. (2009). The influence of tree species, nitrogen fertilization, and soil C to N ratio on gross soil nitrogen transformations. *Soil Sci. Soc. Am. J.,* 73: 638-646.

Christopher L.C.E.W. (2014), Trees and the sacred. http://witcombe.sbc.edu/sacredplaces/trees.html. Accessed on 5th July 2014.

Christopher, M., Tankou, C.M., De Snoo, G.R., De Iongh, H.H. and Persoon, G. (2014). Variation in plant biodiversity across sacred groves and fallows in Western Highlands of Cameroon. *African J. Ecology*, 52: 10-19.

Colding, J., and Folke, C. (2001). Social taboos: "Invisible" systems of local resource management and biological conservation. *Ecological Applications*, 11: 584-600.

DAFF Queensland (2013) [online] available at: http://www.daff.qld.gov.au/__data/assets/pdf_

Dafni, A. (2007). The supernatural characters and powers of sacred trees in the Holy Land. *J. Ethnobiology and Ethnomedicine,* 3: 10.

Dagla, H.R., Paliwal, A. and Shekhawat, N.S. (2007). Oran: A sacred way for biodiversity conservation in Indian Thar Desert. *Current Science*, 93: 279.

Dande, R. (2014). Traditional community management system to climate variability: Case of common property resources in Thar desert of Western Rajasthan. http://admin.indiaenvironmentportal.org.in/files/file/Rucha%20Dande.pdf. Accessed on 28th May 2014.

Day, M.D., Wiley, C.J., Playford, J. and Zalucki, M.P. (2003). Lantana: current management, status and future prospects. Australian Centre for International Agricultural Research, 5: 1-20.

Daye, D.D. and Healey, J.R. (2015). Impacts of land-use change on sacred forests at the landscape scale. *Global Ecology and Conservation*, 3: 349–358.

Deb, D. (2007). Sacred Groves of West Bengal: A Model of Community Forest Management? Dr. Oliver Springate-Baginski, Overseas Development Group, University of East Anglia, Norwich NR4 7TJ, UK.

Deb, D., Deuti, K. and Malhotra, K.C. (1997). Sacred grove relics as bird refugia. *Current Science,* 73: 815-817.

Debabrata, P., Sekhar, B.S. and Palita, S.K. (2015). Floral diversity conservation through sacred groves in Koraput district, Odisha, India: A case study. *International Research Journal of Environment Sciences*, 3(9): 80-86.

Deil, U., Culmsee, H. and Mohamed Berriane, M. (2005). Sacred groves in morocco: a society's conservation of nature for spiritual reasons. *Silva Carelica, Forests, Culture, and Religions,* 49: 185-201.

Derner, J.D. and Schuman, G.E. (2007). Carbon sequestration and rangelands: A synthesis of land management and precipitation effects. *J. Soil and Water Conservation*, 62: 77-85.

Deshmukh, S., Gogate, M.G. and Gupta, A.K. (1998). Sacred groves and biological diversity : Providing pew dimensions to conservation Issue. In: Conserving the sacred for biodiversity management, Ramakrishnan, P.S., Saxena K.G. and Chandrashekara, U.M. (eds.), Oxford and IBH Publishing Co., New Delhi. Pp. 397-414.

Desougi, M.A., Ali, N.A., Adam, A.A. (2016). Effects of soil physical and chemical properties on the distribution of trees in some ecological zones of Zalingei-Darfur, Sudan. *International J. Eng. Res. & Sci.*, 2(2): 131-139.

Dev, A. and De, J.N. (2012). Traditional use of plants against snakebite in Indian subcontinent: a review of the recent literature. *Afr. J. Tradit. Complement. Altern. Med.*, 9: 153–174.

Devagiri, G.M., Kushalappa, C.G., Prakash, N.A., Mohan, G.S., Raghu, H.B. and Satish, B.N. (2012). Payment for carbon sequestration service: a tool for sustainable management of Kodagu landscape in central Western Ghats. *Seronica Journal*, 3: 1-7. http://www.seroindia.org/Article_GMD32.pdf

Dixit, S., Tewari, J.C., Wani, S.P., Vineela, C. Chaurashia, A.K. and Panchal, H.B. (2007). Participatory and conventional biodiversity assessment: creating awareness for better natural resource management. *Annals of Arid Zone*, 46: 197-204.

Dogra, K.S., Kohli, R.K. and Sood, S.K. (2009). An assessment and impact of three invasive species of Himachal Pradesh, India. *International J. Biodiversity and Conservation*, 1: 4-10.

Donovan, G.H., Butry, D.T., Michael, Y.L., Prestemon, J.P., Liebhold, A.M., Gatziolis, D. and Mao, M.Y. (2013). The relationship between trees and human health: evidence from the spread of the emerald ash borer. *American J. Prev. Med.*, 44: 139-145.

Du, H., Zeng, F., Peng, W., Wang, K., Zhang, Z., Liu, L. and Song, T. (2015). Carbon storage in a *Eucalyptus* plantation chronosequence in southern China. *Forests*, 6: 1763-1778.

Eaton, D.A.R., Fenster, C.B., Hereford J., Huang, S. and Ree, R.H. (2012). Floral diversity and community structure in *Pedicularis* (Orobanchaceae). *Ecology,* 93: S182–S194.

Ellison, L. (1960). Influence of grazing on plant succession of Rangelands. *The Botanical Review*, 26: 1–78.

Erfanzadeh, R., Bahrami, B., Motamedi, J. and Petillon, J. (2014). Changes in soil organic matter driven by shifts in co-dominant plant species in a grassland. *Geoderma,* 213: 74-78.

Fallahzade, J. and Hajabbasi, M. A. (2011). Soil organic matter status changes with cultivation of overgrazed pastures in semi-dry West central Iran. *Int. J. Soil Sci.*, 6: 114-123.

FAO. (2010). Global Forest Resources Assessment 2010. Food and Agriculture Organization, Rome. Paper No. 163.

Fernandes, W. (1993). Forest and tribals, informal economy: Dependence and management decision, In: Community and Change in Tribal Society, Indian Institute of Advanced Study, Miri (ed.), Shimla. file/0004/73489/IPA-Mesquite-PP37.pdf Accessed on January 22th 2014.

Follett, R.F., Castellanos, J.Z. and Buenger, E.D. (2005). Carbon dynamics and sequestration in an irrigated Vertisol in Central Mexico. *Soil and Tillage Research*, 83: 148-158.

Fournier, A. and Sanou, L. (2013). Revisiting plant conservation on wooded shrines by a transversal approach: "Core" vs. "satellite" species and anthropological knowledge in

Bwaba land, Burkina Faso. http://agris.fao.org/aos/records/ CN2013200007?output=xml. Retrieved on 4th April 2016.

Fowler-Smith, L. (2009). Hindu tree veneration as a mode of environmental encounter. *Leonardo*, 42: 43–51.

Frazer, J.B. (1915). The Golden Bough. London: McMillian & Co.

Fu, D., Duan, C., Hou, X., Xia, T. and Gao, K. (2009). Patterns and relationships of plant traits, community structural attributes, and eco-hydrological functions during a subtropical secondary succession in central Yunnan (southwest China). *Arch. Biol. Sci., Belgrade*, 61: 741-749.

Gadgil, M. and Berkes, F. (1991). Traditional resource management systems. *Resource Management and Optimization*, 18: 127-141.

Gadgil, M. and Vartak, V.D. (1976). Sacred groves of Western Ghats in India. *Economic Botany*, 30: 152-160.

Gadgil, M., Berkes, F. and Folke, C. (1993). Indigenous knowledge for biodiversity conservation. *A Journal of the Human Environment*, 22: 151-156.

Gaikwad, S.S., Paralikar S.N., Chavan V. and Krishnan S. (2004). Digitizing Indian sacred groves - an information model for web interfaced multimedia database. In: Focus on Sacred Groves and Ethnobotany, Ghate, V., Hema, S. and Ranade, S.S. (eds.), Prisam Publications, Mumbai. Pp. 123-128.

Gandhi, K. and Vardhan, M. (1997). Documentation of people's knowledge and perception about biodiversity and conservation across selected ecosystems and agroecological zones in Rajasthan. Sewa Mandir, Udaipur, Rajasthan.

Gao, Y., Cheng, J., Ma, Z., Zhao, Y. and Su, J. (2014). Carbon storage in biomass, litter, and soil of different plantations in a semiarid temperate region of northwest China. *Annals of Forest Science*, 71: 427-435.

Garg, A. and Singh, V. (2013b). Siddhwari sacred grove in Upper Ganga Ramsar site of Uttar Pradesh. *Current Science*, 105: 1039-1040.

Garg, A. and Singh, V. (2013a). Mandu sacred grove in Upper Ganga Ramsar site, Uttar Pradesh. *Current Science*, 104: 409.

Garg, V.K. (1998), Interaction of tree crops with a sodic soil environment: Potential for rehabilitation of degraded environments. *Land Degradation and Development,* 9: 81-93.

Gartzia-Bengoetxea, N., Gonzalez-Arias, A., Merino, A. and Martinez de Arano, I. (2009). Soil organic matter in soil physical fractions in adjacent semi-natural and cultivated stands in temperate Atlantic forests. *Soil Biol. Biochem.*, 41: 1674-1683.

Gaude, K. and Janarthanam, M.K. (2015). The butterfly (Insecta: Lepidoptera) diversity of four sacred groves of Goa, India. *Journal of Threatened Taxa,* 7(12): 7927–7932

Geesing, D., Al-Khawlani, M. and Abba, M.L. (2004). Management of introduced *Prosopis juliflora* species: can economic exploitation control an invasive species? *Unasylva*, 55: 36-44.

Ghosh, A.K. and Tiwari, K.K. (1984). Biodiversity conservation in North-East India: Problems and Realities. In: Resources Potentials of North-East India Vol. II (Living Resources), Tripathi, R.S. (ed.), Meghalaya Science Society, Shillong. Pp. 105-109.

Giri, J.D., Singh, S.K., Singh, R.S. and Shyampura, R.L. (2007). Carbon stock and its distribution in soils of Ajmer district and management strategies for carbon sequestration. *Agropedology*, 18 (1): 21-32.

Githae, E. W., Gachene, C.K.K. and Njoka, J.T. (2011). Soil physicochemical properties under *Acacia senegal* varieties in the dryland areas of Kenya. *African Journal of Plant Science*, 5(8): 475-482.

Godbole, A., Watv, A., Prabhu, S. and Sarnaik, J. (1998) Role of sacred groves in biodiversity conservation with local people's participation: A case study from Ratnagiri district, Maharashtra. In: Conserving the Sacred for biodiversity management, Ramakrishnan PS, Saxena KG, Chandrashekara UM (eds), Oxford and IBH Publishing Co., New Delhi. Pp. 233-246

Gogkale, Y., Pala, N.A., Negi, A.K., Bhat, J.A. and Todaria, N.P. (2011). Sacred landscapes as repositories of biodiversity: a case study from the Hariyali Devi sacred landscape, Uttarakhand. ISSN: 2067-533X.

GoI (2011). Government of India Census 2011. New Delhi

GOI Monitor (2013). The Orans of Rajasthan: On the knees to nature. http://www.thealternative.in/society/on-the-knees-to-nature/. Accessed on 8th July 2014.

Gokhale, Y. and Pala, N.A. (2011). Ecosystem services in sacred natural sites (ssns) of Uttarakhand: a preliminary survey. *Journal of Biodiversity*, 2: 107-115.

Gokhale,Y. (2000). Sacred conservation tradition in India: An overview. Abstract. National Workshop on Community Strategies on the Management of Natural Resources. Bhopal.

Gold, A.G. and Gujar, B.R. (1989). Of Gods, trees and boundaries: divine conservation in Rajasthan. *Asian Folklore Studies*, 48: 221-229.

Gorshunova, O.V. (2008). Sacred trees of Khodzhi Baror: phytolatry and the cult of female deity in central Asia. *Etnoragraficheskoe Obozrenie*, 1: 71-82.

Gosain, B.G., Negi, G.C., Dhyani, P.P., Bargali, S.S. and Saxena, R. (2015). Ecosystem services of forests: carbon stock in vegetation and soil components in a watershed of Kumaun Himalaya, India. *International Journal of Ecology and Environmental Sciences,* 41(3): 177-188.

Gowda, B. (2006). Sacred Plants. Kalpataru Research Academy, Bangalore.

Gowloog, R.R. 2009. The preservation of indigenous cultural heritage of the Khasis of Meghalaya: Some issues. *Man and Society: A Journal of North East Studies* VI: 95-104.

Gupta, A., Shukla, S., Koradiya, D., Bhavsar, P., Ramji, A., Patel, P. and Taviyad, R. (2000). A cultural and ecological study of sacred groves in Balaram Ambaji and Jessore sanctuary in Banaskantha district of Gujarat. Abstract. National Workshop on Community Strategies on the Management of Natural Resources. Bhopal.

Gupta, B. and Sharma, S. (2014). Estimation of biomass and carbon sequestration of trees in informally protected areas of Rajouri, J&K, India. *It. Res. J. Environment Sci.*, 3: 56-61.

Haberman, D.L. (2013). People trees: worship of trees in northern India. Oxford University Press, New York.

Hamilton, L.S. (2002). Forest and tree conservation through metaphysical constraints. *Forests and Tree Conservation*, 9(3): 57-78.

Hangarge L.M., Kulkarni, D.K., Gaikwad, V.B., Mahajan, D.M. and Chaudhari, N. (2012). Carbon sequestration potential of tree species in Somjaichi Rai (Sacred grove) at Nandghur village, in Bhor region of Pune District, Maharashtra state, India. *Annals of Biological Research*, 3: 342-346.

Hangarge, L.M., Kulkarni, D.K., Gaikwad, V.B. and Mahajan, D.M. (2015). Soil organic carbon (SOC) in selected sacred groves from Bhor region of Western Ghats, Maharashtra. *Asian J. Environ. Sci.*, 10(2): 166-171.

Hansen, E.M. (2014). Forest development and carbon dynamics after mountain pine beetle outbreaks. *Forest Science*, 60(3): 476-488.

Hanski, I. (1998). Metapopulation dynamics, *Nature,* 396: 41–49.

Haregeweyn, N., Tsunekawa, A., Tsubo, M., Meshesha, D. and Melkie, A. (2013). Analysis of the invasion rate, impacts and control measures of *Prosopis juliflora*: a case study of Amibara district, eastern Ethiopia. *Environ. Monit. Assess.*, 185: 7527-7542.

Haridasan, K. and Rao, P. R. (1985). Forest Flora of Meghalya, Vol . 1, Bishen Singh, Dehradun.

Haynes, R.J. (1999). Labile organic matterfractions and aggregate stability under short-time, grass-based leys. *Soil Boil. Biochem.*, 31: 1821-1830.

Hembram, P.C. (1983). Return to the Sacred Groves. In: Tribal Movements in India, Singh, K.S. (ed.), New Delhi.

Hemmat, M.A., Shamekhi, T., Zobeiry, M. and Tabatabaei, M.G. (2012). Sacred places: instances of natural heritage and cultural landscapes in Northern Iran. *Archives Des Sciences,* 65: 605-612.

Hicks, C., Woroniecki, S., Fancourt, M., Bieri, M., Garcia Robles, H., Trumper, K. and Mant, R, (2014). The relationship between biodiversity, carbon storage and the provision of other ecosystem services: critical review for the forestry component of the International Climate Fund. Cambridge, UK.

Holechek, J.L., Gomez, H., Molinar, F. and Galt, D. (1999). Grazing studies: what we've learned. *Rangelands*, 21: 12–16.

Hombegowda, H.C., van Straaten, O., Köhler, M. and Hölscher, D. (2016). On the rebound: soil organic carbon stocks can bounce back to near forest levels when agroforests replace agriculture in southern India. *Soil,* 2: 13-23.

Hooker, J. D. (1897). Flora of British India. 7: 156. L. Reeve & co. London.

Huang, S., Sun, Y-N., Rui, W-Y., Liu, W-R., Zhang, W-J. (2010). Long-term effect of no-tillage on soil organic carbon fractions in a continuous maize cropping systems of northeast China. *Pedosphere*, 20: 285-292.

Hubert, J. (np). Sacred belief and beliefs of sacredness. http://www.columbia.edu/itc/anthropology/schildkrout/6353/client_edit/week9/anth_g6353.pdf. Accessed on 18th July 2014.

Hughes, J.D. (1990). Pan's Travel: Environmental Problems of the Ancient Greek and Romans. Johns Hopkins University Press, Baltimore.

Hughes, J.D. and Chandran, M.D.S. (1998). Sacred groves around the earth: an overview. In: Conserving the Sacred for biodiversity management, Ramakrishnan P.S., Saxena, K.G. and Chandrashekhara U.M. (eds.), Oxford and IBH Publishing Co. Pvt. Ltd., New Delhi. Pp. 69-86.

ICFRE, (2013). Revisit forests types of India. Indian Council of Forestry Research and Education, Dehradun, Uttarakhand

Imam, I., Mazumder, A., Mallick, K., Mitra, S. and Das, P. (2016). Diversity of ground dwelling insect fauna in the sacred groves of Purulia district, West Bengal. *Journal of Entomology and Zoology Studies*, 4(1): 474-478.

Induchoodan, N.C. (1996). Ecological studies on the sacred groves of Kerala. Ph. D. Thesis, Pondicherry University, Pondicherry. Pp. 227.

IPCC (2007). Fourth Assessment Report: Climate Change, 2007 (AR4).

Isbell, F., Calcagno, V., Hector, A., Connolly, J., Harpole, W.S., Reich, P.B., Lorenzen, M.S., Schmid, B., Tilman, D., van Ruijven, J., Weigelt, A., Wilsey, B.J., Zavaletas, E. and Loreau, M. (2011). High plant diversity is needed to maintain ecosystem services. *Nature*, 477: 199-202.

Iyer, S. (2010). Major threats to the Sacred Groves of India. *Eco News*, 16: 27-30.

Jackson, M.L. (1973). Soil Chemical Analysis (II Edition). Prentice Hall of India Private Limited, New Delhi, India

Jain, P. (2011). Dharma and Ecology of Hindu Communities: Sustenance and Sustainability. Ashgate Publishing Ltd, Surrey, U.K.

Jamir, S.A. and Pandey, H.N. (2002). Status of biodiversity in the sacred groves of Jaintia Hills, Meghalaya. *Indian Forester,* 59: 738-744.

Jandl, R., Lindner, M., Vesterdal, L., Bauwens, B., Baritz, R., Hagedorn, F., Johnson, D.W., Minkkinen, K. and Byrne, K. A. (2007How strongly can forest management influence soil carbon sequestration? *Geoderma*, 137: 253-268.

Jaryan, V, Uniyal Gopichand, S.K., Singh, R.D., Lal, B., Kumar, A. and Sharma, V. (2010). Role of traditional conservation practice: highlighting the importance of Shivbari sacred grove in biodiversity conservation. *Environmentalist,* 30: 101-110.

Jayakumar, R. and Nair, K.K.N. (2013). Species Diversity and Tree Regeneration Patterns in Tropical Forests of the Western Ghats, India. *ISRN Ecology*, pp:14. http://dx.doi.org/10.1155/2013/890862

Jayanthi, N., Rajeevan, M., Srivastava, A. K., Devi, S., Roy, Bhowmik, S. K. and Hatwar, H. R., (2006). Monsoon 2006 -A Report, IMD Met Monograph, 4:103.

Jayapal, J., Tangavelou, A.C. and Panneerselvam, A. (2014). Studies on the plant diversity of Muniandavar sacred groves of Thiruvaiyaru, Thanjavur, Tamil Nadu, India. *Hygeia. J. D. Med.,* 6: 48-62.

Jeeva S., Mishra, B.P., Venugopal, N., Kharlukhi, L. and Laloo, R.C. (2006). Traditional knowledge and biodiversity conservation in the sacred groves of Meghalaya. *Indian J. Tradit Knowle*., 5: 563 -568.

Jeeva, S., Mishra, B.P., Venugopal, N. and Laloo, R.C. (2005). Sacred forests: Traditional ecological knowledge heritage in Meghalaya. *J. Scott. Res. Forum*., 1: 93-97.

Jha, M., Vardhan, H., Chatterjee, S., Kumar, K. and Sastry, A.R.K. (1998). Status of Orans (Sacred Groves) in Peepasar and Khejarli villages in Rajasthan. In: Conserving the Sacred for Biodiversity Management. Oxford and IBH, New Delhi.

Johnson, D.R.E. and Coburn, C.A. (2010). Drought stress ecophysiology of shrub and grass functional groups on opposing slope aspects of a temperate grassland valley. *Botany*, 88(9): 850-866

Joshi, P. and Shrivastava, Y. (2000). Drops of nature conservation-sacred grove. *Journal of Human Ecology*, 11: 327-330.

Joshi, S. and Shringi, S.K. (2014). Floristic diversity with special reference to rare and threatened plants of Jawahar Sagar Sanctuary area near Kota Rajasthan. *Biological Forum: an International Journal*, 6: 84-91.

Jost, L. (2006). Entropy and diversity. *Oikos,* 113: 363–375.

Jost, L. (2010). The relation between evenness and diversity. *Diversity*, 2: 207-232.

Jyothi, K.M. and Nameer, P.O. (2015). Birds of sacred groves of northern Kerala, India. *Journal of Threatened Taxa*, 7(15): 8226-8236.

Kakar, K. (2013). Moving to Goa. Penguin UK, Biography & Autobiography. pp: 240.

Kala, C.P. (2005). Ethnomedicinal botany of the Apatani in the eastern Himalayan Region of India. *Journal of Ethnobiology and Ethnomedicine*, 1: 1-12.

Kala, C.P. (2010a). Ethnobotanical and ecological approaches for conservation of medicinal and aromatic plants. *Acta Horticulturae*, 860: 19-26.

Kala, C.P. (2010b). Home gardens and management of key species in the Pachmarhi Biosphere Reserve of India. *Journal of Biodiversity*, 1: 111-117.

Kala, C.P. (2011). Traditional ecological knowledge, sacred groves and conservation of biodiversity in the Pachmarhi Biosphere Reserve of India. *Journal of Environmental Protection*, 2: 967-973.

Kalam, M.A. (1996). Sacred groves in Kodagu district of Karnataka (South India): a socio historical study, Institute Francais de Pondichery, Pondichery, India.

Kandari, L.S., Bisht, V.K., Bhardwaj, M. and Thakur, A.K. (2014). Conservation and management of sacred groves, myths and beliefs of tribal communities: a case study from north-India. *Environmental Systems Research*, 3: 16, doi: 10.1186/540068-014-0016-8.

Kangah-Kesse, L., Attuquayefio, D., Owusu, E. and Gbogbo, F. (2007). Bird species diversity and abundance in the abiriw sacred grove in the eastern region of Ghana. *West Afr. J. Appl. Ecol.*, 11: 38-44.

Kangah-Kesse, L., Attuquayefio, D., Owusu, E. and Gbogbo, F. (2003). Bird species diversity and abundance in the abiriw sacred grove in the eastern region of Ghana.

Kar, S., Pathak, S. and Singh, P. (2015). Existing status of Chilkigarh sacred grove in Midnapore, West Bengal. *Zoo's Print*, XXX(4): 15-17.

Karanth, K.U. 1998. Sacred groves for the 21st century. Seminar 466 -Reconciling the Needs of Man andWildlife in India. Conference Proceedings. IUCN/ SACCat Specialist Group - Digital Cat Library.

Kaur M.R., Malhi, R.K.M. and Kiran, G.S. (2013). Impact of joint forest management strategy on fertility of forest soils. *Bulletin of Environmental and Scientific Research*, 2(2-3): 7-11.

Kemeuze, V.A., Sonwa, D.J., Nkongmeneck, B.A. and Mapongmetsem, P.M. (2016). Sacred groves and biodiversity conservation in semi-arid area of Cameroon: Case study of Diamare plain. Paris, France, United Nations Educational, Scientific and Cultural Organisation (UNESCO).

Kemeuze,V.A., Mapongmetsem, P.M., Sonwa,D.J., Fongnzossie, E. and Nkongmeneck, B.A (2015). Plant diversity and carbon stock in sacred groves of semi-arid areas of Cameroon: case study of Mandara Mountains. *International Journal of Environment*, 4(2): 308-318.

Kent, E.F. (2013). Sacred groves and local gods: religion and environmentalism in south India. Oxford University press, Madison Avenue, New York. Pp: 223.

Khaladkar, R.M., Mahajan, P.N. and Kulkarni, J.R. (2009), Alarming Rise in the Numberand Intensity of Extreme Point Rainfall Events over the Indian Region under Climate Change Scenario, Research Report No. RR-123" August 2009, ISSN 0252-1075., IITM-India

Khamari, S. (2011). Floral Sanctuary: A study of sacred grove of a village in western Orissa. *Orissa Review*, pp: 72-78.

Khandal, D. (2014). Save sacred groves to save our forests. http://www.mydigitalfc.com/leisure-writing/save-sacred-groves-save-our-forests-581. Accessed on 17th July 2014.

Khanwalkar, S. (2005). Rajasthan: Tales of co-existence. In: Community Conserved Areas in India- a directory. Pp: 573-586. http://www.kalpavriksh.org/images/CCA/Directory/Rajasthan_StateChapter.pdf. Accessed on 27th May 2014.

Khiewtam, R.S. and Ramakrishnan, P.S. (1989). Socio-cultural studies at the sacred groves at Cherrapunji and adjoining areas in north-eastern India. *Man in India,* 69: 64-71.

Khoshoo, T.N. (1994). Census of India's Biodiversity: Tasks Ahead. *Current Science,* 67: 577-582.

Khoshoo, T.N. (2008). Environmental Concerns and Strategies. 3rd Edition, APH Publishing, Daryagang, New Delhi.

Khumbongmayum, A.D., Khan, M.L. and Tripathi, R.S. (2004). Sacred groves of Manipur-ideal centres of biodiversity conservation. *Current Science*, 87: 430-433.

Khumbongmayum, A.D., Khan, M.L. and Tripathi, R.S. (2005a). An ethnobotanical study of medicinal plants in the sacred groves of Manipur, northeast India. *Journal of Traditional Ecological Knowledge*, 4(1): 33-38.

Khumbongmayum, A.D., Khan, M.L. and Tripathi, R.S. (2005b). Survival and growth of seedlings of a few tree species in the four sacred groves of Manipur, northeast India. *Current Science*, 88(11): 1781-1788.

Khumbongmayum, A.D., Khan, M.L. and Tripathi, R.S. (2006). Biodiversity conservation in sacred groves of Manipur, northeast India: population structure and regeneration status of woody species. *Biodiversity and Conservation*, 15: 2439-2456.

Khurana, I. (1998). Unnatural decline. *Down to Earth,* 6: 34-38.

King-Oliver, I.E.D., Chitra, V. and Narasimha, D. (1997). Sacred groves: traditional ecological heritage. *Int J. Ecol. Environ. Sci.,* 23: 463-470.

Klopfenstein, S.T., Hirmas, D.R. and Johnson, W.C. (2015). Relationships between soil organic carbon and precipitation along a climosequence in loess-derived soils of the Central Great Plains, USA. *Catena*, 133: 25-34.

Koradia, D. (2000). Sacred groves in dryland sanctuaries. *Honey Bee*, 11: 19-20.

Kosambi, D.D. (1962). Myth and Reality. Bombay: Popular Prakashan.

Kotwal, P.C. and Banerjee, S. (2000). *Biodiversity Conservation in Managed Forests and Protected Areas*. Agribios (India) Publications, Jodhpur. Pp: 217-224.

Kour, K. and Sharma, S. (2014). Diversity and phytosociological analysis of tree species in sacred groves of Vijaypur Block, Samba (J&K). *International Journal of Science and Research*, 3: 859-862.

Krishna, N. (2013). Orans of Rajasthan. http://www.heritageonline.in/2013/10/natural-heritage-discussion-3/. Accessed on 17th July 2014.

Kulkarni, D. K. and Nipunage, D.S. (2009). Floristic diversity and ecological evaluation of 'Dhup-Rahat' sacred grove from Pune district. *Geobios*, 36: 298-302.

Kulkarni, D.K., Nipunage, D.S., Hangarge, L.M. and Kamble, P.B. (2010). Natural heritage of forest conservation in Bhor region of Pune, India. *Asian J. of Environmental Sci.*, 5: 94-98.

Kulkarni, S. and Kulkarni, D. K. (2013). Kalamvihira Sacred grove-A potential tree growth for carbon sequestration in Jawhar taluka of Thane district. *Annals Biological Research*, 4: 119-123.

Kumar, K., Manhas, R.K. and Magotra, R. (2011).The Shankaracharya sacred grove of Srinagar, Kashmir, India. *Current Science*, 101: 262-263.

Lalitha K. P. (2015). An analysis of sacred groves with special reference to Coorg. *International J. Res. Humanities and Social Studies*, 2(10): 38-43.

Laloo, R.C., Kharlukhi, L., Jeeva, S. and Mishra, B.P. (2006). Status of medicinal plants in the disturbed and the undisturbed sacred forests of Meghalaya, northeast India: population structure and regeneration efficacy of some important species. *Current Science*, 90: 225-232.

Lee, C. and Wauchope, S. (2003). The Importance of Sacred Natural Sites for Biodiversity Conservation. Proceedings of the international workshop on the importance of sacred natural sites for biodiversity conservation held in kunming and xishuangbanna biosphere reserve. People's Republic of China.

Leifeld, J. and Kögel, K. (2005). Soil organic matter fractions as early indicators for carbon stock changes under different land-use? *Geoderma*, 124: 143-155.

Li, Q., Zhou, D.W., Jin, Y.H., Wmang, M.L., Song, Y.T. anf Li, G.D. (2014). Effects of fencing on vegetation and soil restoration in degraded alkaline grassland in northeast China. *Journal of Arid Land,* 6(4): 478-487.

Lima, T.T.S., Miranda, I.S. and Vasconcelos, S.S. (2010). Effects of water and nutrient availability on fine root growth in eastern Amazonian forest regrowth, Brazil. *New Phytologist*, 187: 622-630.

Little, P.D. and Brokensha, W. (1987). Local Institutions, Tenure and Resource Management in East Africa. In: Conservation in African People, Policies and Practice, Anderson, D. and Grove, R. (eds.), Cambridge: Cambridge University Press.

Liu, Z., Shao, M. and Wang, Y. (2011). Effect of environmental factors on regional soil organic carbon stocks across the Loess Plateau region, China. *Agriculture, Ecosystem and Environment*, 142: 184-194.

Love, A., Babu, S. and Babu, C.R. (2009). Management of Lantana: an invasive alien weed, in forest ecosystem of India. *Current Science.* 97: 1421-1429.

Lyngwi, N.A. and Joshi, S.R. (2015). Traditional Sacred Groves', an ethnic strategy for conservation of microbial diversity. *Indian Journal of Traditional Knowledge*, 14(3): 474-480.

Lyngwi, N.A., Nongkhlaw, M., Kalita, D. and Joshi, S.R. (2016). Bioprospecting of plant growth promoting bacilli and related genera prevalent in soils of pristine sacred groves: biochemical and molecular approach. PLoS ONE, 11(4): e0152951. doi:10.1371/journal.pone.0152951.

Madeweya, K.H. Oka, H. and Matsumoto, M. (2004). Sustainable Management of Sacred Forests and Their Potential for Eco-Tourism in Zanzibar. *Bulletin of FFPRI*, 3: 33-48.

Magurran, A.E. (1988). Ecological Diversity and Its Measurement, Princeton: Princeton University Press, 110 p.

Mahato, N.K. (2015). Adivasi (indigenous people) perception of landscape: the case of Manbhum. *J. Adivasi and Indigenous Studies,* 2(1): 51-66.

Mahbal, K.B. (2013). India–Farms robbed of water, farmers livelihood. http://www.kractivist.org/tag/irrigation/

Maikhuri, R.K., Nautiyal, S., Rao, K.S. and Saxena, K.G. (1998). Role of medicinal plants in the traditional healthcare systems: A case study from Nanda Devi Biosphere Reserve. *Current Science*, 75(2): 152-157.

Mala, M., Deepa, J. and Saravanakumar, K. (2013). Studies on floristic diversity and ethnomedicinal aspects of sacred grove in Saluppai village, Ariyalur district. *Current Research in Drug Targeting*, 3(1): 1-5.

Malhotra, K.C. (1998). Anthropological dimensions of sacred groves in India: an overview. Pages 423-438, In: Conserving the Sacred for Biodiversity Management, Ramakrishnan, P.S., Saxena, K.G. and Chandra-shekara, U.M. (Eds.), UNESCO and Oxford-IBH Publishing, New Delhi

Malhotra, K.C., Ghokhale, Y., Chatterjee, S. and Srivastava, S. (2001). Cultural and Ecological Dimensions of Sacred Groves in India. INSA, New Delhi.

Malhotra, K.C., Gokhale, Y., Chaterjee, S. and Srivastava, S. (2007). Sacred Groves of India: An overview. Aryan Books, New Delhi. Pp. 170.

Malhotra, K.C., Gokhale, Y., Chatterjee, S. and Srivastava, S. (1999). Sacred Groves in India: An Overview. National Museum of Mankind, Bhopal.

Malhotra, K.C., Stanley, S., Hemam, N.S. and Das, K. (1997). Biodiversity conservation and ethics: Sacred groves and pools, In: Bioethics in Asia, Fujiki, N. and Macer, R.J. (eds.), Eubois Ethics Institute, Japan. Pp. 338- 345.

Malik, V. (2015). Addition to the sacred grove of India. *Journal of Global Biosciences*, 4(7): 2699-2702.

Manikandan, P., Venkatesh, D. R. and Muthuchelian, K. (2011). Conservation and Management of Sacred groves in Theni District, Tamil Nadu, India. *J. Biosci. Res.*, 2: 76-80.

Maru, R.N.and Patel, R.S. (2013). Ethno-botanical survey of sacred groves and sacred plants of jhalod and surrounding areas in Dahod district, Gujarat, India. *Research Journal of Recent Sciences*, 2: 130-135.

McCann, J.C. (1997). The plow and the forest: Narratives of deforestation in Ethiopia, 1840–1992. *Environmen. Hist.,* 2: 138–159.

McGarvey, J.C., Thompson, J.R., Epstein, H.E. and Shugart, Jr, H.H. (2015). Carbon storage in old-growth forests of the Mid-Atlantic: toward better understanding the eastern forest carbon sink. *Ecology*, 96(2): 311–317.

McNeely, J.A., Miller, K.R., Reid, W.R. and Werner, T.B. (1990). Conserving the World's Biological Diversity. IUCN, Gland, Switzerland.

Medhi, P. and Borthakur, S.K. (2013). Sacred groves and sacred plants of the Dimasas of North Cachar Hills of Northeast India. *African Journal of Plant Science*, 7: 67-77.

Meena, D. and Singh, A. (2012). Oran of Rohida: an endangered tree species of Rajasthan. *Current Sci.*, 103: 189.

Meenakshi, B., Chauhan, N.S. and Kak, A. (2011). Dye yielding plants of Himachal Pradesh. *J Econ Taxon Bot*., 35: 429-432.

Mgumia, F.H. and Oba, G. (2003). Potential role of sacred groves in biodiversity conservation in Tanzania. *Environmental Conservation*, 30(3): 259-266.

Mishra, B.P., Tripathi, O.P., Tripathi, R.S. and Pandey, H.N. (2004). Effects of anthropogenic disturbance on plant diversity and community structure of a sacred grove in Meghalaya, northeast India. *Biodiversity and Conservation*, 13: 421–436.

Misra, R. (1968). Ecology Work-Book. Oxford and IBH Publishing Co., New Delhi.

Mitra, A. and Pal, S. (1994). Besieged the forests of the Gods; The sprit of sanctuary. *Down to Earth* (January 31): 21-36.

Mohanan, C.N. and Nair, N.C. (1981). Kunstleria Pram- a new genus record for India and a new species in the genus.Proc. *Ind. Acad. Sci.* B, 90:207-210.

Mohanta, R.K., Adhikari, B.S., Sahu, H.K. and Swain, K.K. (2012). The biological values and conservation status of sacred groves in the Balasore wildlife division, Odisha: a case study. *International J. Conservation Science*, 3(3): 207-216.

Mondal, K. and Roy, M. (2010). Sacred grove: its significance in biodiversity conservation. LAP Lambert Academic Publishing.

Murthy, D.B.N. (2006). Sacred trees. The Hindu, November 3, 2006.

NAEB (1995). Sacred groves of Kurukshetra, Haryana. National Afforestation and Ecodevelopment Board, Ministry of Environment and Forests, Govt. 'of India, New Delhi.

Naidu, M.T., Kumar, O.A. and Venkaiah, M. (2015). Vascular plant diversity in the sacred grove of Modapalli in Visakhapatnam District of Andhra Pradesh, India. *Journal of Threatened Taxa*, 7(10): 7683-7690.

Nair, G.H. (1997). Sacred groves of India – vanishing greenery. *Current Science,* 72(10): 697-698.

Nath, Y.V.S. (1960). Bhils of Ratanmal, Maharaja Sayajirao University. Baroda. Oxford and IBH Publishing Co., New Delhi.

Nieuwhof, A. (2006). Changing landscape and grazing: macroremains from the terp Peins-east, province of Friesland, the Netherlands. *Vegetation History and Archaeobotany*, 15(2): 125-136.

Niraj, A.K. and Gupta, V.K. (2015). Role of sacred grooves in faunal diversity in Alwar district. *Global J. Plant Sciences*, 1(1): http://gjess.co.in/index.php/GJPS/ article/ view/150/0.

Nithyadevi, J. and Sivakumar, R. (2015). Phytosocialogical and ethnomedicinal studies of sacred groves in Konjikuppam village, Cuddalore district, Tamil Nadu. *International Letters of Natural Sciences*, 32: 77-91.

Nwankwo, E. A. and Agboeze, M.U. (2016). Safety issues at selected shrines/sacred groves in eastern Nigeria. *International J. Asian Social Sci.*, 6(1): 80-92.

Ogle, K. and Reynolds, J.F. (2004). Plant responses to precipitation in desert ecosystems: Integrating functional types, pulses, thresholds, and delays. *Oecologia,* 141: 282-294.

Olsson, E.G.A. and Ouattara, S. (2013). Opportunities and Challenges to Capturing the Multiple Potential Benefits of REDD+ in a Traditional Transnational Savanna-Woodland Region in West Africa. *Ambio,* 42(3): 309–319.

Onyekwelu, J.C. and Olusola, J.A. (2014). Role of sacred grove in in-situ biodiversity conservation in rainforest zone of south-western Nigeria. *Journal of Tropical Forest Science*, 26(1): 5–15.

Ormsby, A.A. (2011). The impacts of global and national policy on the management and conversation of sacred groves of India. *Human Ecology,* 39: 783-793.

Ormsby, A.A. (2013). Analysis of local attitudes toward the sacred groves of Meghalaya and Karnataka, India. *Conservation and Society*, 11: 187-197.

Ormsby, A.A. and Bhagwat, S.A. (2010). Sacred forests of India: a strong tradition of community-based natural resource management. *Environmental Conservation,* 37: 320-326.

Ormsby, A.A. and Edelman, C. (2010). Tafi Atome Monkey Sanctuary, Ghana: Community-based ecotourism at a sacred site. In: Sacred natural sites: Conserving nature and culture, Verschuuren, B., Wild, R., McNeely, J. and Oviedo G. (eds.), Earthscan, London. Pp. 233-243.

Ormsby, A.A. and Ismail, S.A. (2015). Cultural and ecological insights into sacred groves: managing timber resources for improved grove conservation. *Forests, Trees and Livelihoods*, 24(4): 244-258.

Ostrom, E. (1998). IFRI Field Manual Version 10.5. 2002. International Forestry Resources and Institutions Research Program Indiana University. Worskshop in Political Theory and Policy Analysis.

Oyelowo, O.J., Aduradola, A.M., Onadeko, S.A. and Agboola, D.A. (2014). Socio-economic contribution of selected sacred groves in south-western Nigeria. *International J. Novel Res. Civil Structural and Earth Sci.*, 1(1): 36-45.

Padmanabhan, P. (2005). Documentation and conservation of small mammals in the sacred groves of Kerala, peninsular India. KFRI Research Report No: 406.

Painter, E.L. and Belsky, A.J. (1993). Application of herbivore optimization theory to rangelands of the western United States. *Ecological Applications*, 3: 2–9.

Pala, N.A., Negi, A.K., Vikrant, K.K., Gokhale, Y. and Todaria, N. P. (2013). Carbon storage in sacred groves-A Study from Chanderbadni sacred grove in Garhwal

Himalaya, Uttarakhand, India. *Journal of Biodiversity and Ecological Sciences*, 3: 75-79.

Pandey, A. (2000). From Sarana (sacred groves) to sustainable natural resources management. Abstract. National Workshops on Community Strategies on the Management of Natural Resources. Bhopal.

Pandey, D.N. (1998a). Ethnoforestry: Local Knowledge for Sustainable Forestry and Livelihood Security. Himanshu Publications, Udaipur.

Pandey, D.N. (1998b). Sacred groves, sacred corridors, sacred gardens and temple forests of Rajasthan. Paper presented in The National Workshop on Sacred Groves at CPR Environmental Education Centre, Chennai.

Pandey, D.N. (1999). Sacred forestry: The case of Rajasthan, India. Sustainable Development International, pp: 1-6. available at: http://www.p2pays.org/ref/40/39748.pdf. Accessed on 25th July 2014.

Pandey, D.N. (2000). Sacred water and sanctified vegetation: tanks and trees in India. http://dlc.dlib.indiana.edu/dlc/bitstream/handle/10535/2384/pandeyd041700.pdf?sequence=1. Accessed on 27th May 2014

Pandit, P.K. and Bhakat, R.K. (2007). Conserving of biodiversity and ethnic culture through sacred groves in Midnapore District, West Bengal, India. *Indian Forester*, 133: 323-344.

Parihar, G.R. (2015). Faunal diversity richness of ecological institute in the village landscape of the Thar eco-region Rajasthan, India. *International J. Eng. Management & Sci.*, 2(6): 6-8.

Patel, H.R. and Patel, R. (2013b). Ethnobotanical plants used by the tribes of R.D.F. Phoshina forest range of Sabarkantha District, North Gujarat, India. *Int. J.Sci. Res.*, 3(2): 1-8.

Patel, H.R. and Patel, R.S. (2012). Study of sacred groves and sacred plants of R.D.F. Poshina Forest Range of Sabarkantha District, North Gujarat, India. *Life sciences Leaflets*, 5: 11-16.

Patel, P.K. and Patel, M.K. (2013a). Sacred groves in conservation of plant biodiversity in Banaskantha district, Gujarat, India. *Recent Research in Science and Technology*, 5: 13-16.

Patel, R., Mahato, A.K.R. and Patel, Y.S. (2014). Study on the floristic diversity of two newly recorded sacred groves from Kachchh district of Gujarat, India. *Indian Journal of Plant Sciences*, 3: 75-81.

Patnaik, S. and Pandey, A. (1998). A study of indigenous community based forest management system: Sarna (sacred groves). In: Conserving the Sacred for Biodiversity Management, Ramakrishnan, P.S., Saxena, K.G. and Chandrasekara, U.M. (eds.), Oxford and IBH, New York, Pp: 315-322.

Pearce, D.W. and Moran, D. (1994). The economic value of biodiversity. Earthscan, London. 24 p.

Pielou, E.C. (1966). The Measurement of Diversity in Different Types of Biological Collections. *J. Theor. Biol.*, 13: 145–163.

Poddar, J. (2014). Sacred groves in Mansarovar cleaned under the eco-school programme. January 31, 2014. http://news.vrindavantoday.org/2014/01/sacred-groves-in-mansarovar-cleaned-under-the-eco-school-programme/. Retrieved on 4th April 2016.

Prasad, K.S. and Raveendran, K. (2012). Endemic plant diversity in selected sacred groves of Kasaragod district, Kerala. *International J. Plant, Animal and Environ. Sci.*, http://www.rroij.com/open-access/endemic-plant-diversity-in-selected-sacred-groves-of-kasaragod-district-kerala-240-244.pdf.php?aid=39699. Retrieved 5th April 2016.

Prasad, M.R., Pati, S., Sarkar, S., Gayen, A., Guin, P. and Mishra, T. (2015). General awareness and perceptions about sacred groves and biodiversity conservation in urban people of Bankura district, West Bengal, India. *Int. Res. J. Environment Sci.*, 4(2): 16-21.

Primack, R. and Higuchi, H. (2006). Climate change and cherry tree blossom festivals in Japan. *Arnoldia,* 65: 15-22. http://people.bu.edu/primack/cherries.pdf. Accessed on 16th July 2014.

Priya, K. and Sharma, S. (2014). Deteriorating condition of sacred groves in Jammu city and its surroundings, J&K. *International Journal of Scientific Research,* 3: 213-215.

Priyanka, N. and Joshi P. K. (2013). A review of Lantana camara studies in India. *International Journal of Scientific and Research Publications,* 3: 1-11.

Pruthi, I. and Burch, W. J. (2009). A socio-ecological study of sacred groves and memorial parks: Cases from USA and India. *J. Environ. Sci. Eng.,* 1: 7-14.

Pundee, S. (2007). Rare trees in forest fragments– will they survive? A study of four climax forest tree species within sacred groves of the Konkan region (northern Western Ghats). Project Report Applied Environmental Research Foundation (AERF), Kothrud, Pune-411029, Maharashtra - INDIA

Pushpagandan, P., Rajendraprasad, M. and Krishnan, P.N. (1998). Sacred groves of Kerala synthesis on the state of the art of knowledge. In: Conserving the Sacred for Biodiversity Management, Ramakrishnan, P.S., Saxena, K.G. and Chandrashekhara, U.M. (eds.), Oxford and IBH Publishing Co. Pvt. Ltd., New Delhi. Pp. 193-209.

Quiroz, D. (2013). The role of Spiritual values. http://www.ciwem.org/news/2013/aug/the-role-of-spiritual-values.aspx

Rahangdale, C.P., Patley, R.K. and Yadav, K.C. (2014). Phytodiversity of ethnomedicinal plants in sacred groves and its traditional uses in Kabirdham District of Chhattisgarh. http://www.academia.edu/3642503/. Accessed on 11th July 2014.

Rajendraprasad, M. (1995). The floristic, structural and functional analysis of sacred groves of Kerala, Ph.D. thesis, University of Kerala, Thiruvananthapuram.

Rajkumar, G., Ravipaul, S., Sivasamy, A., Anuradha, G. and Mohamed, S.R. (2014). Floristic composition and practices on the selected sacred groves of Perambalur distrct, Tamil Nadu. *Int. J. Modn. Res. Revs.,* 2(11): 286-291.

Ramachandran, B. (1999). Significance of Kavu - a note on the sacred grove of Kerala in eco-cultural context. *Journal of Human Ecology,* 10: 285-288.

Ramakrishnan, P.S. (1998). Conserving the sacred for biodiversity: the conceptual framework. In: Conserving the sacred for biodiversity management, Ramakrisnan, P.S., Saxena, K.G. and Chandrashekhara, U.M. (eds.), Oxford and IBH Publishing Co. Pvt. Ltd., New Delhi. Pp. 3-15.

Ramakrishnan, P.S. (2001). Increasing population and declining biological resources in the context of global change and globalization. *Journal of Bioscience,* 26: 465-479.

Ramakrishnan, P.S. (2002). What is Traditional Ecological Knowledge? In: Traditional Ecological Knowledge for Managing the Biosphere Reserve in South and Central Asia, Ramakrishnan, P.S., Rai, R.K., Katwal, R.P.S. and Mehndiratta, S. (eds.), Oxford University Press, New Delhi. Pp. 3-12.

Ramesh, M. and Ishwar, N.M. (2008). Status and distribution of the Indian spiny-tailed lizard *Uromastyx hardwickii* in the Thar Desert, Western Rajasthan. GNAPE Technical Report No T02. Pp. 48.

Ramoliya, P.J., Patel, H.M. and Pandey, A.N. (2004). Effect of salinization of soil on growth and macro-and micro-nutrient accumulation in seedlings of *Acacia catechu* (Mimosaceae). *Annals of Applied Biology,* 144: 321-332

Rampilla, V., Mahammad, K.S. and Kakumanu, B. (2015). Floristic diversity and phyto-sociological studies of Indrakiladri sacred grove in Krishna district, Andhra Pradesh, India. *J. Pharmacy and Biological Sci.*, 10(4): 61-75.

Rani, S.S., Murthy, K.S.R., Goud, P.S.P. and Pullaiah, T. (2003). Tree wealth in the life and economy of the tribes people of Andhra Pradesh, India. *J. Tropical Forest Sci.*, 15(2): 259-278.

Rao, B.R.P. and Sunitha, S. (2011). Medicinal plant resources of Rudrakod sacred grove in Nallamalais, Andhra Pradesh, India. *J. Biodiversity*, 2: 75-89.

Rao, B.R.P., Babu, M.V.S., Reddy, M.S., Reddy, A.M., Rao, V.S., Sunitha, S. and Ganeshaiah, K.N. (2011). Sacred groves in southern Eastern Ghats, India: Are they better managed than forest reserves? *Tropical Ecology*, 52: 79-90.

Rao, M.L.S., Ramakrishna, N. and Saidulu, C.H. (2014). Ethno-veterinary herbal remedies of Gujjars and other folklore communities of Alwar district, Rajasthan, India. *Int. J. Ayur. Pharma Research*, 2: 40-45.

Rathore, M.S. and Shekhawat, N.S. (2012). Ethnobotanical importance of Orans - as a means of conserving biodiversity. *Int. J. Agric. Sci., Res. and Techn.*, 1: 195-200.

Rawat, V.S. (2012). Fine root biomass and soil nutrients in van panchayat forest of Almora district. *Indian Journal of Plant Sciences*, 1: 101-108.

Ray, R., Chandran, M.D.S. and Ramachandra, T.V. (2012). Conservation impacts on sacred forests fragments: a case study from Karnataka, India. *Indian Forester*, 138(6): 248-251.

Ray, R., Chandran, M.D.S. and Ramachandra, T.V. (2014). Biodiversity and ecological assessments of Indian sacred groves. *J. Forestry Research*, 25: 21-28.

Ray, R., Chandran, M.D.S. and Ramachandra, T.V. (2014). Hydrological importance of sacred forest fragments in central western Ghats of India. *Tropical Ecology*, 56(1): 87-89.

Reeder, J.D., Schuman, G.E., Morgan, J.A. and LeCain, D.R. (2004). Response of organic and inorganic carbon and nitrogen to long-term grazing of the short grass steppe. *Environmental Management*, 33: S485-S495

Reisner, M.D., Grace, J. B. Pyke, D.A. and Doescher, P.S. (2013). Conditions favoring *Bromus tectorum* dominance of endangered sagebrush steppe ecosystems. *Journal of Applied Ecology*, 50: 1039–1049

Roy Burman, J.J. (1992). The institution of sacred groves. *Journal of Indian Anthropological Society,* 27: 219-38.

Roy Burman, J.J. (1995). The dynamics of sacred groves. I. *Human Ecol.,* 6: 245-254.

Roy Burman, J.J. (1996). A comparison of sacred groves among the Mahadeo Kolis and Kunbis of Maharashtra. *Indian Anthropologist*, 26: 3746.

Roy Burman, J.J. (2014). Sacred groves: a cultural symbol of tribal political self-assertion. *Journal of Tribal Intellectual Collective India,* 1: 37-47.

Saha, Suranjana, and Coauthors, (2010): The NCEP Climate forecast system reanalysis. *Bull. Amer. Meteor. Soc.*, **91**, 1015–1057.

Sahu, P.K., Kumari, A., Sao, S., Singh, M. and Pandey, P. (2013). Sacred plants and their ethno-botanical importance in central India: a mini review. *Int. J. of Pharm. & Life Sci.* 4: 2910-2914.

Sahu, S.C., Dhal, N.B., Lal, B. and Mohanty, R.M. (2012). Differences in tree species diversity and soil nutrient status in a tropical sacred forest ecosystem on Niyamgiri hill range, Eastern Ghats, India. *J. Mountain Sci.,* 9(4): 492-500.

Sajwan, B.S. and Kala, C.P. (2007). Conservation of medicinal plants: conventional and contemporary strategies, regulations and executions. *Indian Forester*, 133: 484-495.

Samati, H. and Gogoi, R. (2007). Sacred groves of Meghalaya. *Current Science*, 93(10): 1338–1339.

Sambandan, K. and Dhatchanamoorthy, N. (2012). Studies on the Phytodiversity of a Sacred Grove and its Traditional Uses in Karaikal District, U.T. Puducherry. *Journal of Phytology*, 4: 16-21.

Sann, B., Kanzaki, M. and Ohta, S. (2016). Vegetation patterns and species-filtering effects of soil in secondary succession in a tropical dry forest in central Myanmar. *Journal of Tropical Ecology,* 32(2): 116-124.

Sarfo-Mesah, P., Oduro, W., Fredua, E.A. and Amisah, S. (2010). Traditional representations of the natural environment and biodiversity conservation: sacred groves in Ghana. *FEEM Working Paper No. 87.2010* Available at SSRN: http://ssrn.com/abstract=1665065 or http://dx.doi.org/10.2139/ssrn.1665065.

Sariyildiz, T., Savaci, G. and Kravkaz, I.S. (2015). Effects of tree species, stand age and land-use change on soil carbon and nitrogen stock rates in northwestern Turkey. *Journal of Biogeosciences and Forestry,* 9: 165-170.

Sarkar, J. (2011). Ramsar Convention and India. *Current Science*, 101: 1266–1268.

Schenk, H.J. and Jackson, R.B. (2002). Rooting depths, lateral root spreads and below-ground/above-ground allometries of plants in water-limited ecosystems. *Journal Ecology*, 90(3): 480-494.

Sebastiano, T. (2009). Research, protection and evaluation of Sicilian and Mediterranean marine cultural heritage. *Conservation Science- In Cultural Heritage*, 9: 79-112.

Sehgal, J. and Abrol, P. (1994). Soil Degradation in India: Status and Impact, Oxford and IBH Publishing Pvt. Ltd.

Sethi, N. and Viswanath (2003). Sacred grove for fragment's sake. A Down to Earth Supplement. *Down To Earth*. 31st December, 69.

Shah, R. and Patel, R. (2011). Study of traditional worshiping plant of Borsad taluka (Gujarat) India. *Abhinav*, 1: 69-73.

Shannon, C.E. and Weiner, W. (1963). The Mathematical Theory of Communication, Urbana, United States: University of Illinois Press, 177 p.

Sharifi, M., Nadjafi, F., Youssefshahi, H., and Hemmati, Z. (2000). Sacred Grove in western Iran. *Journal of Environmental Studies,* 26: 49-58.

Sharma GP, Raghubanshi AS. 2010. How *Lantana* invades dry deciduous forest: a case study from Vindhyan highlands, India. *Tropical Ecology*, 51(2S): 305–316.

Sharma, B.R. (2000). Sacred groves and their role in social life in Himachal Himalayas. Abstract National Workshops on Community Strategies on the Management of Natural Resources. Bhopal.

Sharma, L. and Upadhyay, D.P. (2014). Biodiversity conservation in Rajasthan: role of a botanist. *Res. J. Pharma., Biol. and Chem. Sci.*, 5: 1037-1044.

Sharma, S. and Devi, R. (2014). Sacred groves: traditional way of conserving plant diversity in block Bhalwal of Jammu District (J&K). *International J. Sci., Res.*, 3(2): 239-242.

Sheridan, M.J. (2009). The Environmental and Social history of African sacred groves: a Tanzanian case study. *African Studies Review*, 52(1): 73-98.

Shetty, B.V. and Singh, V. (1993). Flora of Rajasthan Vol. III. Botanical Survey of India, Calcutta.

Sigu, G.O., Omenda, T.O., Ongugo, P.O. and Opiyo, A. (2001). Sacred groves institutions, rule enforcement and impact on forest condition: the case of Ramogi hill forest reserve, Kenya. https://www.cbd.int/doc/case-studies/for/cs-ecofor-ke-01-en.pdf. Retrieved 5th April 2016.

Sikor, T., Stahl, J., Enters, T., Ribot, J.C., Singh, N. and Sunderlin, W.D. and Wollenberg, L. (2010). REDD-plus, forest people's rights and nested climate governance. *Global Environmental Change,* 20: 423–425.

Silja, V.P., Varma, K.S. and Mohanan, K.V. (2008). Ethnomedicinal plant knowledge of the Mullu kurumatribe of Wayanad district, Kerala. *Indian J. Traditional Knowledge*, 7: 604-612.

Silva, D.M., Batalha, M.A., Cianciaruso, M.V. (2013). Influence of fire history and soil properties on plant species richness and functional diversity in a neotropical savanna. *Acta Bot. Bras.* 27(3): 490-497

Simpson, E. H. (1949). Measurements of Diversity. *Nature*, 163: 683–688.

Singh, A. (2010). Communities manage Orans for protecting their livelihoods. In: Sustaining Participation and scaling up Farmer Empowerment, Radha, T.M., Tafur, J.C., Mertineit, A. and Yap, E. (Eds.), MISEREOR, pp 52-55.

Singh, A. (2011). Oran/Devbani matters trees and tigers conservation by tribes: a case study from Sariska. *Envis Newsletter*, 10: 2-6.

Singh, A. and Sisodia, P. (2003). Oran-A traditional biodiversity management system in Rajasthan. Leisa India. Pp. 19-21. http://www.agriculturesnetwork.org/magazines/india/3-access-to-resources/oran-a-traditional-biodiversity-management-system/at_download/article_pdf. Accessed on 14th July 2014.

Singh, G. (2005). Carbon sequestration under an agri-silvicultural system in the arid region. *Indian Forester*, 131(4): 543-552.

Singh, G. (2008). Assessment of soil carbon stock and dynamics in forest soils of India. http://afri.icfre.org/showdetails87fe.html?pid=16&statusp=d&keepThis=true&TB_iframe=true&height=500&width=600.

Singh, G. (2013). Carbon sequestration during rehabilitation of degraded hills. Pamphlet, AFRI, Jodhpur.

Singh, G. (2014). Studies on carbon sequestration in different forest types of Rajasthan. Project concluding report submitted to ICFRE, Dehradun, India.

Singh, G. (2015b). Studies on the effects of MPOWER proghramme on mitigation and adaptation towards climate change change in western Rajasthan. Project concluding report submitted to MPOWER, Jodhpur. https://www.google.co.in/?gfe_rd=cr&ei=VtayV6zsJ_GK8QfT2pPoCg&gws_rd=ssl#q=Studies+on+the+effects+of+MPOWER+programme+on+mitigation+and+adaptation+towards+Climate+Change+in+western+Rajasthan

Singh, G., Mishra, D., Singh, K. and Parmar, R. (2013). Effects of rainwater harvesting on plant growth, soil water dynamics and herbaceous biomass during rehabilitation of degraded hills in Rajasthan, India. *Forest Ecology and Management*, 310: 612-622

Singh, G.S. and Saxena, K.G. (1998). Sacred groves in the rural landscape: a case study of Shekhala village in Rajasthan. In: Conserving the Sacred for Biodiversity Management, Ramakrishnan, P.S., Saxena, K. G. and Chandrashekara, U.M. (eds.), Science Publishers, New Hampshire/Oxford and IBH New Delhi. Pp. 153-161.

Singh, H., Husain, T. and Agnihotri, P. (2010). Haat kali sacred grove, Central Himalaya, Uttarakhand. *Current Science,* 98: 290.

Singh, K. (2015a). Effects of landuse types on floral diversity and carbon stock in Jodhpur districts of Rajasthan. Ph. D. Thesis, FRI University, Dehradun (Uttarakhand), India.

Singh, K.P., Shukla, A.N. and Singh, J.S. (2010). State level inventory of invasive alien plants, their source regions and use potential. *Current Sci.,* 99(1): 107-114.

Singh, N.K. (2011). A survey of sacred groves of Devipatan region of U.P., Paper presented at the National Conference on 'Forest Biodiversity Earth's Living Treasure'. UP State Biodiversity Board Souvenirs, Pp. 153 -157.

Singh, R.K.R. and Garg, A. (2014). The translocated Kurdi Angod sacred site –a conservatory of RET plants of the Western Ghats. *Current Science,* 106(1): 11-12.

Singh, Y.P., Sharma, D.K., Singh, G., Nayak, A.K., Mishra, V.K. and Singh, R. (2008). Alternate land use management for sodic soils. Tech. Bull. No. 2/2008. Central Soil Salinity Research Institute, Regional Research Station, Lucknow-226005 (INDIA), pp. 16.

Sinha, B. and Maikhuri, R.K. (1998). Conservation through socio-cultural-religious practices in Garhwal Himalaya: A case study of Haryali sacred site. A P H Publication Corporation, New Delhi. Pp. 289-299.

Sisodia, V. and Malhotra, K.C. (1963). Terracotta Motifs among the Warlis of Dharampur. *East and West,* 14: 97-105.

Skinner, A.K., Lunt, I.D., Spooner, P. and Mcintyre, S. (2009). The effect of soil compaction on germination and early growth of eucalyptus albens and an exotic annual grass. *Austral Ecol.*, 34: 698-704.

Sochaczewski, P. (1998). Life reserves. Arbor Vitae (IUCN/WWF), April, 14.

Sokal, R.R. and Rohlf, F.J. (1981). Biometry. 2nd edn. New York, USA: W.H. Freeman.

Soliveres, S. and Eldridge, D.J. (2014). Do changes in grazing pressure and the degree of shrub encroachment alter the effects of individual shrubs on understorey plant communities and soil function? *Functional Ecology*, 28(2): 530-537.

Sponsel, L.E. (2008). Sacred places and biodiversity conservation. http://www.eoearth.org/view/article/155815/. Retreived on 5th April 2016.

Srivastava, M.K. (1994). Hill Korwa: Past, present and potential. Shri Mudran and Publication, Raipur.

Stephenson, N.L., Das, A.J., Condit, R., Russo, S.E. et al. (2014). Rate of tree carbon accumulation increases continuously with tree size. *Nature,* doi:10.1038/nature12914 (2014).

Sukumaran, S. and Raj, A.D.S. (2010). Medicinal plants scared groves in Kanyakumari district, Southern Western Ghats. *Indian J. Trad, Knowl,* 9: 294-299.

Sundarapandian, S.M. and Subbiah, S. (2015). Diversity and tree population structure of tropical dry evergreen forests in Sivagangai district of Tamil Nadu, India. *Tropical Plant Research,* 2(1): 36–46.

Sundriyal, R.C., Sharma, E.E. and Raj, S.C. (1994). Tree structure regeneration and woody biomass removal in a subtropical forest of Mamlay watershed in Sikkim Himalaya. *Vegetation,* 113: 53-63.

Swamy, P.S., Sundarpandian, S.M. and Chandrasekaharan, S. (1998). Sacred groves of Tamil Nadu. In: Conserving the sacred for biodiversity management, Ramakrishnan, P.S., Saxena, K.G. and Chandra-sekhara, U.M. (eds.), UNESCO and Oxford-IBH Publishing, New Delhi. Pp. 357-361.

Takahashi, Y., Veríssimo, D., MacMillan, D.C. and Godbole, A. (2012). Stakeholder perceptions of potential flagship species for the sacred groves of the north Western Ghats, India. *Human dimensions of Wildlife: An International Journal*, 17: 257-269.

Tegegn, G. G. (2008). Experiences on Prosopis management case of Afar region. FARM-Africa, London.

Tewari, S. and Rao, P. (2015). Ecosystem conservation through sacred groves. http://www.krishisanskriti.org/vol_image/11Sep201505095226.pdf. Retrieved on 23rd March 2016.

Thomey, M.L., Ford, P.L., Reeves, M.C., Finch, D.M., Litvak, M.E., Collins, S.L. (2014). Review of climate change impacts on future carbon stores and management of warm deserts of the United States. Gen. Tech. Rep. RMRS-GTR-316. Fort Collins, CO: U.S. Department of Agriculture, Forest Service, Rocky Mountain Research Station. 26 p.

Tiwari, B.K., Tynsong, H., Lynrah, M.M., Lapasam, E., Deb, S. and Sharma, D. (2013). Institutional arrangement and typology of community forests of Meghalaya, Mizoram and Nagaland of North-East India. *Journal of Forestry Research*, 24: 179-186.

Tiwari, U.K., Bank S.K. and Tripathi, R. (1998). Sacred groves of Meghalaya, In: Conserving the sacred for biodiversity management, Ramakrishnan, P.S., Saxena, K.G. and Chandrashekara, U.M. (eds.), Oxford and IBH Publishing Co., New Delhi. Pp. 253-262.

Tokarev, S. (1989). History of Religion. Progress Publishers, Moscow.

Tripathi, G., Deora, R. and Singh, G. (2013). The influence of litter quality and micro-habitat on litter decomposition and soil properties in a silvopasture system. *Acta Oecologica*, 50: 40-50.

Tripathi, N., Singh, R.S., Bakhori, B., Dalal, C., Parmar, D. and Mishra, B. (2013). The world's only inland mangrove in sacred grove of Kachchh, India, is at risk. *Current Science,* 105(8): 1053-1055.

Tripathi, R.S., Tiwari, B.K. and Barik, S.K. (1995). *Sacred Groves of Meghalaya: Status and Strategy for Their Conservation.* NAEB, NEHU, Shillong. Pp. 112-125.

Tripathi, S. and Shukla, R.P. (2007). Effect of clipping and grazing on various vegetat-ional parameters of grassland communities of Gorakhpur, Uttar Pradesh. *Tropical Ecol.*, 48(1): 61-70.

Trouse, A., Jr. and Humbert, R. (1961). Some effects of soil compaction on the development of sugar cane roots. *Soil Sci.*, 91: 208–217.

Ugwu, C.O.T. (2014). The demise of the African god/s: Fallacy of reality. An inaugural lecture. Nsukka: University of Nigeria.

UNDP (2008). Human development profile. Jhalawar district. Institute of Development Studies, Jaipur.

Unnikrishnan, E. (1990). Part played by the Sacred Groves in Local Environments, Centre for Science and Environment, New Delhi.

Unnikrishnan, V. (1995). Sacred groves of North Kerala: an ecofolklore study (in Malayalam). Jeevarekha, Thrissur.

UN-REDD+ (2012). Retrieved 22nd March 2016, from http://reddpluspartnership.org/en/.

Vågen, T.-G. and Winowiecki, L.A. (2013). Mapping of soil organic carbon stocks for spatially explicit assessments of climate change mitigation potential. *Environmental Research Letter*, 8(1): 7 page, doi:10.1088/1748-9326/8/1/015011.

Vartak, V.D., Kumbhojkar, M.S. and Nipuge, D.S. (1987). Sacred groves in tribal areas of Western Ghats: treasure trove of medicinal plants. *B. Medi. Ethno. Bot. Res.*, 8: 77-78.

Vasan, S. and Kumar, S. (2006). Situating conserving communities in their place: political economy of Kullu *Devban. Conservation Soc.*, 4: 325-346.

Vatsyayan, K. (1992). Ecology and Indian Myth. *India International Centre Quarterly*, 19: 157-180.

Venkanna, K., Mandal, U.K., Raju, A.J.S., Sharma, K.L., Adake, R.V., Pushpanjali, Reddy, B.S., Masane, R.N., Venkatravamma, K. and Babu, B.P. (2014). Carbon stocks in major soil types and land-use systems in semiarid tropical region of southern India. *Current Science*, 106: 604-611.

Verschuuren, B., McNeely, J., Oviedo, G. and Wild, R. (2012). Sacred natural sites: conserving nature and culture. Earthscan, London. 336p.

Vidyarthi, L. P. and Rai, B. K. (1977). Tribal culture of India.Concept Publishers, Delhi.

Viji, C. (1995). The sacred groves. The Hindu, 5 September.

Vinothkumar, D., Murugavelu, S. and Kethsy Prabhavathy, A. (2011). Phytosociological and ethenobotanical studies of sacred groves in Udukottai District, Tamil Nadu, India. *Asian J. Exp. Biol. Sci.*, 2: 306-315.

Vipat, A. and Bharucha, E. (2014). Sacred groves: the consequence of traditional management. *Journal of Anthropology*, Volume 2014, Article ID 595314, 8 pages, http://dx.doi.org/ 10.1155/2014/595314.

Wagh, V.V. and Jain. A.K. (2015). Invasive alien flora of Jhabua district, Madhya Pradesh, India. *International J. Biodiversity and Conservation*, 7(4): 227-237.

Waghchaure, C.K., Tetali, P., Gunale, V.R., Antia, N.H. and Birdi, T.J. (2006). Sacred groves of parinche valley of Pune district of Maharashtra, India and their importance. *Anthropology and Medicine*, 13: 55-76.

Walkely, A. and Black, I.A. (1934). An examination of Degtjareff methods for determining soil organic matter and a proposed modification of the chromic acid titration method. *Soil Science*, 37: 29-38.

Wang, X., Wang, J. and Zhang, J. (2012). Comparisons of Three Methods for Organic and Inorganic Carbon in Calcareous Soils of Northwestern China. PLoS ONE, 7(8): e44334. doi:10.1371/journal.pone.0044334

Warrier, K.C.S., Kunhikannan, C. and Sasidharan, K.R. (2015). Endangering sacred groves of a non forested region in Kerala, India and strategies for their conservation. *Indian Forester*, 141(8): 832-837.

Wei, Y., Li, M., Chen, H., Lewis, B.J., Yu, D., Zhou, L., et al. (2013). Variation in carbon storage and its distribution by stand age and forest type in boreal and temperate forests in northeastern China. PLoS ONE, 8(8): e72201. doi:10.1371/ journal.pone.0072201

Wild, R.W. and McLeod, C.M. (eds.) (2008). Sacred natural sites: guidelines for protected area managers. Gland: IUCN.

Willatt, S.T. and Pullar, D.M. (1984). Changes in soil physical properties under grazed pastures. *Australian J. Soil Research,* 22(3): 343-348.

WWF- Andhra Pradesh, (1996). Sacred Groves of Andhra Pradesh. World Wide Fund for Nature, Andhra Pradesh State Office, Hyderabad.

Xaxa, V. (1991). Orans: Religion, Custom and Environment. In: Indigenous Vision, Sen, G. (ed.), Saga Publication, New Delhi, pp. 101-109.

Xu, W., Liu, L., He, T., Cao, M., Sha, L., Hu, Y., Li, Q. and Li, J. (2016). Soil properties drive a negative correlation between species diversity and genetic diversity in a tropical seasonal rainforest. *Scientific Reports*, 6, Article number: 20652, doi:10.1038/srep20652

Xu, W., Liu, W., Yang, W., Tang, C. and Blank, D. (2012). Rhombomys opimus contribution to the "fertile island" effect of tamarisk mounds in Junggar Basin. *Ecological Research,* 27: 775-781.

Xuluc-Tolosa, F.J., Vester, H.F.M., Ramírez-Marcial, N., Castellanos-Albores, J. and Lawrence, D. (2003). Leaf litter decomposition of tree species in three successional phases of tropical dry secondary forest in Campeche, Mexico. *Forest Ecology and Management,* 174: 401–412.

Yadav, S., Yadav, J.P., Arya, V. and Panghal, M. (2010). Sacred groves in conservation of plant biodiversity in Mahendergarh district of Haryana. *Indian J. Traditional Knowledge*, 9: 693-700.

Yelvattimath, G.P. and Kotresha, K. (2011). Phytodiversity studies in Sri Ramathirth sacred grove, Halasi, Khanapur taluk, Belgavi district, Karnataka. *Life sciences Leaflets*, 18: 670- 683.

Yousefi, M., Tavili, A., Jafari, M., Zarechahouki, M.A. (2008). Relationships between dominant species and soil chemical properties in garmsar region. *Pajouhesh-va-Sazandegi*, 21(3): 161-168.

Yuan, H. and Hou, F. (2015). Grazing intensity and soil depth effects on soil properties in alpine meadow pastures of Qilian Mountain in northwest China. *Acta Agriculturae Scandinavica*, 65(3): 222-232.

Acronyms

AFRI	:	Arid Forest Research Institute
ACZ	:	Agro-Climatic Zone
AER	:	Agro-Ecological Region
ANOVA	:	Analysis of Variance
BD	:	Bulk density
Bigha	:	Unit of area equivalent to 0.163 hectare
BSI	:	Botanical Survey of India
cm	:	Centimeter
D	:	Species dominance or Simpson's Index
DMRT	:	Duncan Multiple Range Tests
dSm^{-1}	:	Deci Siemens per metre
e'	:	Species evenness
ENS	:	Effective number of species
FAO	:	Food and Agriculture Organization
g	:	gram
G	:	Gravel fraction
GOI	:	Government of India
GOR	:	Government of Rajasthan
H'	:	Shanon:Weiner index of diversity
ha	:	Hectare
ICAR	:	Indian Council of Agricultural Research
ICFRE	:	Indian Council of Forestry Research & Education
IPCC	:	Intergovernmental Panel on Climate Change
IVI	:	Importance value index
IUCN	:	International Union for Conservation of Nature
JBF	:	Jal Bhagirathi Foundation
km	:	Kilometer
M	:	Molarity (gram molecular weight of solute in 1 liter of solvent)
m	:	Meter
Mg	:	Mega gram

mm	:	Millimeter
NH_4-N	:	Ammonium nitrogen
NO_3-N	:	Nitrate nitrogen
NGO	:	Non Government organization
NTFP	:	Non Timber Forest Produces
PCCF	:	Principal Chief Conservator of Forests
PO_4-P	:	Phosphate phosphorus
R	:	Species richness
RD	:	Relative density
RF	:	Relative frequency
RDom	:	Relative dominance
REDD	:	Reducing Emissions from Deforestation and Forest Degradation
SE	:	Standard Error
SFD	:	State Forest Department
SOC	:	Soil Organic Carbon
Sq	:	Square
SOC	:	Soil organic carbon
SIC	:	Soil inorganic carbon
SRI	:	Simpson's Reciprocal Index (i.e., 1/D)
TGA	:	Total geographical area
TNOSP	:	Number of tree species
Tree30	:	Trees of girth greater than 30 cm at breast height
Tree10	:	Trees of girth 10 to 30 cm at breast height
UNCBD	:	United Nations Convention on Biological Diversity
UNFCCC	:	United Nations Framework Convention on Climate Change
V	:	Volume
WWF	:	World Wildlife Fund

Annexure I

Tree and shrub species recorded in the sacred groves of different districts in Rajasthan.

SNo.	Species name	Habit	Family
1.	*Acacia auriculiformis* A. Cunn. ex Benth.	Tree	Fabaceae
2.	*Acacia catechu* Willd.	Tree	Fabaceae
3.	*Acacia leucophloea* (Roxb.) Willd.	Tree	Fabaceae
4.	*Acacia nilotica* L.	Tree	Fabaceae
5.	*Acacia nilotica* var. *cupressiformis* (J. L. Stewart) Ali & Faruqi	Tree	Fabaceae
6.	*Acacia pennata* (L.) Willd.	Tree	Fabaceae
7.	*Acacia senegal* (L.)Willd.	Tree	Fabaceae
8.	*Acacia tortilis* (Forssk.) Hayne	Tree	Fabaceae
9.	*Adansonia digitata* L.	Tree	Bombacaceae
10.	*Adina cordifolia* (Roxb.) Brandis	Tree	Rubiaceae
11.	*Aegle marmelos* (Linn.) Corr.	Tree	Rutaceae
12.	*Ailanthus excelsa* Roxb.	Tree	Simaroubaceae
13.	*Albizia lebbeck* (L.) Benth	Tree	Fabaceae
14.	*Albizia procera* (Roxb.) Benth.	Tree	Fabaceae
15.	*Anogeissus latifolia* (DC.) Wallich ex Guill. & Perr.	Tree	Combretaceae
16.	*Anogeissus pendula* Edgew.	Tree	Combretaceae
17.	*Anogeissus sericea* (Brandis.)	Tree	Combretaceae
18.	*Anogeissus sericea* ssp. *nummularia* Brandis	Tree	Combretaceae
19.	*Azadirachta indica* A. Juss.	Tree	Meliaceae
20.	*Balanites aegyptiaca* (Linn.) Del.	Tree	Balanitaceae
21.	*Bauhinia racemosa* Lam.	Tree	Fabaceae
22.	*Bombex ceiba* L.	Tree	Bombacaceae
23.	*Boswellia serrata* Roxb. ex Colebr.	Tree	Burseraceae
24.	*Bridelia retusa* (L.) Spreng.	Tree	Euphorbiaceae
25.	*Butea monosperma* (Lam.)	Tree	Fabaceae
26.	*Capparis decidua* (Forsk.) Edgew.	Tree	Capparaceae
27.	*Cassia fistula* L.	Tree	Fabaceae
28.	*Cassia siema* Lam.	Tree	Fabaceae

SNo.	Species name	Habit	Family
29.	*Cocos nucifera* L.	Tree	Arecaceae
30.	*Cordia dichotoma* G. Forst.	Tree	Boraginaceae
31.	*Cordia gharaf* (Forssk.) Ehrenb. ex Asch.	Tree	Boraginaceae
32.	*Cordia myxa* L.	Tree	Boraginaceae
33.	*Crataeva religiosa* Forst. f.	Tree	Capparaceae
34.	*Crataeva nurvala* Buch. Ham.	Tree	Capparaceae
35.	*Dalbergia sissoo* Roxb.	Tree	Fabaceae
36.	*Delonix regia* Raf.	Tree	Fabaceae
37.	*Diospyros cordifolia* Roxb.	Tree	Ebenaceae
38.	*Diospyros melanoxylon* Roxb.	Tree	Ebenaceae
39.	*Dolichandrone falcata* (Wall.ex DC.) Seem	Tree	Bignoniaceae
40.	*Ehretia aspera* Willd.	Tree	Boraginaceae
41.	*Ehretia laevis* Roxb.	Tree	Boraginaceae
42.	*Elaeocarpus ganitrus* (Roxb.)	Tree	Elaeocarpaceae
43.	*Emblica officinalis* Gaertn.	Tree	Euphorbiaceae
44.	*Eucalyptus camaldulensis* Dehn	Tree	Myrtaceae
45.	*Feronia limonia* Linn.	Tree	Rutaceae
46.	*Ficus benghalensis* Linn.	Tree	Moraceae
47.	*Ficus glomerata* Roxb.	Tree	Moraceae
48.	*Ficus infectoria* Roxb.	Tree	Moraceae
49.	*Ficus religiosa* (L.)	Tree	Moraceae
50.	*Grewia tiliifolia* Vahl	Tree	Tiliaceae
51.	*Guazuma ulmifolia* Lam.	Tree	Sterculiaceae
52.	*Holarrhina antidysenterica* Wall. Ex A. DC.	Tree	Apocynaceae
53.	*Holoptelea integrifolia* Roxb.	Tree	Ulmaceae
54.	*Lannea coromendelliaca* (Houtt.) Merr.	Tree	Anacardiaceae
55.	*Lawsonia inermis* Linn.	Tree	Lythraceae
56.	*Leucaena leucocephala* (Lam.) De Wit.	Tree	Fabaceae
57.	*Madhuca indica* J.Gmelin	Tree	Sapotaceae
58.	*Mangifera indica* Linn.	Tree	Anacardiaceae
59.	*Maytenus emarginata* (Willd.)	Tree	Celastraceae
60.	*Miliusa tomentosa* (Roxb.) J. Sinclair	Tree	Annonaceae
61.	*Millingtonia hortensis* Linn.	Tree	Bignoniaceae
62.	*Mitragyna parviflora*, (Roxb.) Korth	Tree	Rubiaceae
63.	*Morinda tinctoria* (Roxb.)	Tree	Rubiaceae
64.	*Moringa concanensis* nimmo ex Dalz. & Gibson	Tree	Moringaceae
65.	*Morus alba* L.	Tree	Moraceae
66.	*Ougeinia oojeinensis* (Roxb.)	Tree	Fabaceae
67.	*Phoenix sylvestris* Roxb.	Tree	Arecaceae
68.	*Pithecellobium dulce* (Roxb.) Benth.	Tree	Fabaceae
69.	*Polyalthia longifolia* (Sonn.) Thwaites	Tree	Annonaceae
70.	*Pongamia pinnata* L.	Tree	Fabaceae
71.	*Prosopis cineraria* (L.) Druce.	Tree	Fabaceae

SNo.	Species name	Habit	Family
72.	*Prosopis juliflora* (Sw) DC	Tree	Fabaceae
73.	*Putranjiva roxburghii* Wall.	Tree	Putranjivaceae
74.	*Randia dumentorum* Lam.	Tree	Rubiaceae
75.	*Randia dumetorum* (Retz.) Poiret	Tree	Rubiaceae
76.	*Salvadora oleoides* Decne	Tree	Salvadoraceae
77.	*Salvadora persica* L.	Tree	Salvadoraceae
78.	*Santallum album* Linn.	Tree	Santalaceae
79.	*Sapindus mukorossi* Gaertn.	Tree	Sapindaceae
80.	*Soymida febrifuga* (Roxb.) A.Juss.	Tree	Meliaceae
81.	*Sterculia urense* Roxb.	Tree	Sterculiaceae
82.	*Syzygium cumini* (L.)	Tree	Myrtaceae
83.	*Tamarandis indica* Linn.	Tree	Fabaceae
84.	*Tamarix aphylla* (L.) H. Karst.	Tree	Tamaricaceae
85.	*Tecomella undulata* D.Don	Tree	Bignoniaceae
86.	*Tectona grandis* L.f.	Tree	Verbenaceae
87.	*Terminallia arjuna* (DC.) Wight & Arn	Tree	Combretaceae
88.	*Terminallia bellerica* (Gaertn.) Roxb.	Tree	Combretaceae
89.	*Thespesia populnea* Linn	Tree	Malvaceae
90.	*Wrightia tinctoria* (Roxb.) R. Br.	Tree	Apocynaceae
91.	*Zizyphus mauritiana* Lam.	Tree	Rhamnaceae
92.	*Zizyphus rotundifolia* Lam.	Tree	Rhamnaceae
93.	*Acacia jacquemontii* Benth	Shrub	Fabaceae
94.	*Adhatoda vasica* Nees	Shrub	Acanthaceae
95.	*Aegave Americana* L.	Shrub	Agavaceae
96.	*Aerva persica* (Burm. f.) Merrill. in Philipp.	Shrub	Amaranthaceae
97.	*Aerva pseudotomenosa* Blatt. & Hallb	Shrub	Amaranthaceae
98.	*Annona squamosa* L.	Shrub	Annonaceae
99.	*Aristolochia bracteata* Retz.	Shrub	Aristolochiaceae
100.	*Barleria prionitis* L.	Shrub	Acanthaceae
101.	*Cadaba fruticosa* (L.) Druce	Shrub	Capparaceae
102.	*Callygonum polygonides* L.	Shrub	Polygonaceae
103.	*Calotropis procera* (Ait.) Ait.f	Shrub	Asclepiadaceae
104.	*Capparis sepiera* L.	Shrub	Capparaceae
105.	*Carissa carandas* Linn	Shrub	Apocynaceae
106.	*Cassia auriculata* L.	Shrub	Fabaceae
107.	*Clerodendrum phlomidis* Linn.	Shrub	Verbinaceae
108.	*Commiphora wightii* (Am) Bhand,	Shrub	Burseraceae
109.	*Crotolaria burhia* Buch.-Ham.	Shrub	Fabaceae
110.	*Dichrostachys cinerea* (L.) Wight & Arn	Shrub	Fabaceae
111.	*Dyerophytum indicum* Kuntze	Shrub	Plumbaginaceae
112.	*Euphorbia caducifolia* Haines	Shrub	Euphorbiaceae
113.	*Euphorbia neriifolia* L.	Shrub	Euphorbiaceae
114.	*Flacortia indica* (Burm. f.)	Shrub	Flacourtiaceae

SNo.	Species name	Habit	Family
115.	*Grewia flavescens* Juss.	Shrub	Tiliaceae
116.	*Grewia tenax* (Forssk.)	Shrub	Tiliaceae
117.	*Grewia villosa* Willd.	Shrub	Tiliaceae
118.	*Gymnosporea montana* (Roth) Bemth	Shrub	Celeastraceae
119.	*Haloxylon salicornicum* (Moq.) Bunge ex Boiss	Shrub	Chenopodiaceae
120.	*Jatropha gossipifolium* L.	Shrub	Euphorbiaceae
121.	*Jatropha curcus L.*	Shrub	Euphorbiaceae
122.	*Lantana camara* L.	Shrub	Verbenaceae
123.	*Leptadenia pyrotechnica* (Forssk.) Decne	Shrub	Asclepiadaceae
124.	*Lycium barbarum* L.	Shrub	Solanaceae
125.	*Maerua oblongifolia* (Forssk.) A. Rich.	Shrub	Capparaceae
126.	*Nerium odoratum* Lam	Shrub	Apocynaceae
127.	*Opuntia dillenii* (Ker-Gawl) Haw.	Shrub	Cactaceae
128.	*Pendanus odoratissinus*	Shrub	Pandanaceae
129.	*Rhus mysorensis* G.Don	Shrub	Anacardiaceae
130.	*Securengia virosa* Roxb. ex Willd.	Shrub	Euphorbiaceae
131.	*Zizyphus nummularia* (Burm. f.) Wight et Arn.	Shrub	Rhamnaceae

Annexure II

Diversity variables of trees of girth class >30 cm and 10-30 cm observed in sacred groves of different districts of Rajasthan.

SN	Name of Oran	District	Tree of >30 cm girth					Trees of 10-30 cm girth				
			pop	Species	H'	ENS	1/D	pop	Species	H'	ENS	1/D
1	Devnarayan Ji Ka Oran	Ajmer	63	5	0.32	1.38	1.14	6	1	0.00	1.00	1.00
2	Bheruji Ka Oran	Ajmer	59	5	0.46	1.59	1.23	9	2	0.69	1.99	1.98
3	Bhuna Bai ji ka Oran	Ajmer	64	3	0.27	1.31	1.13	43	1	0.00	1.00	1.00
4	Manaldan ji Todli Ka Oran	Alwar	80	10	1.64	5.17	3.49	22	5	1.42	4.16	3.67
5	Hanuman Ji Ka Oran	Alwar	98	8	1.43	4.19	2.77	12	3	0.82	2.28	1.95
6	Jharna Ji Ka Oran	Alwar	75	9	1.78	5.90	4.52	23	7	1.78	5.90	5.34
7	Rupu Ka Vas Oran	Alwar	272	2	0.16	1.17	1.08	1	1	0.00	1.00	1.00
8	Shiv Ji Maharaj Ka Oran	Alwar	43	7	1.70	5.48	4.68	18	7	1.82	6.15	5.59
9	Dagia Bheruji Ka Oran	Banswara	54	5	0.32	1.38	1.14	373	5	0.83	2.29	1.92
10	Prathurinath ji Ka Oran	Banswara	47	6	0.87	2.39	1.66	-	-	0.00	1.00	
11	Sai Baba Ka Mandir	Banswara	50	5	1.01	2.73	2.13	3	2	0.64	1.89	1.80
12	Devnarayan Ji Ka Oran	Baran	52	9	1.41	4.09	2.58	-	-	0.00	1.00	
13	Musain Mata Ji Ka Oran	Baran	61	8	1.44	4.22	3.05	63	7	1.29	3.63	2.78
14	Kala ji Ka Bagh (Nag)	Baran	141	10	1.75	5.77	4.59	48	2	0.56	1.75	1.60
15	Baukal Mata ji Ka Oran	Barmer	27	3	0.85	2.34	1.98	34	2	0.30	1.35	1.19
16	Viratra Mata Ka Oran	Barmer	27	3	0.50	1.65	1.35	-	-	0.00	1.00	
17	Mata ji Ka Oran	Barmer	20	1	0.00	1.00	1.00	-	-	0.00	1.00	
18	Khuwas Mata Ji Ka Oran	Barmer	26	4	1.03	2.81	2.37	-	-	0.00	1.00	
19	Mama ji Ka Selenadi	Barmer	41	4	1.26	3.54	3.25	1	1	0.00	1.00	1.00
20	Ghanshsyam Baba Ka Ashram	Bharatpur	237	12	1.91	6.75	0.81	34	9	2.00	7.37	6.49
21	Adi Badri Dham	Bharatpur	64	9	1.76	5.84	4.49	3	1	0.00	1.00	1.00
22	Sawai Bhoj/ Devnarayan Mandir	Bhilwara	30	8	1.82	6.16	4.64	6	4	1.24	3.46	3.00
23	Dennaryan Ji Ki Bani	Bhilwara	150	4	0.45	1.56	1.31	-	-	0.00	1.00	
24	Devnaryan	Bhilwara	35	7	1.58	4.84	3.64	-	-	0.00	1.00	
25	Goga Ji Ka Oran	Bikaner	36	3	0.92	2.51	2.34	1	1	0.00	1.00	1.00

SN	Name of Oran	District	Tree of >30 cm girth					Trees of 10-30 cm girth				
			pop	Species	H'	ENS	1/D	pop	Species	H'	ENS	1/D
26	Karni Mata Ji ka Oran	Bikaner	65	5	1.38	3.96	2.13	14	3	0.76	2.14	1.78
27	Karni Mata Ji Ka Oran	Bikaner	70	1	0.00	1.00	1.00	-	-	0.00	1.00	
28	Siya mata ji Ki Bani	Bundi	114	4	1.27	3.54	3.27	18	2	0.69	2.00	2.00
29	Gangrawal Ji Maharaj Ki Vani	Bundi	115	3	1.02	2.78	3.18	38	3	0.71	2.03	1.69
30	Parbetheshwar Mahadeo ki Bani	Bundi	68	10	2.12	8.33	10.78	32	9	1.71	5.51	5.95
31	Devnarayan ji Ka Oran	Chittorgarh	76	7	0.51	1.67	1.25	11	1	0.00	1.00	1.00
32	Rishi Mangri Oran	Chittorgarh	92	3	0.33	1.39	1.19	16	5	1.24	3.47	2.67
33	Nahar Singhji Mata ka Oran	Chittorgarh	173	3	0.18	1.20	1.07	21	3	0.85	2.33	2.17
34	Junjan Dodaji Ka Oran	Churu	101	4	0.54	1.72	1.33	2	1	0.00	1.00	1.00
35	Gogaji Ka Oran	Churu	79	4	1.09	2.97	2.36	3	1	0.00	1.00	1.00
36	Gosai Ji Ka Oran	Churu	53	5	0.54	1.71	0.56	57	1	0.00	1.00	1.00
37	Kaludas Baba Ki Dungri	Dausa	263	2	0.14	1.15	1.06	11	4	1.12	3.07	2.57
38	Devnarayna Ji Ka Oran	Dausa	107	5	0.61	1.83	1.37	22	3	0.91	2.48	2.12
39	Sawanria ji Seth Dham	Dausa	207	3	0.58	1.78	1.55	4	1	0.00	1.00	1.00
40	Bali Baba Ki Dargah	Dholpur	47	5	0.62	1.85	1.41	44	4	0.96	2.62	3.55
41	MuchKund	Dholpur	30	6	1.23	3.43	2.57	71	7	1.30	3.67	2.71
42	Mahadeo Ji Ka Oran	Dholpur	24	4	1.21	3.34	3.45	4	3	1.04	2.83	5.33
43	Govindguru Chhari Magri	Dungarpur	80	3	0.18	1.20	1.08	1	1	0.00	1.00	1.00
44	Bhuneshwar Mahadev Ka Oran	Dungarpur	189	9	1.14	3.13	2.15	92	6	0.57	1.77	1.31
45	Neelkanth Mahadev	Dunagrpur	61	6	0.71	2.02	1.46	5	2	0.50	1.65	1.47
46	Bhadra Kali Mata Ji Ka Oran	Hanumangarh	99	7	1.49	4.42	3.18	5	1	0.00	1.00	1.00
47	Chisti Peer ki Dargah	Hanumangarh	48	5	0.40	1.50	1.19	6	1	0.00	1.00	1.00
48	Guru Gorakhnath Ji Ka Oran	Hanumangarh	30	7	1.45	4.27	3.00	6	5	1.56	4.76	4.50
49	Kalimata ji ki Bani	Jaipur	29	6	1.17	3.23	2.56	4	3	1.04	2.83	2.67
50	Kapil Muni Ashram	Jaipur	51	11	1.72	5.57	3.83	3	2	0.64	1.89	1.80
51	Shiv Dungri	Jaipur	59	4	0.70	2.02	1.62	12	5	1.31	3.72	3.00
52	Ramdevji ka Oran	Jaisalmer	37	4	1.12	3.08	2.81	13	3	1.06	2.88	2.77
53	Bhadariya ji Ka Oran (old)	Jaisalmer	27	2	0.16	1.17	1.08	17	1	0.00	1.00	1.00
54	Malani Oran	Jaisalmer	36	1	0.00	1.00	1.00	12	2	0.29	1.33	1.18
55	Bellane Oran	Jaisalmer	14	2	0.60	1.82	1.69	2	2	0.69	2.00	2.00
56	Dharneswar Mahadeo ka Oran	Jalore	149	6	0.49	1.63	1.27	23	2	0.46	1.58	1.41
57	Dudeshwar Mahadeo Ka Oran	Jalore	140	4	0.91	2.48	1.89	5	2	0.50	1.65	1.47
58	Jhootana Mama ji Ka Oran	Jalore	116	3	0.81	2.25	1.89	11	2	0.30	1.35	1.20
59	Kalla ji Ka Oran	Jhalawar	34	6	1.30	3.68	2.82	34	3	0.87	2.38	2.21
60	Bapulala Smirti Van	Jhalawar	89	3	0.74	2.10	2.03	100	2	0.68	1.98	1.96
61	Devnarayan Ji Ka Asthan	Jhalawar	63	7	1.15	3.16	2.46	14	1	0.00	1.00	1.00

SN	Name of Oran	District	Tree of >30 cm girth					Trees of 10-30 cm girth				
			pop	Species	H'	ENS	1/D	pop	Species	H'	ENS	1/D
62	Shiv Mandir Kelkheura	Jhalawar	51	4	0.83	2.30	1.78	13	2	0.67	1.95	1.90
63	Bhairav Ji Ka Bag	Jhalawar	106	10	1.62	5.03	3.61	29	6	1.52	4.59	3.64
64	Bihari ji Ka Mandir	Jhunjunnu	60	3	0.77	2.16	1.95	1	1	0.00	1.00	1.00
65	Prithviraj Ki Bani	Jhunjunnu	38	3	0.63	1.88	1.59	-	-	0.00	1.00	
66	Panchmir Pir Ki Bani	Jhunjunnu	57	5	0.89	2.43	1.76	8	2	0.69	2.00	2.00
67	Kolu Pabuji Ka Oran	Jodhpur	37	2	0.04	1.04	1.08	-	-	0.00	1.00	
68	Karni Mata Ji Oran	Jodhpur	37	4	0.90	2.47	2.23	32	4	0.99	2.70	2.38
69	Ramdeo Ji ka Oran	Jodhpur	25	2	0.55	1.74	1.57	7	3	1.08	2.94	2.88
70	Pabuji Ka Oran	Jodhpur	15	1	0.00	1.00	1.00	19	1	0.00	1.00	1.00
71	Joleyali Oran	Jodhpur	33	2	0.23	1.26	1.13	-	-	0.00	1.00	
72	Jardabala Ji Ka Oran	Karauli	27	11	2.17	8.75	7.36	-	-	0.00	1.00	
73	Barbasni Mata Ji Ka Oran	Karauli	65	4	0.58	1.79	1.42	-	-	0.00	1.00	
74	Devnarayan Ji Ki Banni	Karauli	34	1	0.00	1.00	1.00	-	-	0.00	1.00	
75	Dardevi Mata ji Ka Oran	Kota	113	15	2.18	8.84	5.90	18	4	1.04	2.82	2.45
76	Devnarayan Bhagwan Ka Oran	Kota	74	10	1.50	4.50	2.81	1	1	0.00	1.00	1.00
77	Devnarayan Ji Ka Oran	Kota	100	3	0.64	1.89	1.51	14	2	0.26	1.29	1.15
78	Devnarayan Ji Ka Oran	Kota	80	8	1.59	4.91	4.00	13	3	0.90	2.46	2.25
79	Dhokad Wali Mata ji	Kota	72	4	0.57	1.77	1.41	5	2	0.50	1.65	1.47
80	Devnarayan Ji Ka Oran	Kota	193	8	1.09	2.97	2.01	86	7	1.42	4.13	3.35
81	Hanuman Ji ka Oran	Kota	98	2	0.06	1.06	1.02	20	3	0.94	2.57	2.38
82	Devnarayan Ji ka Oran	Kota	120	7	1.47	4.35	3.34	41	4	1.16	3.20	3.01
83	Kalaji-Gora Ji Oran	Kota	71	10	1.16	3.19	2.00	5	2	0.50	1.65	1.47
84	Bheruji Ka Oran	Nagour	23	5	1.27	3.57	3.06	-	-	0.00	1.00	
85	Bhairon Ji Ka Oran	Nagour	105	5	1.09	2.98	2.50	22	4	1.00	2.71	2.40
86	Gosai Ji Ka Oran	Nagour	54	3	1.00	2.70	2.45	21	1	0.00	1.00	1.00
87	Sathika Jogmaya Ji Ka Oran	Nagour	31	3	0.61	1.85	1.49	1	1	0.00	1.00	1.00
88	Laba ji Maharaj Ka Oran	Pali	24	2	0.17	1.19	1.09	25	2	0.44	1.55	1.37
89	Mama ji Ka Oran	Pali	134	4	0.74	2.09	1.58	118	4	0.75	2.11	1.92
90	Devnarayanji Ka Oran	Pali	23	3	0.95	2.59	2.33	152	2	0.22	1.25	1.13
91	Ashapura Mata ji ka Oran	Pali	71	5	0.99	2.69	1.96	47	2	0.10	1.11	1.04
92	Mata ji Ka Oran (Mada)	Pali	16	6	1.47	4.36	3.28	124	6	0.99	2.69	2.12
93	Dhundhlimata Ki Banni	Pratapgarh	207	16	1.56	4.76	2.70	298	8	1.26	3.53	2.70
94	Bijasau mata ji Bani	Pratapgarh	44	6	1.30	3.68	2.86	22	3	0.37	1.44	1.20
95	Chandimata ji Ka Oran	Pratapgarh	34	6	1.41	4.11	3.27	28	2	0.15	1.17	1.07
96	Devnarayan ji Ki bani	Pratapgarh	51	2	0.43	1.54	1.36	15	2	0.58	1.79	1.64
97	Devnarayan ji Ki Bani	Rajsamand	100	3	0.15	1.17	1.06	2	2	0.69	2.00	2.00
98	Bheru Nath ji Bani	Rajsamand	110	9	1.44	4.24	3.13	3	1	0.00	1.00	1.00
99	Chamunda Mata ji Ki Bani	Rajsamand	93	2	0.56	1.75	1.59	1	1	0.00	1.00	1.00
100	Haumanji ki Dungri	Sawaimadhopur	39	3	0.77	2.17	2.01	6	1	0.00	1.00	1.00
101	Jasram Baba Ji Ki Bani	Sawaimadhopur	31	8	1.81	6.12	5.14	6	5	1.56	4.76	4.50

SN	Name of Oran	District	Tree of >30 cm girth					Trees of 10-30 cm girth				
			pop	Species	H'	ENS	1/D	pop	Species	H'	ENS	1/D
102	Peer Baba Ki Dargah	Sawaimadhopur	82	6	0.74	2.09	1.57	14	2	0.68	1.98	1.96
103	Bheruji Ki Bani	Sikar	42	4	1.23	3.43	3.25	11	2	0.66	1.93	1.86
104	Kheri Walai Balaji ka Oran	Sikar	79	5	1.49	4.44	3.98	6	3	1.01	2.75	2.57
105	Satimata Ji Ka Mandir	Sikar	46	2	0.69	1.99	1.98	38	1	0.00	1.00	1.00
106	Sidheswar Mahadeo Mandir	Sirohi	160	5	1.04	2.82	2.20	37	3	1.06	2.87	2.99
107	Vorada Hanuman ji Ka Oran	Sirohi	148	3	0.76	2.15	1.84	3	1	0.00	1.00	1.00
108	Magriwali Mata ji Ka Oran	Sirohi	100	7	1.06	2.88	1.87	19	3	0.81	2.25	1.99
109	Gosai Ji Ka Jod	SriGanganagar	99	5	0.54	1.71	1.31	34	2	0.13	1.14	1.06
110	Baba Gadanpur Ki Dargah	SriGanganagar	48	4	1.16	3.18	2.61	13	2	0.67	1.95	1.90
111	Miyan Mama Peer Baba	Sri Ganganagar	117	4	0.88	2.41	1.97	12	3	0.57	1.76	1.41
112	Baba Rahupeer Dargah	Sri Ganganagar	55	5	1.17	3.22	2.51	9	2	0.53	1.70	1.53
113	Mata Ji Ka Darra	Tonk	104	2	0.64	1.89	1.81	131	2	0.04	1.05	1.02
114	Devnarayan Ka Asthan	Tonk	203	7	0.49	1.64	1.24	20	5	1.01	2.74	1.94
115	Devmaharaj Ka Asthan	Tonk	158	6	0.30	1.35	1.12	1	1	0.00	1.00	1.00
116	Pava Bab ji Oran	Udaipur	104	16	1.82	6.20	3.49	20	10	2.08	8.02	6.67
117	Hanuman Ji Ka Oran	Udaipur	37	2	0.21	1.23	1.11	-	-	0.0	1.00	
118	Shringi Rishi Ji Ka Oran	Udaipur	65	8	1.65	5.22	3.95	7	1	0.00	1.00	1.00
119	Bhanvar Mata ji Ki Bani	Udaipur	125	8	0.78	2.17	1.59	53	4	0.90	2.47	2.08
120	Kali Pahari Oran	Udaipur	77	4	0.80	2.22	1.72	28	2	0.26	1.29	1.15
121	Kundeshwar Mahadevji Oran	Udaipur	69	13	1.76	5.83	3.52	9	6	1.68	5.35	4.76
122	Tarkeshwar Mahadev Oran	Udaipur	76	16	2.35	10.47	6.75	23	5	0.90	2.45	1.76
123	Ubeshwar ji Oran	Udaipur	64	12	1.91	6.72	4.65	19	7	1.57	4.82	3.72

H': Shannon-Wiener diversity index; ENS: effective species number and 1/D: Simpson's reciprocal index.

Annexure III

Diversity variables of shrubs and tree saplings (gith <10 to 3 cm) observed in sacred groves of different districts of Rajasthan.

SN	Name of Oran	District	Shrubs					Trees sapling (<10 cm girth)				
			pop	Species	H'	ENS	1/D	pop	Species	H'	ENS	1/D
1	Devnarayan Ji Ka Oran	Ajmer	2	2	0.69	2.00	2.00	3	1	0.00	1.00	1.00
2	Bheruji Ka Oran	Ajmer	19	5	1.55	4.70	2.96	4	1	0.00	1.00	1.00
3	Bhuna Bai ji ka Oran	Ajmer	14	5	1.39	4.00	2.93	3	1	0.00	1.00	1.00
4	Manaldan ji Todli Ka Oran	Alwar	27	5	0.83	2.28	1.62	1	1	0.00	1.00	1.00
5	Hanuman Ji Ka Oran	Alwar	38	5	1.26	3.54	3.19	-	-	0.00	1.00	-
6	Jharna Ji Ka Oran	Alwar	39	5	1.17	3.24	2.48	-	-	0.00	1.00	-
7	Rupu Ka Vas Oran	Alwar	1	1	0.00	1.00	1.00	-	-	0.00	1.00	-
8	Shiv Ji Maharaj Ka Oran	Alwar	31	10	1.95	7.00	4.98	1	1	0.00	1.00	1.00
9	Dagia Bheruji Ka Oran	Banswara	11	3	0.93	2.55	2.37	-	-	0.00	1.00	-
10	Prathurinath ji Ka Oran	Banswara	-	-	0.00	1.00	-	14	2	0.60	1.82	1.69
11	Sai Baba Ka Mandir	Banswara	7	4	1.28	3.59	3.27	21	4	1.21	3.36	2.96
12	Devnarayan Ji Ka Oran	Baran	7	5	1.48	4.37	3.77	5	1	0.00	1.00	1.00
13	Musain Mata Ji Ka Oran	Baran	10	4	1.09	2.97	2.38	50	5	1.00	2.72	2.04
14	Kala ji Ka Bagh (Nag)	Baran	18	1	0.00	1.00	1.00	5	2	0.67	1.96	1.92
15	Baukal Mata ji Ka Oran	Barmer	9	2	0.64	1.89	1.80	78	2	0.69	1.99	1.99
16	Viratra Mata Ka Oran	Barmer	64	8	1.76	5.79	4.73	-	-	0.00	1.00	-
17	Mata ji Ka Oran	Barmer	24	4	1.27	3.55	3.31	5	1	0.00	1.00	1.00
18	Khuwas Mata Ji Ka Oran	Barmer	33	3	1.09	2.97	2.94	3	2	0.64	1.89	1.80
19	Mama ji Ka Selenadi	Barmer	4	1	0.00	1.00	1.00	-	-	0.00	1.00	-
20	Ghanshsyam Baba Ka Ashram	Bharatpur	-	-	0.00	1.00	-	-	-	0.00	1.00	-
21	Adi Badri Dham	Bharatpur	5	1	0.00	1.00	1.00	-	-	0.00	1.00	-
22	Sawai Bhoj/ Devnarayan Mandir	Bhilwara	11	5	1.47	4.34	3.90	8	1	0.00	1.00	1.00
23	Dennaryan Ji Ki Bani	Bhilwara	-	-	0.00	1.00		-	-	0.00	1.00	-
24	Devnaryan	Bhilwara	1	1	0.00	1.00	1.00	3	1	0.00	1.00	1.00
25	Goga Ji Ka Oran	Bikaner	1	1	0.00	1.00	1.00	-	-	0.00	1.00	-

SN	Name of Oran	District	Shrubs					Trees sapling (<10 cm girth)				
			pop	Species	H'	ENS	1/D	pop	Species	H'	ENS	1/D
26	Karni Mata Ji ka Oran	Bikaner	17	1	0.00	1.00	1.00	34	1	0.00	1.00	1.00
27	Karni Mata Ji Ka Oran	Bikaner	83	5	1.12	3.06	2.50	-	-	0.00	1.00	-
28	Siya mata ji Ki Bani	Bundi	-	-	0.00	1.00		-	-	0.00	1.00	-
29	Gangrawal Ji Maharaj Ki Vani	Bundi	8	3	0.96	2.60	2.33	14	1	0.00	1.00	1.00
30	Parbetheshwar Mahadeo Ki Bani	Bundi	13	4	1.03	2.81	2.20	-	-	0.00	1.00	-
31	Devnarayan ji Ka Oran	Chittorgarh	2	2	0.69	2.00	2.00	-	-	0.00	1.00	-
32	Rishi Mangri Oran	Chittorgarh	11	3	0.86	2.36	2.05	13	2	0.54	1.72	1.55
33	Nahar Singhji Mata ka Oran	Chittorgarh	6	2	0.45	1.57	1.38	10	5	1.42	4.13	-
34	Junjan Dodaji Ka Oran	Churu	31	4	1.20	3.32	2.94	-	-	0.00	1.00	
35	Gogaji Ka Oran	Churu	7	1	0.00	1.00	1.00	-	-	0.00	1.00	
36	Gosai Ji Ka Oran	Churu	36	3	0.71	2.03	1.79	-	-	0.00	1.00	
37	Kaludas Baba Ki Dungri	Dausa	-	-	0.00	1.00		8	1	0.00	1.00	1.00
38	Devnarayna Ji Ka Oran	Dausa	60	6	1.37	3.94	3.14	-	-	0.00	1.00	
39	Sawanria ji Seth Dham	Dausa	8	1	0.00	1.00	1.00	2	1	0.00	1.00	1.00
40	Bali Baba Ki Dargah	Dholpur	105	3	1.08	2.95	2.90	-	-	0.00	1.00	
41	MuchKund	Dholpur	31	4	1.16	3.18	2.79	-	-	0.00	1.00	
42	Mahadeo Ji Ka Oran	Dholpur	29	3	1.09	2.99	2.97	-	-	0.00	1.00	
43	Govindguru Chhari Magri	Dungarpur	1	1	0.00	1.00	1.00	1	1	0.00	1.00	1.00
44	Bhuneshwar Mahadev Ka Oran	Dungarpur	17	7	1.71	5.54	4.59	3	1	0.00	1.00	1.00
45	Neelkanth Mahadev	Dunagrpur	10	3	1.03	2.80	2.63	48	1	0.00	1.00	1.00
46	Bhadra Kali Mata Ji Ka Oran	Hanumangarh	-	-	0.00	1.00		5	2	0.50	1.65	1.47
47	Chisti Peer ki Dargah	Hanumangarh	34	3	0.55	1.73	1.43	7	1	0.00	1.00	1.00
48	Guru Gorakhnath Ji Ka Oran	Hanumangarh	1	1	0.00	1.00	1.00	-	-	0.00	1.00	
49	Kalimata ji ki Bani	Jaipur	3	1	0.00	1.00	1.00	6	1	0.00	1.00	1.00
50	Kapil Muni Ashram	Jaipur	10	3	1.03	2.80	2.63	-	-	0.00	1.00	
51	Shiv Dungri	Jaipur	1	1	0.00	1.00	1.00	5	1	0.00	1.00	1.00
52	Ramdevji ka Oran	Jaisalmer	40	2	0.53	1.70	1.54	4	1	0.00	1.00	1.00
53	Bhadariya ji Ka Oran (old)	Jaisalmer	-	-	0.00	1.00		-	-	0	1.00	
54	Malani Oran	Jaisalmer	197	5	0.87	2.40	1.97	-	-	0	1.00	
55	Bellane Oran	Jaisalmer	31	2	0.67	1.95	1.90	1	1	0.00	1.00	1.00
56	Dharneswar Mahadeo ka Oran	Jalore	26	3	0.47	1.60	1.45	-	-	0.00	1.00	
57	Dudeshwar Mahadeo Ka Oran	Jalore	14	3	0.13	1.14	1.10	9	1	0.00	1.00	1.00
58	Jhootana Mama ji Ka Oran	Jalore	3	1	0.15	1.16	1.12	10	1	0.00	1.00	1.00
59	Kalla ji Ka Oran	Jhalawar	-	-	0.00	1.00		-	-	0.00	1.00	
60	Bapulala Smirti Van	Jhalawar	-	-	0.00	1.00		-	-	0.00	1.00	
61	Devnarayan Ji Ka Asthan	Jhalawar	5	1	0.00	1.00	1.00	5	1	0.00	1.00	1.00

SN	Name of Oran	District	Shrubs					Trees sapling (<10 cm girth)				
			pop	Species	H'	ENS	1/D	pop	Species	H'	ENS	1/D
62	Shiv Mandir Kelkheura	Jhalawar	-	-	0.00	1.00		-	-	0.00	1.00	
63	Bhairav Ji Ka Bag	Jhalawar	19	2	0.34	1.40	1.23	19	2	0.34	1.40	1.23
64	Bihari ji Ka Mandir	Jhunjunnu	-	-	0.00	1.00		-	-	0.00	1.00	
65	Prithviraj Ki Bani	Jhunjunnu	12	1	0.00	1.00	1.00	-	-	0.00	1.00	
66	Panchmir Pir Ki Bani	Jhunjunnu	25	2	0.59	1.81	1.68	-	-	0.00	1.00	
67	Kolu Pabuji Ka Oran	Jodhpur	-	-	0.00	1.00		-	-	0.00	1.00	
68	Karni Mata Ji Oran	Jodhpur	7	5	1.55	4.71	4.45	5	1	0.00	1.00	1.00
69	Ramdeo Ji ka Oran	Jodhpur	6	1	0.00	1.00	1.00	4	1	0.00	1.00	1.00
70	Pabuji Ka Oran	Jodhpur	-	-	0.00	1.00		19	1	0.00	1.00	1.00
71	Joleyali Oran	Jodhpur	-	-	0.00	1.00		5	1	0.00	1.00	1.00
72	Jardabala Ji Ka Oran	Karauli	34	4	1.05	2.85	2.48	-	-	0.00	1.00	
73	Barbasni Mata Ji Ka Oran	Karauli	-	-	0.00	1.00		15	1	0.00	1.00	1.00
74	Devnarayan Ji Ki Banni	Karauli	12	2	0.29	1.33	1.18	-	-	0.00	1.00	
75	Dardevi Mata ji Ka Oran	Kota	12	3	0.82	2.28	1.95	24	2	0.45	1.57	1.38
76	Devnarayan Bhagwan Ka Oran	Kota	7	1	0.00	1.00	1.00	14	2	0.60	1.82	1.69
77	Devnarayan Ji Ka Oran	Kota	-	-	0.00	1.00		23	2	0.52	1.69	1.52
78	Devnarayan Ji Ka Oran	Kota	25	3	1.05	2.86	2.73	-	-	0.00	1.00	
79	Dhokad Wali Mata ji	Kota	9	1	0.00	1.00	1.00	-	-	0.00	1.00	
80	Devnarayan Ji Ka Oran	Kota	44	4	1.28	3.60	3.29	35	1	0.00	1.00	1.00
81	Hanuman Ji ka Oran	Kota	20	3	0.73	2.08	1.68	23	2	0.52	1.69	1.52
82	Devnarayan Ji ka Oran	Kota	10	1	0.00	1.00	1.00	47	2	0.52	1.68	1.50
83	Kalaji-Gora Ji Oran	Kota	9	3	0.85	2.34	1.98	-	-	0.00	1.00	
84	Bheruji Ka Oran	Nagour	54	7	1.75	5.77	5.12	-	-	0.00	1.00	
85	Bhairon Ji Ka Oran	Nagour	37	3	0.64	1.90	1.61	14	1	0.00	1.00	1.00
86	Gosai Ji Ka Oran	Nagour	2	1	0.00	1.00	1.00	-	-	0.00	1.00	
87	Sathika Jogmaya Ji Ka Oran	Nagour	86	5	1.37	3.94	3.48	-	-	0.00	1.00	
88	Laba ji Maharaj Ka Oran	Pali	-	-	0.00	1.00		20	2	0.65	1.91	1.83
89	Mama ji Ka Oran	Pali	7	2	0.13	1.14	1.96	5	1	0.00	1.00	1.00
90	Devnarayanji Ka Oran	Pali	1	1	0.00	1.00	1.00	2	1	0.00	1.00	1.00
91	Ashapura Mata ji ka Oran	Pali	31	2	0.51	1.67	1.49	35	2	0.41	1.51	1.32
92	Mata ji Ka Oran (Mada)	Pali	15	4	0.36	1.43	1.45	4	2	0.56	1.75	1.60
93	Dhundhlimata Ki Banni	Pratapgarh	7	2	0.60	1.82	1.69	5	1	0.00	1.00	1.00
94	Bijasau mata ji Bani	Pratapgarh	3	1	0.00	1.00	1.00	-	-	0.00	1.00	
95	Chandimata ji Ka Oran	Pratapgarh	11	1	0.00	1.00	1.00	-	-	0.00	1.00	
96	Devnarayan ji Ki bani	Pratapgarh	8	2	0.38	1.46	1.28	-	-	0.00	1.00	
97	Devnarayan ji Ki Bani	Rajsamand	4	2	0.56	1.75	1.60	4	2	0.69	2.00	2.00
98	Bheru Nath ji Bani	Rajsamand	4	2	0.56	1.75	1.60	7	1	0.00	1.00	1.00
99	Chamunda Mata ji Ki Bani	Rajsamand	14	1	0.00	1.00	1.00	3	1	0.00	1.00	1.00
100	Haumanji ki Dungri	Sawaimadhopur	4	1	0.00	1.00	1.00	3	1	0.00	1.00	1.00
101	Jasram Baba Ji Ki Bani	Sawaimadhopur	11	3	0.76	2.14	1.75	-	-	0.00	1.00	

SN	Name of Oran	District	Shrubs					Trees sapling (<10 cm girth)				
			pop	Species	H'	ENS	1/D	pop	Species	H'	ENS	1/D
102	Peer Baba Ki Dargah	Sawaimadhopur	5	1	0.00	1.00	1.00	-	-	0.00	1.00	
103	Bheruji Ki Bani	Sikar	18	3	0.79	2.20	1.91	-	-	0.00	1.00	
104	Kheri Walai Balaji ka Oran	Sikar	33	2	0.67	1.96	1.91	-	-	0.00	1.00	
105	Satimata Ji Ka Mandir	Sikar	2	1	0.00	1.00	1.00	-	-	0.00	1.00	
106	Sidheswar Mahadeo Mandir	Sirohi	26	2	0.53	1.70	1.61	19	2	0.66	1.93	1.87
107	Vorada Hanuman ji Ka Oran	Sirohi	6	1	0.00	1.00	1.00	7	1	0.00	1.00	1.00
108	Magriwali Mata ji Ka Oran	Sirohi	5	1	0.00	1.00	1.00	11	2	0.69	1.99	1.98
109	Gosai Ji Ka Jod	SriGanganagar	33	2	0.67	1.96	1.91	-	-	0.00	1.00	
110	Baba Gadanpur Ki Dargah	SriGanganagar	-	-	0.00	1.00		16	2.00	0.38	1.46	1.28
111	Miyan Mama Peer Baba	Sri Ganganagar	4	2	0.69	2.00	2.00	-	-	0.00	1.00	
112	Baba Rahupeer Dargah	Sri Ganganagar	21	1	0.00	1.00	1.00	14	2.00	0.60	1.82	1.69
113	Mata Ji Ka Darra	Tonk	1	1	0.00	1.00	1.00	5	1	0.00	1.00	1.00
114	Devnarayan Ka Asthan	Tonk	3	1	0.00	1.00	1.00	12	2	0.64	1.89	1.80
115	Devmaharaj Ka Asthan	Tonk	0	0	0.00	1.00		5	1	0.00	1.00	1.00
116	Pava Bab ji Oran	Udaipur	96	3	0.98	2.66	2.49	14	4	1.30	3.67	3.38
117	Hanuman Ji Ka Oran	Udaipur	-	-	0.00	1.00		-	-	0.00	1.00	
118	Shringi Rishi Ji Ka Oran	Udaipur	18	5	1.43	4.17	3.52	-	-	0.00	1.00	
119	Bhanvar Mata ji Ki Bani	Udaipur	27	5	1.06	2.88	2.36	27	1	0.00	1.00	1.00
120	Kali Pahari Oran	Udaipur	-	-	0.00	1.00		-	-	0.00	1.00	
121	Kundeshwar Mahadevji Oran	Udaipur	3	2	0.64	1.89	1.80	-	-	0.00	1.00	
122	Tarkeshwar Mahadev Oran	Udaipur	58	5	1.03	2.81	2.40	62	6	1.55	4.70	4.01
123	Ubeshwar ji Oran	Udaipur	111	8	1.23	3.41	2.36	4	2	0.56	1.75	1.60

H': Shannon-Wiener diversity index; ENS: effective species number and 1/D: Simpson's reciprocal index.

Annexure IV

Dominance of trees of >30 cm girth, 10-30 c girth, shrubs and tree saplings in different sacred groves of Rajasthan.

SN	Name of Oran	Trees of >30 cm girth		Trees of 10-30 cm girth		Shrubs		Saplings	
		Species	IVI	Species	IVI	Species	IVI	Species	IVI
1	Devnarayan Ji Ka Oran	*P. juliflora*	194	*P. juliflora*	300	*L. camara*	170	*P. juliflora*	300
2	Bheruji Ka Oran	*M. emarginata*	231	*P. juliflora*	168	*A. jacqmontii*	117	*P. juliflora*	300
3	Bhuna Bai ji ka Oran	*P. juliflora*	246	*P. juliflora*	300	*C. procera*	125	*P. juliflora*	300
4	Manaldan ji Todli Ka Oran	*A. nilotica*	145	*M. emarginata*	116	*Z. nummularia*	205	*H. integrifolia*	300
5	Hanuman Ji Ka Oran	*S. oleoides*	142	*H. integrifolia*	165	*C. phlomidis*	107	-	-
6	Jharna Ji Ka Oran	*P. juliflora*	92	*M. emarginata*	68	*Z. nummularia*	150	-	-
7	Rupu Ka Vas Oran	*A. pendula*	248	*A. pendula*	300	*A. lanata*	300	-	-
8	Shiv Ji Maharaj Ka Oran	*A. leucophloea*	110	*P. juliflora*	76	*A. vasica*	73	*M. emarginata*	300
9	Dagia Bheruji Ka Oran	*T. grandis*	150	*T. grandis*	178	*L. camara*	161	-	-
10	Prathurinath ji Ka Oran	*T. grandis*	160	-	-	-	-	*T. grandis*	209
11	Sai Baba Ka Mandir	*T. grandis*	131	*H. integrifolia*	171	*L. camara*	114	*H. antidysent-erica*	108
12	Devnarayan Ji Ka Oran	*B. monosperma*	133	-	-	*Z. nummularia*	129	*B. monosperma*	300
13	Musain Mata Ji Ka Oran	*A. pendula*	164	*A. pendula*	114	*F. indica*	175	*H. antidysen-terica*	132
14	Kala ji Ka Bagh (Nag)	*B. monosperma*	95	*P. juliflora*	204	*C. sepiera*	300	*B. monosperma*	186
15	Baukal Mata ji Ka Oran	*T. undulata*	181	*T. undulata*	245	*A. persica*	249	*T. undulata*	160
16	Viratra Mata Ka Oran	*M. emarginata*	224	-	-	*Z. nummularia*	92	-	-
17	Mata ji Ka Oran	*C. decidua*	300	*C. decidua*	300	*Z. nummularia*	127	*C. decidua*	300
18	Khuwas Mata Ji Ka Oran	*S. oleoides*	174	-	-	*D. cineraria*	138	*S. oleoides*	178
19	Mama ji Ka Selenadi	*P. cineraria*	101	*No tree*	0.00	*Z. nummularia*	300	-	-
20	Ghanshsyam Baba Ka Ashram	*S. oleoides*	224	*A. pendula*	61	-	-	-	-
21	Adi Badri Dham	*M. parviflora*	74	*C. dichotoma*	300	*Z. nummularia*	300	-	-
22	Sawai Bhoj/Devnarayan Mandir	*P. juliflora*	88	*C. decidua*	139	*E. nerifolia*	113	*P. juliflora*	300

SN	Name of Oran	Trees of >30 cm girth		Trees of 10-30 cm girth		Shrubs		Saplings	
		Species	IVI	Species	IVI	Species	IVI	Species	IVI
23	Dennaryan Ji Ki Bani	*A. pendula*	225	-	-	-	-	-	-
24	Devnaryan	*A. leucophloea*	98	-	-	*S. virosa*	300	*P. juliflora*	300
25	Goga Ji Ka Oran	*Z. rotundifolia*	140	*P. juliflora*	300	-	-	-	-
26	Karni Mata Ji ka Oran	*Z. rotundifolia*	154.79	*A. tortlis*	210	*L. barbarum*	300	*A. tortlis*	300
27	Karni Mata Ji Ka Oran	*Z. rotundifolia*	300	-	-	*L. pyrotech-nica*	141	-	-
28	Siya mata ji Ki Bani	*A. pendula*	230	*P. juliflora*	181	-	-	-	-
29	Gangrawal Ji Maharaj Ki Vani	*A. pendula*	220	*A. pendula*	201	*A. vasica*	171	*P. juliflora*	300
30	Parbatheshwar Mahadeo ki Bani	*M. parviflora*	151	*B. aegyptiaca*	68	*D. cineraria*	113	-	-
31	Devnarayan ji Ka Oran	*A. pendula*	210	*A. pendula*	300	*M. oblongifolia*	150	-	-
32	Rishi Mangri Oran	*A. pendula*	229	*A. pendula*	144	*D. cineraria*	148	*A. pendula*	222
33	Nahar Singhji Mata ka Oran	*A. pendula*	277	*A. pendula*	162	*L. camara*	220	*H. integrifolia*	119
34	Junjan Dodaji Ka Oran	*M. emarginata*	223	*M. emarginata*	300	*C. procera*	108	-	-
35	Gogaji Ka Oran	*M. emarginata*	162	*A. tortlis*	300	*L. pyrotech-nica*	300	-	-
36	Gosai Ji Ka Oran	*S. oleoides*	331	*A. tortlis*	300	*Z. nummularia*	219	-	-
37	Kaludas Baba Ki Dungri	*A. pendula*	265	*A. pendula*	146	-	-	*H. integrifolia*	300
38	Devnarayna Ji Ka Oran	*A. nilotica*	205	*A. nilotica*	188	*A. vasica*	117	-	-
39	Sawanria ji Seth Dham	*H. integrifolia*	231	*H. integrifolia*	300	*A. vasica*	300	*H. integrifolia*	300
40	Bali Baba Ki Dargah	*P. juliflora*	154	*H. integrifolia*	150	*Z. nummularia*	145	-	-
41	MuchKund	*A. pendula*	152	*A. pendula*	165	*C. procera*	131	-	-
42	Mahadeo Ji Ka Oran	*F. bengalensis*	132	*A. indica*	75	*Z. nummularia*	112	-	-
43	Govindguru Chhari Magri	*P. juliflora*	264	*P. juliflora*	300	*E. laevis*	300	*P. juliflora*	300
44	Bhuneshwar Mahadev Ka Oran	*A. pendula*	148	*A. pendula*	214	*D. cineraria*	98	*A. pendula*	300
45	Neelkanth Mahadev	*P. juliflora*	182	*P. juliflora*	241	*J. curcus*	151	*P. juliflora*	300
46	Bhadra Kali Mata Ji Ka Oran	*P. juliflora*	105.22	*P. juliflora*	300.00	*No shrub*	0.00	*P. juliflora*	237
47	Chisti Peer ki Dargah	*C. decidua*	232	*C. decidua*	300	*Z. nummularia*	244	*D. sissoo*	300
48	Guru Gorakhnath Ji Ka Oran	*S. oleoides*	167	*A. indica*	80	*Z. nummularia*	300	-	-
49	Kalimata ji ki Bani	*A. tortlis*	129	*P. juliflora*	137	*C. sepiera*	300	*P. juliflora*	300
50	Kapil Muni Ashram	*A. indica*	93	*A. indica*	192	*C. procera*	138	-	-
51	Shiv Dungri	*P. juliflora*	178	*P. juliflora*	138	*Z. nummularia*	300	*P. juliflora*	300
52	Ramdevji ka Oran	*S. oleoides*	134	*S. oleoides*	142	*C. burhia*	170	*A. tortlis*	300
53	Bhadariya ji Ka Oran (old)	*Z. rotundifolia*	265	*Z. rotundifolia*	300	-	-	-	-
54	Malani Oran	*C. decidua*	300.00	*P. juliflora*	243	*C. procera*	210	-	-
55	Bellane Oran	*C. decidua*	201.91	*A. senegal*	179	*Z. nummularia*	196	*A. tortlis*	300
56	Dharneswar Mahadeo ka Oran	*T. undulata*	233	*T. undulata*	240	*A. jacqmontii*	151	-	-

SN	Name of Oran	Trees of >30 cm girth		Trees of 10-30 cm girth		Shrubs		Saplings	
		Species	IVI	Species	IVI	Species	IVI	Species	IVI
57	Dudeshwar Mahadeo Ka Oran	*S. oleoides*	193	*P. juliflora*	208	*A. jacqmontii*	203	*P. juliflora*	300
58	Jhootana Mama ji Ka Oran	*S. oleoides*	198	*P. juliflora*	264	*C. auriculata*	300	*P. juliflora*	300
59	Kalla ji Ka Oran	*A. leucophloea*	155	*P. juliflora*	154	-	-	*A. nilotica*	300
60	Bapulala Smirti Van	*A. pendula*	193	*D. melanox-ylon*	174	-	-	*D. melanoxylon*	300
61	Devnarayan Ji Ka Asthan	*A. leucophloea*	148	*P. juliflora*	300	*C. procera*	300	*A. nilotica*	103
62	Shiv Mandir Kelkheura	*D. melanoxylon*	161	*D. melanoxy-lon*	175	-	-	-	-
63	Bhairav Ji Ka Bag	*B. monosperma*	100	*M. parviflora*	94	*C. sepiera*	262	*M. parviflora*	104
64	Bihari ji Ka Mandir	*S. oleoides*	160	*S. oleoides*	300	-	-	-	-
65	Prithviraj Ki Bani	*P. cineraria*	218	-	-	*C. procera*	300		-
66	Panchmir Pir Ki Bani	*A. tortlis*	125	*A. tortlis*	174	*Z. nummularia*	237	-	-
67	Kolu Pabuji Ka Oran	*C. decidua*	240	-	-	-	-	-	-
68	Karni Mata Ji Oran	*C. decidua*	139	*P. juliflora*	154	*Z. nummularia*	82	*P. juliflora*	300
69	Ramdeo Ji ka Oran	*P. cineraria*	235	*P. juliflora*	121	*C. procera*	300	*P. juliflora*	300
70	Pabuji Ka Oran	*P. juliflora*	300	*P. juliflora*	300	-	-	*P. juliflora*	300
71	Joleyali Oran	*P. cineraria*	267	-	-	-	-	*P. juliflora*	300
72	Jardabala Ji Ka Oran	*E. officinalis*	71	-	-	*C. procera*	150	-	-
73	Barbasni Mata Ji Ka Oran	*A. pendula*	215	-	-	-	-	*A. pendula*	300
74	Devnarayan Ji Ki Banni	*A. pendula*	300	-	-	*D. cineraria*	265	-	-
75	Dardevi Mata ji Ka Oran	*T.arjuna*	101	*P. juliflora*	153	*Z. nummularia*	196	*P. juliflora*	234
76	Devnarayan Bhagwan Ka Oran	*A. catechu*	120	*P. juliflora*	300	*D. cineraria*	300	*P. juliflora*	222
77	Devnarayan Ji Ka Oran	*E. camaldule-nsis*	208	*P. juliflora*	264	-	-	*P. juliflora*	225
78	Devnarayan Ji Ka Oran	*A. nilotica*	92	*P. juliflora*	166	*D. cineraria*	177	-	-
79	Dhokad Wali Mata ji	*A. pendula*	211	*A. pendula*	228	*C. sepiera*	300	-	-
80	Devnarayan Ji Ka Oran	*A. catechu*	163	*C. decidua*	111	*D. cineraria*	127	*B. monosperma*	300
81	Hanuman Ji ka Oran	*B. monosperma*	271	*B. monospe-rma*	152	*Z. nummularia*	202	*B. monosperma*	230
82	Devnarayan Ji ka Oran	*P. juliflora*	102	*B. aegyptiaca*	102	*C. sepiera*	300	*B. aegyptiaca*	232
83	Kalaji-Gora Ji Oran	*A. pendula*	158	*A. pendula*	190	*Z. nummularia*	165	-	-
84	Bheruji Ka Oran	*P. cineraria*	111	-	-	*A. jacqmontii*	135	-	-
85	Bhairon Ji Ka Oran	*P. juliflora*	133	*P. juliflora*	157	*L. barbarum*	202	*P. juliflora*	300
86	Gosai Ji Ka Oran	*C. decidua*	114	*C. decidua*	300	*Z. nummularia*	300	-	-
87	Sathika Jogmaya Ji Ka Oran	*C. decidua*	194	*C. decidua*	300	*Z. nummularia*	168	-	-
88	Laba ji Maharaj Ka Oran	*A. sericea*	279	*A. tortlis*	235	-	-	*A. tortlis*	180
89	Mama ji Ka Oran	*P. juliflora*	126	*P. juliflora*	170	*C. auriculata*	263	*C. decidua*	300
90	Devnarayanji Ka Oran	*S. oleoides*	196	*P. juliflora*	249	*C. auriculata*	300	*P. juliflora*	300
91	Ashapura Mata ji ka Oran	*S. oleoides*	199	*P. juliflora*	280	*C. sepiera*	228	*P. juliflora*	286

SN	Name of Oran	Trees of >30 cm girth		Trees of 10-30 cm girth		Shrubs		Saplings	
		Species	IVI	Species	IVI	Species	IVI	Species	IVI
92	Mata ji Ka Oran (Mada)	*A. leucophloea*	149	*P. juliflora*	160	*C. auriculata*	210	*P. juliflora*	169
93	Dhundhlimata Ki Banni	*T. grandis*	132	*W. tinctoria*	126	*F. indica*	208	*W.tinctoria*	300
94	Bijasau mata ji Bani	*T. grandis*	133	*H. integrifolia*	254	*L. camara*	300	-	-
95	Chandimata ji Ka Oran	*T.arjuna*	105	*B. monosperma*	274	*L. camara*	300	-	-
96	Devnarayan ji Ki bani	*T. grandis*	227	*T. grandis*	209	*L. camara*	250	-	-
97	Devnarayan ji Ki Bani	*A. pendula*	253	*A. pendula*	171	*D. cineraria*	194	*A. pendula*	153
98	Bheru Nath ji Bani	*A. pendula*	130	*P. juliflora*	300	*G. tenax*	197	*P. juliflora*	300
99	Chamunda Mata ji Ki Bani	*A. pendula*	216	*A. pendula*	300	*R. mysurensis*	300	*A. pendula*	300
100	Haumanji ki Dungri	*P. juliflora*	141	*A. pendula*	300	*C. procera*	300	*A. nilotica*	300
101	Jasram Baba Ji Ki Bani	*M. parviflora*	87	*P. juliflora*	91	*C. phlomidis*	221	-	-
102	Peer Baba Ki Dargah	*A. pendula*	212	*P. juliflora*	131	*C. sepiera*	300	-	-
103	Bheruji Ki Bani	*P. cineraria*	125	*A. tortlis*	193	*C. procera*	191	-	-
104	Kheri Walai Balaji ka Oran	*C. decidua*	104	*C. decidua*	145	*Z. nummularia*	171	-	-
105	Satimata Ji Ka Mandir	*P. cineraria*	184	*C. decidua*	300	*C. procera*	300	-	-
106	Sidheswar Mahadeo Mandir	*A. senegal*	134	*A. senegal*	112	*Z. nummularia*	183	*P. juliflora*	168
107	Vorada Hanuman ji Ka Oran	*S. oleoides*	197	*P. juliflora*	300	*C. auriculata*	300	*P. juliflora*	300
108	Magriwali Mata ji Ka Oran	*A. pendula*	172	*A. pendula*	200	*G. tenax*	300	*A. pendula*	158
109	Gosai Ji Ka Jod	*C. decidua*	197	*C. decidua*	280	*Z. nummularia*	188	-	-
110	Baba Gadanpur Ki Dargah	*C. decidua*	99	*P. juliflora*	192	*No shrub*	0.00	*P. juliflora*	257
111	Miyan Mama Peer Baba	*A. nilotica*	172	*C. decidua*	234	*C. procera*	169	-	-
112	Baba Rahupeer Dargah	*S. oleoides*	180	*P. juliflora*	204	*Z. nummularia*	300	*P. juliflora*	221
113	Mata Ji Ka Darra	*P. juliflora*	152	*P. juliflora*	281	*C. sepiera*	300	*P. juliflora*	300
114	Devnarayan Ka Asthan	*A. pendula*	207	*A. pendula*	199	*C. sepiera*	300	*A. pendula*	217
115	Devmaharaj Ka Asthan	*A. pendula*	223	*P. juliflora*	300	-	-	*P. juliflora*	300
116	Pava Bab ji Oran	*W. tinctoria*	112	*W. tinctoria*	64	*G. villosa*	132	*T. bellerica*	87
117	Hanuman Ji Ka Oran	*F. bengalensis*	277	-	-	-	-	-	-
118	Shringi Rishi Ji Ka Oran	*A. senegal*	114	*A. senegal*	300	*G. tenax*	119	-	-
119	Bhanvar Mata ji Ki Bani	*A. pendula*	188	*A. pendula*	181	*E. caducifolia*	163	*A. pendula*	300
120	Kali Pahari Oran	*P. juliflora*	180	*P. juliflora*	272	-	-	-	-
121	Kundeshwar Mahadevji Oran	*F. bengalensis*	141	*S. album*	79	*C. sepiera*	183	-	-
122	Tarkeshwar Mahadev Oran	*P. roxburghii*	53	*P. roxburghii*	157	*A. squamosa*	168	*H. integrifolia*	100
123	Ubeshwar ji Oran	*E. camaldulensis*	83	*B. monosperma*	112	*L. camara*	136	*D. cordifolia*	209

Annexure V

Common fauna recorded in sacred groves of Rajasthan through people interactions.

SNo.	Zoological name	Common name	Animal type
1.	*Acridotheres ginginianus*	Bank Myna	Bird
2.	*Acridotheres tristis*	Myna	Bird
3.	*Agma agilis*	Agma	Reptile
4.	*Anhinga anhinga*	Snake bird	Bird
5.	*Aquila hastate*	Eagle	Bird
6.	*Athene brama*	Spotted owlet	Bird
7.	*Axis axis*	Deer	Mammal
8.	*Bos taurus*	Cow	Mammal
9.	*Boselaphus tragocamelus*	Blue Bull	Mammal
10.	*Canis aureus auveus*	Jackal	Mammal
11.	*Canis lupus pallipes*	Wolf	Mammal
12.	*Chamaeleo zeylanicus*	Chameleon	Reptile
13.	*Columba livia*	Rock Pigeon	Bird
14.	*Columba livia domestica*	Pigeon	Bird
15.	*Corvus macrorhynchos*	Jungle crow	Bird
16.	*Corvus splendens*	Crow	Bird
17.	*Coturnix coturnix*	Grey Quil	Bird
18.	*Cuon alpinus*	Wild dog	Mammal
19.	*Echis carinatus*	Viper	Reptile
20.	*Ardea alba*	Egret	Bird
21.	*Eryx conicus*	Sand boa	Reptile
22.	*Eryx johnii*	Earth snake	Reptile
23.	*Eudynamys scolopaceus*	Koel	Bird
24.	*Felis chaus*	Wild cat	Mammal
25.	*Felis silvetri oranata*	Desert Cat	Mammal
26.	*Francolinus pondicerianus*	Grey partridge	Bird
27.	*Funambulus palmarum*	Squirrel	Mammal

SNo.	Zoological name	Common name	Animal type
28.	*Funambulus pennati*	Five-striped	Mammal
29.	*Gazella gazella*	Chinkara	Mammal
30.	*Gyps bengalensis*	White vulture	Bird
31.	*Hemidactylus flarviviridis*	House gecko	Reptile
32.	*Hemiechinus auritus callaris*	Hedghog	Mammal
33.	*Herpestes edwardsi*	Mongoose	Mammal
34.	*Hyaena hyaena*	Hyena	Mammal
35.	*Hystrix indica indica*	Porcupine	Mammal
36.	*Lanius exubitor*	Grey shrike	Bird
37.	*Leptocoma zeylonica*	Sun-bird	Bird
38.	*Lepus nigricollis*	Common hare	Mammal
39.	*Macaca fascicularis*	Monkey	Mammal
40.	*Melursus ursinus*	Bear	Mammal
41.	*Meriones hurrianae*	Indian desert gerbil	Mammal
42.	*Ophiomorus tridactylus*	Sand fish	Reptile
43.	*Ophiophagus hannah*	King Kobra	Reptile
44.	*Oryctolagus cuniculus*	Rabbit	Mammal
45.	*Panthera pardus*	Panther	Mammal
46.	*Passer domesticus*	Sparrow	Bird
47.	*Pavo cristatus*	Pea fowl	Bird
48.	*Picnonotus cafer*	Red-vented bulbul	Bird
49.	*Psittacula eupatria eupatria*	Alexandrine parakeet	Bird
50.	*Psittacula krameri*	Parrot	Bird
51.	*Pterocles exustus*	Indian sand goose	Bird
52.	*Pteropus giganteus*	Fruit bat	Mammal
53.	*Rattus norvegicus*	Rat	Mammal
54.	*Rusa unicolor*	Samber deer	Mammal
55.	*Semnopithecus entellus*	Hanuman Langoor	Mammal
56.	*Streptopelia capicola*	*Ring*-necked *dove*	Bird
57.	*Streptopelia decocto*	Spotted dove	Bird
58.	*Sus scrofa*	Wild pig	Mammal
59.	*Tetera indica*	Indian gerbil	Mammal
60.	*Turdoides caudatus*	Common babbler	Bird
61.	*Uromastix hardwickii*	Indian spiny-tailed	Reptile
62.	*Varanus bengalensis*	Varanus-Pata goh	Reptile
63.	*Varanus griseus*	Varanus-chandan goh	Reptile
64.	*Vulpes bengalehsis*	Indian Fox	Mammal
65.	*Vulpes vulpes pusilla*	Desert Fox	Mammal